# ADVANCES IN AGROFORESTRY

# ADVANCES IN AGROFORESTRY

*By*

**Prof. L.K. Jha**

*Head, Deptt. of Forestry*
School of Agril Sciences and Forestry
NEHU, Mizoram Campus
Aizawl - 796007.
INDIA

**APH PUBLISHING CORPORATION**
**ANSARI ROAD DARYA GANJ,**
**NEW DELHI - 110 002**

*Published by*
S.B. Nangia
**A P H Publishing Corporation**
4435-36/7, Ansari Road, Daryaganj
New Delhi 110002
Ph.: 23274050
E-mail : aphbooks@gmail.com

2023



Rs. 4995/-

*Printed at*
**Balaji Offset**
Navin Shahdara, Delhi 110032

*Dedicated to*
*My beloved wife*
***Mrs. Punam Roy***

and

*My Dearest Children*
***Manish, Madhu and Anu***

# Foreword

Agroforestry is considered "alternate agriculture" by several precepts. Although agroforestry has been practised throughout the world in some form or the others. Since very long time, its status as a separate scientific discipline is of recent origin. With the increase in population of both cattle and human kind in geometrical progression and the area of land being finite, there is no alternative but to obtain from the same unit of land food, fibre and fuel and feed with a view to meeting the future needs. In other words, there is an urgency to evolve an integrated land use pattern which the agroforestry aims at.

Agroforestry is a nascent science and it will take time for the crystallisation of the concept through the process of carefully conducted experiments, analyses of the existing data more maticulously than had been done so far and application of scientific principles and practices. Like any other application oriented scientific discipline, agroforestry also needs blending of the scientific principles of agriculture, forestry, agrostology and animal husbandry, etc. Unfortunately such an integrated approach is lacking both at the institutional and individual level. Consequently, instead of having an integrated approach, agroforestry has been viewed and interpreted by many from a narrower angle, restricting their approaches with a bias to the line of specialisation of the individual institution or scientist. This has resulted in a myriads of approaches. Confusion has ruled supreme.

The present book written by Prof. Lalit Kumar Jha, Head, Department of Forestry, North-East Hill University is a result of painstaking collection and collation of the scientific data that exist in numerous journals, periodicals and a few books for a number of years. The book has attempted to depict a connected picture of agroforestry as a separate scientific descipline, though applied in nature. In this task the author has achieved commendable success.

The chapters cover a very wide range of subjects which are connected with the agroforestry. At the end of each chapter, reference to the literature has been given. The value of the book has been enhanced by the inclusion of a large numbers of illustrations, flow-charts, photographs etc.

The book will be useful both as a text-book for the students of the Agroforestry as well as a compendium of the scientists. The prose is quite simple and readable without any ambiguity in meaning. For a scientific trait this is good achievement.

The book I am sure, will receive acclaim and will be avidly read by students, teachers, scientists, planners, administrators, forestry, non-government organisations and voluntary agencies who will find the book very useful.

**Prof. P.K. Sen-Sarma**

Residence:
Flat No. 1, 2nd Floor
West Block
302, N.S.C. Bose Road
Calcutta - 700047
(India)

# Preface

Agroforestry as a new applied science is of recent origin. An efficient agroforestry system would aim at systematically developing integrated land use system and practices where the positive interaction between trees and crops is encouraged and maximized.

There has been a long felt need by the scientists, teachers and students for a comprehensive book on agroforestry in India. One interested in acquiring knowledge about the agroforestry has to refer to journals, periodicals and a few foreign books. Being aware of this lacuna, since, I joined as Incharge of All India Co-ordinated Project on Agroforestry, ICAR, Regional Centre, Faculty of Forestry, B.A.U., Ranchi in the year 1987, it has been my ambition to write a book on agroforestry and present title is out come of my efforts.

The book deals with various aspect of agroforestry and covers syllabus of B.Sc. and M.Sc. agroforestry. It is based on various research journals, reports and edited volumes published in India as well as in foreign countries. I am sure, it will prove very useful to the students, scientists, foresters and extension workers. The reference given at the end of each chapter would help the reader to acquire additional knowledge on the concern topic.

I wish to acknowledge my profound thanks to the authors of the papers and reports as well as editors of the books, that form the basis of this volume. I am thankful to foresters, scientists of different Research Institute and Universities for their help in the preparation of manuscript by providing reprints, reports, photoplates etc. I am grateful to them.

I am indebted to Prof. A.K. Sinha, Chief Scientist, Department of Agril. Engineering, Birsa Agril. University, Ranchi; Prof. P.K. Sen-Sarma, Emeritus Scientist, Social Forestry Support Office, W.L.D.B., Govt. of India, NEHU, Shillong; Sri P.P. Sen, I.F.S. (Rtd.), Purulia; Dr. R.P. Tiwari, Head, Geology Department, Pachhunga University College, Aizawl and Prof. K.N. Jha, Head, Economics Department, Koshi College, Khagaria, Bhagalpur

University for their kind permission to reproduce their work related to social and agroforestry.

I wish to express my sincere indebtedness to Prof. P.K. Sen-Sarma, Emeritus Scientist, Social Forestry Support Office, NEHU, who in spite of his busy schedule, spared his valuable time to read out the preliminary draft of the manuscript. Words fail me to convey my deep sense of gratitude to him for writing an apt foreword for the book.

During the preparation of manuscript I got inspiration specially from my parents (Prof. K.N. Jha and Smt. Khusi Devi) and father-in-laws, (Mr. K.N. Roy, Chartered Accountant and Smt. Manju Devi). For many favours, I am grateful to my wife Mrs. Punam Roy, children Manish, Madhu and Anu, and brothers Sri Keshab Kumar Jha, Dr. Anil Kumar Jha, Sri Sunil Kr. Jha and Sri Pramil Kumar Jha. I am also grateful to my uncle Sri Krishna Chadra Jha (a practical agroforester) and aunt Smt. Karpoor Devi. Infact my uncle gave me basic training in the field of agrihorticulture and silvipastoral system.

Lastly, but not the least, I am indebted to the Ashish Publishing House, Delhi for their efforts to bring out the publication in time.

I will highly appreciate suggestions for improvement of this volume.

**LALIT KUMAR JHA**

# Contents

# *List of Charts*

# *List of Illustrations*

# 1

# *Agroforestry Defined*

Agroforestry is neither an invention nor a new concept. The practice of combining tree species with crops had been practised in some form or the other in most of the countries. However, agroforestry as a new applied science is of recent origin. With the increase in population of both cattle and human beings in geometrical progression and the land area remaining finite, there is no other alternative but to grow food, fodder, feed and fibre in an integrated manner in the same unit of land. Consequently, agroforestry is progressively emerging as a new discipline of human knowledge. The recent emphasis on agroforestry which includes the concept of co-existence of farm and forest, is the outcome of the change of people's attitude born out of dire necessities.

To make the subject more explicit it is necessary to define forestry first and than social and agroforestry. At one time forestry was defined as "the theory and practice of all that consistute the creation, conservation and scientific management of forest and rational utilisation of this resources." Scientific management includes forest management on the principle of sustained production of goods and services. Srivastava (1987) defined, based on the prèmise of the facts stands above as "the scientific management of forest for continuous production of renewable and socially most acceptable goods and such services as are not subject to the principle of exclusion". Renewable socially acceptable goods primarily constitute fuelwood, leaf fodder, fencing material, timber, bamboo and minor forest products like resin, lac, honey, birds, insects, etc.

The community service rendered by the forest constitutes watershed management, flood control, climatic regulation, waste treatment and recycling, carbon dioxide absorption, production of oxygen, conservation of environmental quality, maintenance of habitat of wild life, etc.

It may be mentioned that the forester mainly worked in isolation and assumed the role of policing to conserve the forest. However, since 1968 a change of attitude has drawned among the professional foresters who are now very keen to involve people's participation in the creation and management of forest. An epoch making decision was taken by the Government of India through the notification of June, 1990 in which peoples participation in the management of degraded forests has been accepted as a cardinal principle. This change in attitude has been brought about by the fact that without people's participation and active involvement degraded forests cannot be conserved and maintained under the broad concept of social forestry which was difined for the first time by Westoby in the IX-Commonwealth Forestry Congress (1968) as "Social Forestry is a Forestry which aims at producing flow of protection and recreational benefits for the community". It was a revolutionary step as it aimed at the establishment of a happy relationship between forest and society.

National Commission on Agriculture in its report (1976) has enlarged the scope of the social forestry in view of the socio-economic imperatives of forests for the rural community as well as in the management of forest resources. Consequently the NCA (1976) have included the following under the ambit of social forestry.

(A) **Farm Forestry** (Agroforestry)

Raising rows of trees on bunds or boundaries of the farm and individual trees in private agricultural lands.

(B) **Extension Forestry**

(*a*) Mixed forestry comprising creation of fuel, fodder, fruit and woodlots in the village common lands, government wastelands and Panchayat lands.

(*b*) Raising of shelter belts and wind breaks.

(*c*) Planting of trees on road, canal and rail sides.

(*d*) Reafforestation of degraded forest lands situated in and around human habitats which lost their floristic and faunistic glory because of anthropogenic intervention and biotic interferences.

(*e*) Recreation forestry.

(*f*) Urban forestry.

This clearly proves that social forestry is an umbrella subject and we can define it differently, depending on the objectives of the programmes.

Tiwari (1983) has included under social forestry several components that may be defined in the Indian context as the science and art of growing trees and other vegetation on all lands available for the purpose mainly outside traditional forest areas, with voluntary and close involvement of the people and more or less integrated with other operations, resulting in a balanced and complimentary land use so as to providing a wide range of goods and services to the individual as well as to the society.

On the basis of above discussion general definition of social forestry can be formulated as "Plantation out side the traditional forest area (marginal, submarginal, private land and community land) and in degraded forest lands (provided above categories of land are not available) by the involvement of individuals preferably below the poverty line or socially and economically depressed people, with a view to meeting their requirements in respect of crops, legumes, tubers, fuel, fodder, small timber, etc.

Agroforestry is, by implication, a component of social forestry and is an integral part of the social forestry system so as to make it possible to increase the production of food, crops, legumes, etc. in the same unit of land.

It may be worthy of mentioning that the system of raising trees along with agricultural crops has been in vogue in Forests Department for over 100 years (Taungya cultivation). The usual practice has been to lease out the forest lands to the cultivators to raise cereal crop after the annual coupe has been felled, exploited and replanted. In lieu, the beneficiaries are expected to protect the tree seedlings and saplings.

It would be worth mentioning here that in the year 1937 "The Hindu", a national Newspaper of India included the features, in its 6th November issue as mentioned below.

'A scheme known as agriforestry, which was a combination of agriculture and forestry, was adopted at the meeting of the Central Fodder and Grazing Committee of the Imperical Council of Agricultural Research (present Indian Council of Agricultural Research) which met at New Delhi'. The committee passed the following resolutions :-

"The committee strongly approves the scheme put by the United Provinces (present Uttar Pradesh) Forest Department for a preliminary survey to examine the possibilities of creating fuel and fodder plantations in the plains district (forestry-cum-agriculture known as the Taungya in the United Provinces) has proved in forest areas most beneficial to the cultivator as a means of supplying poles, firewood and fodder, as also giving good land for cultivation and at the same time allowing regulated forest development. The possibility of applying this system to lands remote from forests and private lands is now to be explored".

There is a school of thought which is of the opinion that apporting a land to grow annuals and perennial separately on separate parts of the same land is "agroforestry" which is in fact, highly, erroneous, as such an use either concurrently or separately are nothing but another way of practising agriculture and forestry in lands that are mutually exclusive and negates the fundamental principles of agroforestry which primarily aim at wedding seasonal agricultural crops with annual in the same unit of land in such a manner that their roles are mutualistic and complementary. Agroforestry is thus an integrated landuse planning system following the principle of generating multiple resources from the same unit of lands (Sinha, 1985).

Regarding the definition of agroforestry, several investigators across the world have defined it differently the commonality of basic theme notwithstanding.

Torres (1983) has stated that agroforestry can be defined as the deliberate combination of trees with crop plantation or pastures, or both, in an effort to optimise the use of accessible resources to satisfy the objectives of the producer in a sustainable way. Lungdrean and Raintree (1983) have opined that agroforestry is a collective name for land use systems and technology where woody perennials (trees, shrubs, palms, bamboos, etc.) are deli-

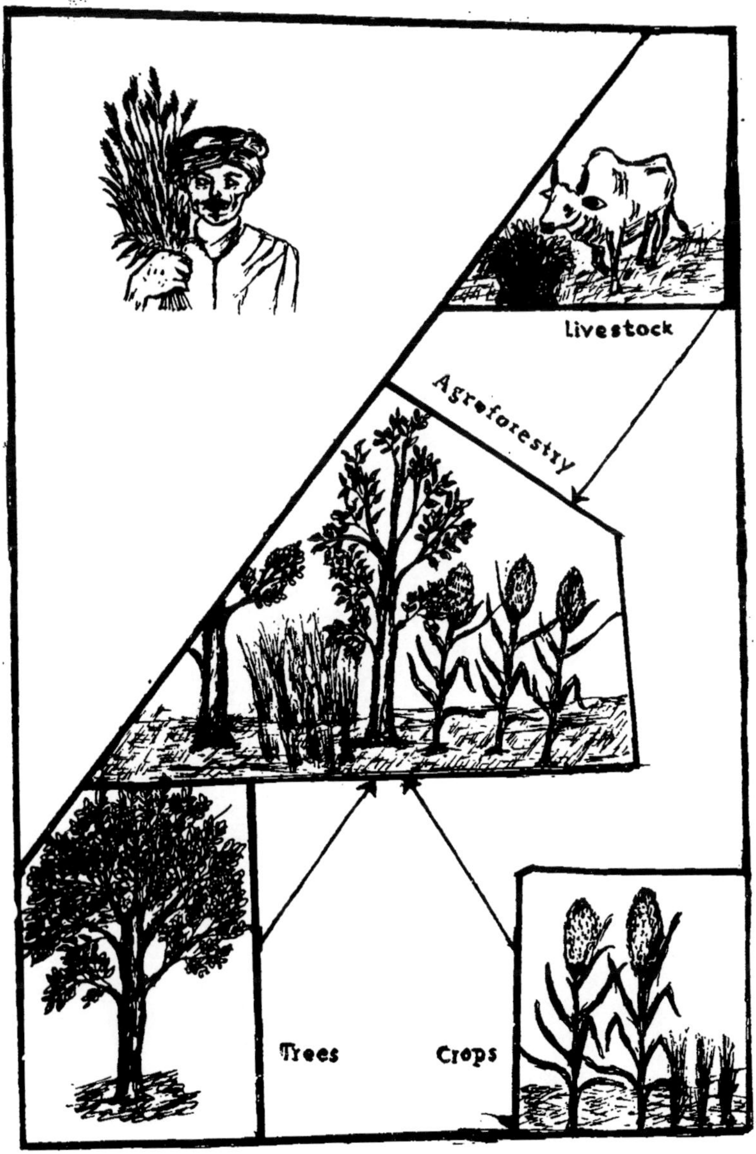


Fig. 1 : Components of agroforestry system.

berately used on the same land management unit as agriculture crops and/or animals, either on the same form of spatial arrangement or on ecological sequence. In agroforestry systems there are both ecological and economical interactions between the different components.

These definition however, outline the broad boundaries of agroforestry and the typical characteristics of the systems :–

* Agroforestry normally involves two or more species of plants (or plants and animals) at least one of which is a woody perennial ;
* An agroforestry system has two or more outputs ;
* The cycle of an agroforestry system is always more than one year.
* Even the most simple agroforestry system is more complex ecologically, than a monocropping system.

According to Walt (1989) "agroforestry has two dimensions. One is the way of thinking. It means looking for ways to maximize the production from land using a mixture of wood fibre, wildlife recreation and farm cropes, both plant and animal, to make the biggest buck possible consistent with keeping the land productive. The second dimension echoes the first. It is the actual management of land according to this ethic."

Rao (1989) has mentioned that in the broadest sense, the term "agroforestry" encompasses any and all techniques that attempt to establish or maintain both forest' tree and agricultural production on the same piece of land. According to the more strictly scientific definition of agroforestry, all forms of agroforestry are characterised by :-

* The deliberate introduction of woody perennials on the same unit of land as agricultural crops and/or animal, either in some form of spatial mixture or temporal sequence ;
* There must be a significant interaction (positive and/or negative) between the woody and non woody components of the system, either ecological and/or economical.

A more precise definition of agroforestry is :

"a land use that involves deliberate retention, introduction or mixture of trees or other Woody perennials in crop/animal production fields to benefit from the resultant ecological and economic interaction (Rao & MacDicken, 1991)."

An efficient agroforestry system would aims at systematically developing integrated land use systems and practices where the positive interaction between trees and crops is encouraged and maximized. This seeks to achieve a more productive, sustainable and diversified output from the land than is possible with the conventional mono-cropping system.

Westly (1990) very correctly had pointed out that researchers might also define agroforestry practices differently depending on the focus of their work. For instance, an agronomist interested in the effects of nitrogen fixing trees on the surrounding vegetation inevitably looks at an agroforestry practice differently from an economist interested in the potential cash income derived from selling of poles, once those trees are harvested. Finally, there is the question of scale. We can look at agroforestry from the perspective of individual trees and plants, of fields, of farms, of local geographic units such as watersheds or of broad ecological regions. [Readers can go through the paper written by Andriesse (1978), King (1978 and 1979), Huxley (1983), Foley and Barnard (1984), Bentley (1985), Rocheleau and Raintree (1986), Labelle (1987), Gordon and Bentley (1990), Young (1990) for more informations.]

## REFERENCES


Andriesse, J.P. (1978). From shifting cultivation to agroforestry or permanent agriculture, Proc ; 50th symposium, *Tropical Agril. Bull ;* 303, Deptt. of Agricultural Research, Amsterdam, pp. 35-43.

Bentley, W.R. (1985). *Agroforestry : A strategy for research and action in India,* Discussion paper No. 17, The Ford Foundation, N. Delhi, p. 10.

Foley, G. and Barnard, G. (1984). Farm and community forestry, *Tech. Rep.,* 3 Earthscan, London.

Ghotz, H.L. Ed. (1987). *Agroforestry : realities, possibilities and potentials,* Dredrecht (The Netherlands) : Martinus Nijhoff in cooperation with ICRAF, pp. 1-227.

Gordon, J.C. and Bentley, W.R. (1990). *A hand book on the management of agroforestry Research,* Winrock International, U.S.A. and South Asia Books, pp. 1-72.

Huxley, P.A. (1983). The role of trees in agroforestry. Some comments, In (Huxley, P.A., ed.), *Plant research in agroforestry,* Nairobi, pp. 257-70.

Huxley P.A. (1983). Comments on agroforestry classification with special reference to plant aspect. In : (Huxley, P.A., ed.), *Plant research in agroforestry,* Nairobi, ICRAF, pp. 161-171.

King, K.F.S. (1978). *Agroforestry,* paper presented in the 50th Agril. Conf. Amsterdam.

King, K.F.S. (1979). Keynote address to the National seminar on Agroforestry, India, delivered on May 16-18, Shillong (unpublished).

Labelle, R. (1987). Agroforestry : General concept, early work and current initiatives a review of the literature, pp. 1-62. (Cross Ref).

Lungdrean, B. and Raintree, J.B. (1983). Sustained agroforestry, *Agril. research for development potential and challenges in Asia,* The Hague, ISNAR, pp. 1-25 (Cross Ref).

National Commission on Agriculture (1976). Govt. of India, part IX (Forestry).

Rao, Y.S. (1989). Why, What, How and Where agroforestry in the Asia pacific region, *Forest News* (Tiger paper) Thailand, pp. 1-9.

Rao, Y.S. and MacDicken, K.G. (1991). Foreword, In : Agroforestry in Asia and the Pacific, Mellick, W., Rao, Y.S. and MacDicken, K.G. (eds.), RAPA Publication, pp. 1-2.

Rocheleau, D.E. and Raintree, J.B. (1986). Agroforestry and the future of food production in developing countries, *Impact of Science on Society,* 142 : 127-141.

Sinha, B.N. (1985). Role of agroforestry in soil and water conservation, In. (Muhammad, S., ed.), *Proceeding, Social forestry workshop,* B.A.U., pp. 90-99.

Srivastava, J.L. (1987). *Defining social forestry,* Ensine Printers, Gumla, pp. 1-10.

Srivastava, V.C. (1985). Agroforestry concept, progress, In : (Muhammad, S., ed.), *Social forestry workshop proceeding,* B.A.U., Ranchi, pp. 97-105.

The Hindu (1937). A national news paper of India, Dated 6th November (Cross Ref.).

Tiwari, K.M. (1983). Social Forestry as an aid to environmental conservation, *Van Vigyan,* Vol. 121, No. 1 & 2.

Tiwari, K.M. (1983). *Social Forestry in India,* Natraj Publishers, Dehradun, pp. 1-285.

Torres, F. (1983). Agroforestry : Concepts and practices. In : *Agroforestry system for small scale farmers,* Hockstra, D.A. and Kuguru. F.M., Eds. Proc. ICRAF/BAT, workshop, Nairobi, Sept. 1982, ICRAF, Nairobi, pp. 27-42.

Torres, F. (1985). *Net work of generation of agroforestry technology in Africa,* ICRAF, working paper 31, Nairobi, Kenya, pp. 28.

Walt, H.R. (1989). Weneed agroforestry in a big way, Workshop on *Agroforestry in California : planning for the twenty-first centuary,* Deptt. of Forestry and Resources Management University of California, Berkeley, pp. 1-3.

Westoby, J.C. (1968). Changing objectives of forest management address to the *Ninth Common Wealth Forestry Conference,* New Delhi, 1968.

Westly, B. Sidney, (1990). Defining agroforestry technologies, A.F. databank, *Agroforestry today,* Vol. 2(1), p. 21.

Young, A. (1990). Agroforestry, environment and sustainability, *Outlook on Agriculture,* 19 : 155-160.

2

# Types of Agroforestry System

Agroforestry systems are more complex than agriculture or forestry when practised individually. These systems improve the productivity of plants and animals and it is based on principle of optimum utilisation of land. The systems are very complex. In a few types of agroforestry systems, more than four components interact. All these are based upon maximum utilisation of the resources to increase the economic benefits. Most of the systems is related to each other. Hence, it is a very difficult task to make clear cut distinction among the various types of agroforestry system. The types of agroforestry system mentioned here are based upon its components and functions.

### Agrisilviculture

In this system tree species are grown and managed in the farmland along with agricultural crops. The aim is to increase the overall yield of the land. This system is also based on the principle of sustained yield. Woody plants and the crops interact both ecologically and economically. Interaction may be negative or positive. This system can be of great value, if trees, and crops are appropriately well managed according to their requirements. This can solve the problem of shortage of food, fuelwood, timber and conserve soil moisture contents and also ameliorate the harsh climatic condition.

### Silvipastoral System

Under silvipastoral system of land management, improved pasture species are grown along with tree species. Here selection of tree species could be either for timber alone or for dual purpose i.e. fuel cum fodder. Grasses or grass legume mixtures are grown along with tree species simultaneously on the same unit of land. Combining of trees, grasses and legumes also helps to conserve soil moisture and improve the fertility status of soil.

**Fig. 2 : Agrisilviculture**

This system may be grouped into two sub-types :-

(*a*) Silvipastoral system with control and supervised grazing of domestic ruminants.

(*b*) Silvipastoral system in which domesticated ruminants are not permitted for grazing. Forages are utilised only for stall feeding. Some workers are of the opinion that this system also falls under agrisilviculture (King, 1979).

Fig. 3 : Silviagriculture: Kulthi grown along with *Melia azedarach* at Rarha Village, Ranchi. (*Courtesy* : Faculty of Forestry, B.A.U., Ranchi).

Fig. 4 : Silviagriculture : Saflower grown along with *Azadirachta indica* Kumharia Village, Ranchi. (*Courtesy* : Faculty of Forestry, B.A.U., Ranchi).

**Fig. 5 :** Silvipastoral : Grass grown along with *Acacia auriculiformis.*

### Agrisilvipastoral System

This system is the result of the union between silvipastoral and agrisilviculture systems. Under this system, the same unit of land is managed to get agricultural and forest crops where farmers can also rear animals.

### Agrihorticulture System

If agricultural crops are grown along with fruit trees, the system is referred to as agrihorticulture system. This land management system aims at production of both agricultural crops and fruits. This practice could continue till 5 or 6 years or till canopy of fruit trees becomes fully closed. This system makes soil congenial for the rapid growth of fruit trees. However, if the fruit trees are planted at wider spacing, agriculture can continue simultaneously too.

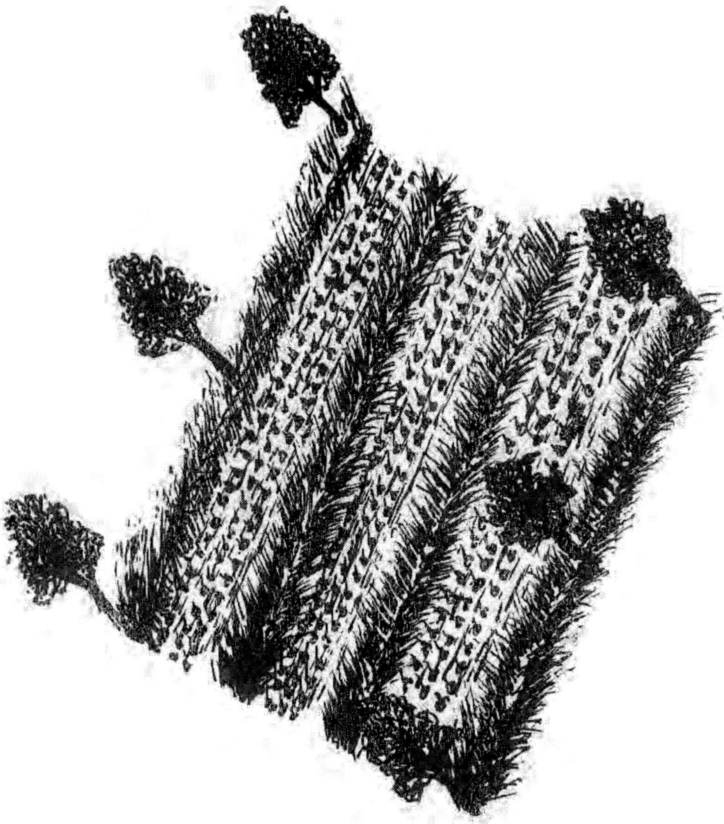

Fig. 6 : Agrisilvipastoral.

**Silvihorticulture System**

This system is similar to agrisilviculture system. This system is based on the principle of sustained yield. Tree species are managed to get timber, fuelwood etc., and horticultural crops are grown in the interspace. Thorough judicious introduction of horticultural crops additional benefit in the form of fruits and vegetables are achieved. Thus, the farmer can have the dual benefit.

**Fig. 7 :** Silviagrihorticulture : Mustard grown in mango orchard and sissoo trees on the border area at Tardhia Village, Madhubani (Bihar).

**Fig. 8 :** Silvihorticulture System

**Silvihortipastoral System**

A combination of tree species, horticultural crops, grasses are practised in this system. The three combinations are based on the principle that each of its component draw nutrients from different layers of soil. This system is more productive than the pure agricultural crops.

**Silviagrisericulture System**

This is a very complex system of agroforestry. In this system crops/vegetables are grown along with tree species (Silk host plants). After three years, tree leaves are used for rearing of silkworm. If tree species is *Terminalia arjuna,* tassar worms are reared in the field itself. The larval excreta are good manure for the crops/vegetables. Silk is the end product. The pruned branches of the tree species serve to supplement domestic energy need.

**Silviagrilacculture System**

In this system crops/vegetables are grown along with lac host plants. After maturity of the host plants, lac insects are reared. *Moghania macrophylla* which is a woody shrub is the ideal lac host for incorporation in this system. Being a shrub, it does not interfere in any manner the growth of agricultural crop. Lac insects being sessile cannot migrate to the crop plants. This system is very common in the Chotanagpur plateau of Bihar. Farmer also grows fodder along with lac host plant. This may be called as silvipastoral-lac culture system.

**Silviagriapiculture System/Hortisilviagri-apiculture System**

The land is managed for concurrent production of flowers, crops and honey. In this system trees and crops are judiciously selected so that worker bees in course of ramblings, never miss to collect nectar in their honey stomach and pollen in their pollen basket. There should not be dearth of flower. Beside honey, farmers can also collect fuelwood, small timber and agricultural crops from the same unit of land. In Bihar tribal farmers are practicing this system. Honey is named and sold after the name of nourishing trees (*e.g.* Karanj Honey). In several parts of North East citrus trees are grown and citrus honey is a prized commodity. The system can be called horti-silvi-agri-apiculture system. Flowering plants often

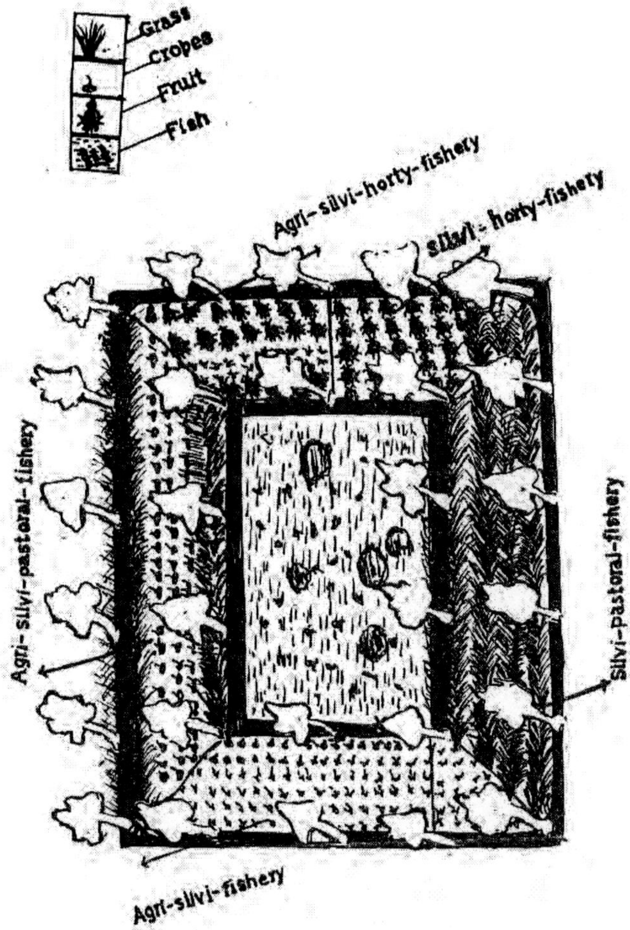


Fig. 9 : Silviagripastoral and Fisheries system.

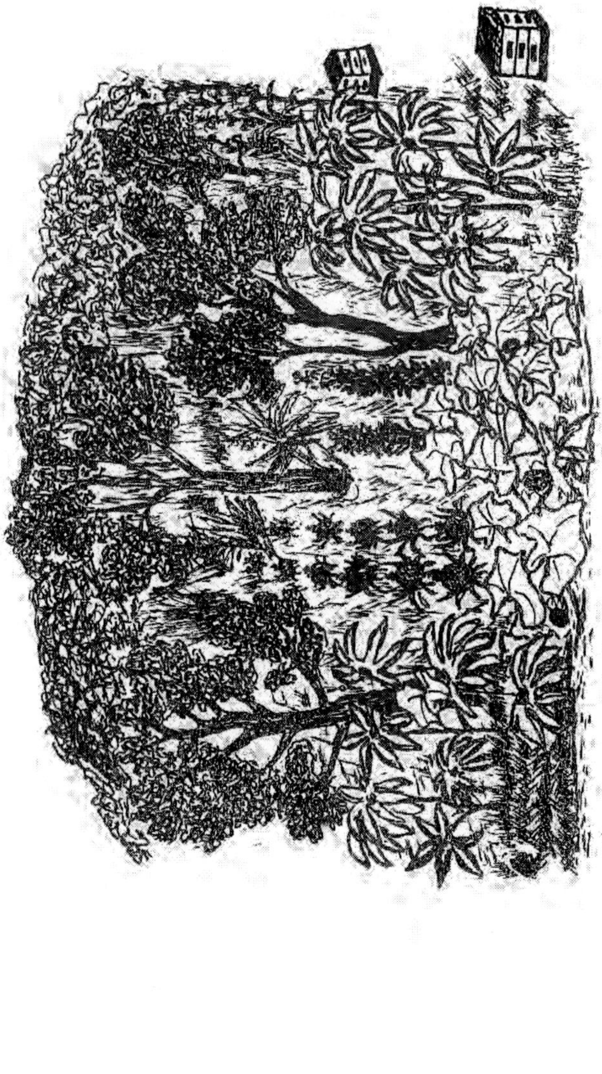

Fig. 10 : Multistoried agroforestery system.

**Figs. 11 and 12 :** Multistoried agroforestry system (Mizoram)

favour increase of parasites and predators of crop pests and thus an antiregulatory biocontrol system operates in this system.

**Silviagripastoral and Fisheries System**

This may be considered as most complex system which integrates all the four components in a harmonious manner. In North Bihar and West Bengal this system is prevalent. Fruit trees *Mangifera indica* and *Litchi chinensis* in North Bihar and *Cocus nucifera* and *Areca catechu* in West Bengal are planted on the terraced land around the water tanks, ponds etc. When fruit trees are planted at wider space, crops/vegetables are grown in the interspace. When trees cause shade on the floor, some shade tolerent crops/vegetables/spices are selected. The ponds are used for breeding fish and Makhana (*Euryale ferox*) cultivation. The fallen leaves of tree species, Makhana leaves etc. enrich the pond nutrients for fishes. This system is characterized by the long chain of material-energy flow within an ecosystem.

**Multistoried Agroforestry System**

This complex agroforest is managed by the combination between cultural practices and respects the natural processes of vegetation production and reproduction. It represents a profitable production system and constitutes an efficient buffer between villages and forest. It is a good model of association between integration of forest resources and cultivation of cash crops/ vegetables etc. in the form of a sustainable and flexible system (Michon, *et al.* 1986).

Agroforestry systems mentioned above are based on the principle of full utilization of natural resources. However, research base for many agroforestry systems is either weak or nothing is practically known. Thus, there is ample scope for further research by strengthening existing research base.

## REFERENCES

King, K.F.S. (1979). Key note address to the National Seminar on Agroforestry, India, delivered on May, 16 (Un. Pub).

Michon, G., Mary, F. and Bompard, J (1986). Multistoried agroforestry garden system in West Sumatra Indonesia, *Agroforestry System* 4 : 315-338.

3

# *Gains of Agroforestry System : An Overview*

A combination of appropriately selected woody components and herbaceous crops essentially contributes to both productivity and sustainability of the farming system on marginal and submarginal lands by increased production of organic matter, maintaining soil fertility, reduction of water and wind erosion and creating a micro-climate favourable for associated crops and livestock. Further, integration of farmers into tree management processes has a salubrious effect in making the farmers conscious of the importance of the tree cover for the sustainability of the farm production. (see Fig. 13).

Appropriate agroforestry system have the potential to control erosion, maintain soil organic matter and physical properties, augment nitrogen fixation and promote efficient nutrient cycling (Young, 1989).

He has also suggested that 'appropriate' means that the systems are adopted to local environmental conditions, are well designed and properly managed, and meet the needs of the farmers. When agroforestry is successfully implemented, it can generate several positive environmental impact. In fact, it is difficult to discuss direct and indirect benefit of agroforestry within a few pages. The main benefits of appropriate agroforestry system are as follows :-

**Soil Conservation**

In general all good-land has to be protected from erosion for continued high productivity and economic stability of the country. In our country nearly 44 per cent of the total land area is subjected to severe erosion and it requires urgent soil and water conservation measures. Verma (1984) has reported that about 6000 million

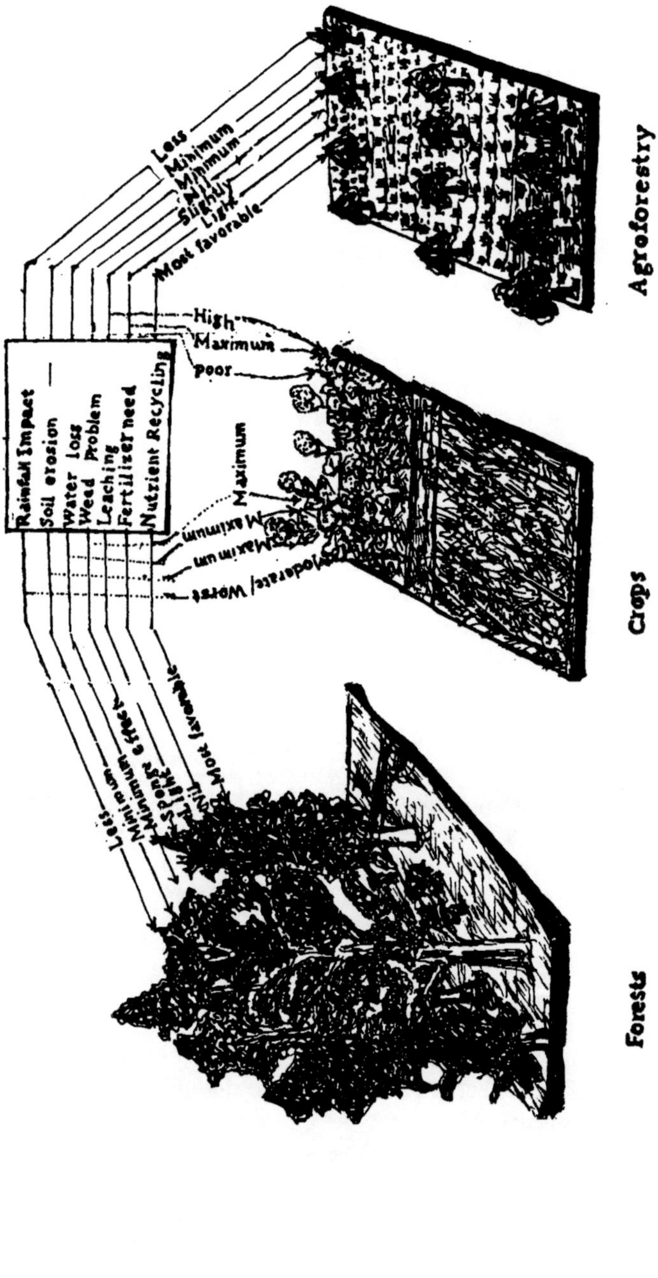


**Fig. 13 :** Gains of Agroforestry system (comparison)

tonnes of soil is lost every year because of erosion, depriving the country of about six million tonnes of vital plant nutrients. According to study made by CSWCRTI, Dehra Dun, annual average loss of top soil is approximately 16 tonnes/ha, far above the permissible limits of around 4 tonnes/ha. Soil conservation includes the application of the science of agronomy, engineering and forestry up to a stage that the land becomes in balance with the complexity of nature *i.e.* the loss of soil by erosion is less than the soil formed so that the land remains productive on a sustained basis.

Agricultural crops occupy the fields for a part of the year and during rest period the fields remain vacant. Strong summer winds blow away quite a substantial quantity of top soil, and another substantial portion of soil is carried away by the first rains. Agroforestry system plays an important role in protecting land from the process of erosion. This system may be considered as long lasting as well as recurring measure with a view to preventing the ultimate deterioration of the land, reducing the soil loss and maintaining the soil productivity on a sustained basis.

Tree species under agroforestry system prevent the soil from being blown away by reducing the velocity of wind at the ground surface. Young (1989) has stated that trees are capable of erosion control. This is achieved in two ways—with trees acting as barriers or as cover. The barrier function is the conventional approach to erosion control by checking run-off of water, and suspended sediment. The cover function involves reducing rain drop impact and run-off by increasing soil cover, with living or dead plant material spatially mixed agroforestry system, such as home gardens or plantation crop combination, achieve erosion control largely through the maintenance of a ground surface litter cover.

According to Reynolds *et al.* (1988), on slopping land, when leguminous trees species are planted along the contours, the trees limit soil erosion. Tree foliage (prunings) can be used as mulch and fertilizer for food crops, and/or for animal feed. Alley farming combines cropping and fallow period, separated under traditional farming system. Chatterjee and Maiti (1979) reviewed the role of grasses in soil conservation with particular reference to eastern India. Species composition of grasses and their management played an important role in controlling run-off, soil loss and water infiltration rates. Mishra *et al.* (1975) reported that with appropriate soil and water conservation measures the rate of sediment could be reduced from 80 tonnes/ha/annum to 6-7 tonnes/ha/annum within 5-6 years of treatment. Singh and Singh (1980) have pointed out that among all the farming system the highest amount of soil loss is observed in shifting cultivation plots and lowest recorded in the bamboo forest in Meghalaya. He has shown that erosion rates on 50-60 per cent slopes are 147 tonnes of soil loss per hectare under the first year of shifting cultivation, 170 tonnes/ha in the second year's shifting cultivation, and only 9 tonnes/ha in a natural bamboo forest.

The above problem can be tackled only by developing alternative models of agroforestry that would simulate the traditional practices. On sloping sites, trees and fillar crops (within rows of trees) have to be established on the contour line. If lands are with multiple slopes, most practical solution is to plant one row on the contour line and the remaining rows should be parallel to the first line. According to Rao and Roger (1990) contour planting alone may not solve erosion problem where land slopes steeply and rainfall is intense. Such sites may call for the construction of contour or graded bunds with grassed water ways, or some plots may require ridge and furrow or tied ridges.

It is apparent that soil erosion and water loss are the most critical phenomena throughout the nation. This requires management of the land surface and vegetation gradually so as to conserve more moisture and soil. It will be worthwhile to mention here that change of farming practices from traditional rotations to continuous row cropping, have accelerated soil erosion in recent years. This problem can be tackled only by developing appropriate agroforestry model for each agroclimatic zone. Thus, combined intervention of agronomy, forestry and engineering will minimise the soil erosion and help land remaining productive on a sustained basis.

**Land slides**

Only few reference on the possible role of agroforestry in minimization of land slides are available. The best protection against land slides is given by mixed forest and grasses. This requires careful selection of tree species, grasses etc. Besides this we should also consider the elevation, aspect of the site and nature of soil. We should develop economically feasible and environmentally sound agroforestry models. Farmers can plant timber tree species/fruit tree species at proper spacing and the space between the line of trees may be utilised by hedgerows as fillars. Between hedgerows farmers can grow a combination of permanent, semipermanent and annual crops. The deep rooted trees planted along contours, and hedges along the cover crops (permanent, semipermanent) help in soil stabilization and reduce land slide problems. Weeding requires much care which should be confined only around the plants until the hedgerows are established properly to hold the soil in place.

**Surface run-off and conservation of water**

High rainfall, rolling topography, heavy deforestation, misutilization of grazing lands and faulty methods of cultivation have contributed to heavy soil loss due to increase in surface run-off. Soil disturbances increase the surface run-off of atmospheric water. Surface and sheet erosion is the result of intensified surface run-off of water and in general it remains disturbed, uniformly on the gentle slopes.

Rill erosion is also induced by the surface run-off of water. This is pronounced on steeper slopes. Ultimately gullies develop.

According to Zonn (1986) deforestation intensifies silting of rivers, and increased silt carries away larger quantities of nutrients along with soil. The degree of increase in the run-off and the damage to soil are found to be drastically higher in ploughed land than in meadows and on steeper slopes than on more gentle ones.

If we consider the problems of North Eastern States, it is observed that due to wrong soil tillage practice in shifting cultivation areas, run-off and soil erosion increased beyond the limit of control. In these states luxuriant vegetation makes people to think that the soil is extremely rich. But this impression is misleading, because these soils are poor and heavy rain can destroy them within few months or few years (A.F.C.L., 1989).

The main reason of degradation of soils in Indian hills is the monoculture farming system. Farmer's never consider about the biological equilibrium and soils get finally degraded.

In the dryland agriculture, crops depend upon availability of reserve supply of soil moisture. Emphasis is given on the effective conservation and efficient use of rain water. This problem can best be tackled by the use of land according to capability and treatment according to needs.

To augment and conserve moisture, measures may be adopted to increase the infiltration rate, and decrease in evaporation from soil. Proper agroforestry models may have answer.

In the alley cropping system light showers are absorbed readily in the plough furrows and helps in moisture conservation. In case of heavy rainfall, this practice reduces the velocity of run-off and checks erosion. Soil is susceptible to erosion when fields are bare of vegetation. The cover crop of agroforestry system solves the problem by checking soil erosion and adding organic matter to the soil. The cover crops causes reduction in run-off and erosion and increased infiltration of water.

Agroforestry system keeps land almost continuously covered, thus improves the infiltration rate of soil, reduce soil erosion and build up the fertility status of soil. Besides the above, intercrops or tree foliage may be utilised as mulch. The mulch acts as a surface barrier, prevents evaporation, suppresses weed growth and decreases soil temparature. The best mulch will absorb little moisture and permit the rain water to move downward rapidly into the soil. This help in improving the soil structure and farmers can do this work themselves at a nominal cost.

**Nutrient status**

In traditional farming system farmer uses bush fallow period. In this system, soil nutrient status is gradually resorted during the fallow period.

According to ICRAF Annual report (1989) the short term fertility effects of trees are associated with decomposition of leaf litter, while long term effects are related to root decomposition and improvements in soil physical properties. In a tree/crop interface experiment at Machakos, *Cassia siamea* trees were grown in replicated plots for four years, while the adjoining plots were kept under crops. The same fertilizer regime was applied to all the plots throughout the period. When the trees were felled, all woody material were removed from the site, including the stumps. Maize was then planted over the whole area. Although, soil analysis on the two sets of plots did not reveal any consistent differences, subsequently a maize crop was considerably better on land previously under the trees.

In another study at Machakos, both annual crops and newly planted *Leucaena* hedges showed superior growth on land occupied during the previous four years by hedges of *Leucaena, Cassia* and *Gliricidia,* compared with adjacent areas previously under crops.

In these instances, the beneficial effects were of short term nature and could have been due to several possible causes. These might include the decomposition of leaf litter, fine tree roots, reduced export of nutrients during the period under trees or soil disturbance during tree removal.

According to Noordwijk and Dommergues (1990) root of nitrogen-fixing trees have more nodules, when nitrogen fixation takes place and when they are in close contact with roots of non-nitrogen fixing plants. This increased nodulation may lead to the direct transfer of nitrogen to the non-nodulatory plants. In southern China, abundant nodules were found under eucalyptus roots in a mixed eucalyptus—casuarina plantation (Dommergues, 1987). On close inspection, the nodules were found to be attached to the casurina roots.

Shannon, *et al.* (1990) reported that Alley cropping research in Zaire began in the early 1980s, but when the National Legume programme of Zaire began its work in 1985, no data was available to show whether alley cropping gives higher crop yields or maintains soil productivity better than traditional farming methods. A long term trial was started in January, 1986 at Gandajika station (7° s latitude, 780 m elevation), on a site which had previously contained Cassava. Rainfall average 1400 mm between the late August and mid-May, permitting 2 maize crops in a year. *Leucaena leucocephala* grows very well in these soils (pH 5 to 6) and was chosen as the hedgerows species. Treatment factors were presence or absence of fertilizer. Initially, recommended fertilizer at the rate of 64 kg N and 46 kg $P_2O_5$/ha were used to avoid nutrient imbalance, 30 kg $K_2O$/ha was added as potassium sulphate, because K-deficiency symptoms were observed on maize leaves. Alley cropping without fertilizer resulted in stable but low maize yields over the 3 and a half years of the trial yields, without alley cropping yield appeared to decline after the third season, even with fertilizer application. During the first two season, alley cropping did not benefit the maize crops because of insufficient prunings which were used for mulch. During the third season, the central rows of maize within the alley produced more than the respective rows in plots without hedgerows, but the gain was negated by poor yields in maize in the rows adjacent to hedgerows. From the fourth season onwards, maize with alley cropping yielded consistently and significantly higher than without alley cropping.

Fertilization increased yields in all seasons, both with an without alley cropping. Nevertheless, fertilizer alone was not able to halt the apparent decline in yield from the third season onwards in plots without alley cropping. The highest yield was obtained from the combination of alley cropping with a low rate of fertilizer application.

The result suggests that alley cropping alone maintain soil productivity, but integration of alley cropping with moderate fertilizer application may be the best solution to maintain yield stability and increased productivity for farmers in south-central-zone of Africa.

Konwar, *et al.* (1988) reported that in silvipasture system *Leucaena* seedlings were planted in contour furrows spaced 7.5 m apart, at an inter-row spacing of 2 m. The inter-row spaces were sown with two pasture species, *Cenchrus ciliaris* and *Stylosanthus hamata,* each on 50% of the land. The pasture was harvested every year, while *Leucaena* was allowed to grow as a tree for pole and wood. He also pointed out that soil fertility under silvipasture system showed progressive increase in organic carbon content of the soil. The rate of increase was more in case of stylo system than the cenchrus, which could be attributed to higher production in the farming system. There were slight decrease in P-content which might be due to continuous removal by the pastures, without fertilization. The K-content of the soil remained almost constant. Since these soils are rich in K, no visible K-depletion was expected.

In alley cropping system, hedgerows are cut and utilised as mulch or green manure. The nutrients are released from the prunings which are more readialy available to the crops when incorporated into the soil as green manure. The addition of organic matter helps in improving soil physical characteristics.

In alley cropping system, two alley widths, 3.6 m and 7.8 m (paired rows of 60 cm) were tested for two types of *Leucaena* lopping utilisation, lopping for mulching (to supply N—to arable crop) during crop season and for fodder/fuelwood purpose during off-season, and lopping for fodder purpose throughout the year. Four crops viz. Sorghum, Pearl millet, Caster and Pigeonpea were tested. A low cutting height of 10-15 cm was adopted for *Leucaena.* The arable crops received 30 kg $P_2O_5$/ha and 50% of the recommended N *i.e.* 20 kg/ha.

The yield of arable crop increased especially in 7.8 m alley width in the mulch system. The mulch system supplied—N to the arable crop and hence the yield increased (Konwar, *et al.* 1988). They have suggested that this system is quite useful when the farmers want to reduce the requirement of the organic fertilizer- N, by substituting the same with the organic- N, generated in situ by the alley cropping system.

The leguminous shrubs and trees have great influence in building up soil nitrogen, soil organic matter and some of them also help in improving other soil nutrients through leaf litters. The influence of different trees under silvipastoral system after 10-year plantation as compared to non-tree situation (open) on soil fertility built up and water retention properties showed a favourable trend (Hazra, 1990). This indicates a gain of available soil nitrogen with tree species in degraded soils after 10 years. The water retention properties and organic matter component of soil also greatly improved over the years with trees.

**Soil organic matter**

The strongest indirect evidence for the capacity of agroforestry systems to maintain soil organic matter is the high organic matter content of soils under natural forests. This is coupled with observation of the decline in soil organic matter when land is cleared for agriculture and subsequent recovery during the fallow period of shifting cultivation. It is possible to draw up balance sheet, comparing the probable rate of soil organic matter decline under crop production with the increases that can realistically be expected under agroforestry systems. This approach has been tested by computer modelling, using observed data on biomass and litter production. This effect is gradually reduced if prunings are removed for live stock feed, although the tree root residues continue to contribute to soil organic matter and there can, of course, be further benefit if farmyard manure is returned to arable land (Young, 1989).

Pathak and Gupta (1987) have reported organic matter addition through leaf litter in 2-year old *Leucaena* plantation to be 5.6 t/ha annually, which improved tilth, cation exchange capacity, water holding capacity, bulk density, brought down soil pH from alkaline to normal and improved the yield of successive crops.

**Soil Physical Properties**

Soil is three dimensional body, consisting of solid, liquid and gaseous phases. Under suitable environmental condition soil supports and nourishes plants. It is important to mention here that soil physical properties are not less important than nutrient levels for crop production. Now we have sufficient evidence that agroforestry systems maintain physical properties of soil. Agroforestry system also promotes increased activity of soil fauna.

The proper growth of roots and their functioning depends upon many factors like, soil environment, resistance to root extension, water supply, aeration, etc. If soil is fine textured root growth is restricted by deficient aeration. In dry condition, root extension stops due to deficiency of water, and finally absorbing capacity of roots reduced.

Above problems can be solved under agroforestry systems in which land is kept under regular cultural (ploughing, hoeing and weeding) operations, thus, maintaining proper level of oxygen and carbon dioxide in soil which also plays important role for the root growth. The trees are coppiced regularly, so that roots of trees die back, thus helping in improving physical properties of soil.

**Root Competition and Interaction**

Generally, growths of root system are restricted when grown in competitions with other systems. Shoot growths are less infleunced in comparison to root growth. Grass inhibits the root growth of trees by increasing carbon dioxide and depleting oxygen level in the soil (Howard, 1925). Main reason for root competition and interaction is allelopathic effect of tree root system, leaf system and competition for the nutrient, minerals and waters.

Competition at the root zone between trees and crops in an agroforestry system can be eliminated/minimised by selecting appropriate models. In India during last decade many agroforestry centres have conducted research at different agricultural zones to find out suitable management practices to provide a positive interaction between trees/shrubs and their associate food crops and live stock. Presently many models are already suggested by the scientists, which minimise the negative intraction between trees and crops (tree species are so chosen as to be complementary for understory production) and helps in increasing productivity of the field.

**Micro-meteorological parameters**

Micro meteorological parameters plays crucial role in deciding the suitability of particular agroforestry system in any agroclimatic zone. The most relevant parameters are interception of rainfall and radiation by canopy structure, relative humidity and soil temperature.

**Rainfall interception**

Ramkrishna (1984) has observed that in *Acacia tortilis* plantation of 13 and 7 years of age and in *Holoptelia integrifolia* plant cover (8 years old) at Jodhpur, the rainfall interception varied between 23 to 33 per cent and 14 to 19 per cent respectively under the two *Acacia tortilis* plantations while it was only 3 to 8 per cent under *Holoptelia integrifolia* plant cover under moderate to heavy rainfall condition (20 to 35 mm), which can be attributed to the canopy shape and density of plant cover in case of *Acacia tortilis*.

Under agroforestry system we need not worry much about the rainfall interception, because we select tree species having following characteristics : multipurpose, fast growing, deep rooted with narrow root zone, ability to grow back after cuttings, light canopy to allow sunlight penetration etc. The tree foliage are used as mulch, green manure, and or for animal feed. The selected tree species remain productive under regular pruning. The rainfall interception is less and this system allows improved crop production.

**Relative humidity**

Relative humidity also plays important role in growing agricultural crop beneath the tree crop. Ramkrishna and Sastri (1977) have observed that in *Acacia tortilis* (7 year old plantation) and guar crop at Jodhpur mean daily relative humidity values beneath the tree canopy during the active cropping season of guar was 7 per cent more than that in the open and we know that increase in R.H. decreases the evaporation as well as transpiration rate. By judicious management of tree canopy we can maintain relative humidity of the particular site.

**Temperature**

If crops are grown alone land remains fallow after harvesting. Temperature of exposed soil increases. This influences the growth of future crops. Under agroforestry system, trees maintain the soil temperature even during fallow period. Mann and Muthana (1984) have observed the micro climate under *A. tortilis.* The mean daily maximum soil temperature beneath these trees were lower by as much as 10 to 16 °C in the top soil zone (10 - 15 cm) and 4 to 5 °C upto 30 cm depth than those recorded in the open indicating better soil thermal regime under the tree plantations for the inter cropping system. Hazra and Patil (1986) have assessed the micrometeorological condition under silvipastoral system. The air temperature under the tree cover were lower (36.8 - 37.8 °C) as against non-tree open cover (38.4 °C). The leaf temperature of grasses under different tree covers were 31.5 °C to 32.2 °C, whereas it was 33.8 °C in open.

The leaf temperature is believed to be a good indicator of the relative photorespiratory process which would ultimately be the determinant in net carbohydrate accumulation in plants (Dutta *et al.,* 1980). The higher air-leaf temperature difference is also an indication of higher net photosynthesis as it is resultant of lower canopy temperature and consequently lower photo-respiratory processes. The lower radiant energy availability under the trees reduces the air temperature and also the underneath grass temperature.

Hazra and Patil have stated that the biomass production is a physiological process. It is largely controlled by the micro-meteorological parameters particularly radiation availability and to some extent by air temperature difference and relative humidity.

**Radiant Energy**

Productivity under agroforestry system is also influenced by total energy received on the canopy and the radiation available to the intercrops. The tree species moderate the radiation infiltration quite substantially and infiltrating radiation level is greately influenced by the tree species and the type of canopy structure and coverage by the trees (Hazra, 1990). Mann and Muthana (CAZRI Monograph series-23) stated that net radiant energy recorded underneath the *Acacia tortilis* tree cover during August revealed that it remained very poor from 14 to 30 and 15 to 24 per cent, respectively, of the energy received in the open area. Similar trend was also recorded during September. The poor germination and ultimate failure of *Cenchrus ciliaris* grass establishment underneath the tree cover in spite of abundant soil moisture availability during kharif season are attributable to poor radiant energy availability.

It is apparent that radiant energy has influence over the productivity under agroforestry system. This problem is tackled by canopy manipulation through lopping, pruning, etc.

**Pest and Disease Suppression**

Insects and diseases play a very important role in the food grain productivity. The pest and disease situation varies with the nature of the cropping system. Monoculture plantation favours the multiplication of pests/disease causing organism. Agroforestry system reduce the incidence of major pests and diseases due to diversity in the plant species. Singh, P. and Singh, S. (1987) have opined that ecosystems in agroforestry systems are so variable that management practices will need to be oriented depending upon the local conditions, pollution hazards, economical considerations, available expertise etc. Pests and diseases can also be best managed by selection of entomologically and pathologically desirable mixture of trees and agricultural crops. Under agroforestry system their is minimum chance of pest outbreak due to crop diversity.

### Generation of Steady Income

Agroforestry is increasingly accepted and promoted as an alternate approach to land use which generates steady income for subsistence farmers.

Large areas of uplands which are only partially cultivated can be made to yield high return per unit area under a combination of suitable agricultural crops legume/fodder with tree crops along with suitable water management technique. Muhammad (1984) has stated that agroforestry opens a big scope for not only increasing production of wood (Long term) but also provides increased production of cereals, fodder, oil seeds, legumes, vegetables etc. (short term), and generate steady income for the farmers.

### Socio-Economic Condition

On account of progressive increase of population of man-kinds and his domesticated ruminants, demand for cereals, vegetables, fruits, milk, meat, fuelwood, timber has increased enormously. Present system of land use cannot fulfil these needs. This has necessitated adoption of integrated land use system to meet the growing needs. Proper introduction of agroforestry practices may sustain the crop productivity of a given land. This programmes is labour intensive, and creates new employment opportunities among the agricultural labour and small/marginal cultivators. Their income may be largely raised by this programme and thus it can help in the formation of harmonious social system in the village.

### Environmental Conservation

Agroforestry is closely linked with environmental amelioration and improved quality of life through socio-economic upliftment. The relationship between ecosystem, agroforestry and environment are so intimately interlinked as to preclude a simplistic understanding. To take the environment alone, the agroforestry when extensively and successfully implemented can generate several positive environmental impacts like improvement in hydrological balance, improvement of physical properties of soil, favourable to improved infiltration, retention capacity and indepth percolation of surface run-off and reduction of sedimentation of rivers, streams etc., recycling of nutrients, creation of favourable microclimate condition conducive to higher food production, maintaining balance in oxygen-carbon dioxide, atmospheric temperature and relative humidity.

## REFERENCES

A.F.C.L. (1989) *Status report on Dhaleswari Catchment,* Mizoram, sponsored by Department of Agriculture, Govt. of Mizoram.

Bredero, F. (1987) *Vetiver grass, method of vegetative soil and moisture conservation*, World Bank Pub., New Delhi, pp. 1-69.

Chatterjee, B.N. & Maiti, S. (1974) Role of grasses in soil conservation in eastern India, *Soil Conservation Digest,* Vol. 2(1), pp. 15-23.

Dommergues, Y.R. (1987) The role of biological nitrogen fixation in agroforestry, In : *Agroforestry : a decade of development,* Steppler H.A. and Nair, P.K.R. Eds., Nairobi, ICRAF, pp. 245-71.

Dutta, T.R., Patil, B.D. and Hazra, C.R. (1980) Environmental analysis on dryland crops of sorghum and cowpea (Pure and mixed), *Annals of arid zone,* Vol. 19(4), pp. 433-435.

Greenfield, J.G. (1987) *Vetiger grass. A method of vegetative soil and moisture conservation,* Fran Bredera, Eds., P.S. Press Service Pvt., Ltd., New Delhi, pp. 2-4 (C.R)

Hazra, C.R. and Patil, B.D. (1986) Forage production under silvipastoral system, light and temperature relation, *Ind. J. Range, Manage,* Vol. 7(1), pp. 33-36.

Hazra, C.R. (1990) Soil and climatological studies in relation to agroforestry research, *1st NARP. Training Programme,* organised by IGFRI, Jhanshi.

Howard, A. (1925). The origin of alkali land, *Agril. J. India,* 20 : 461-629 (C.R.)

I.C.R.A.F. (1989) Annual Report, Narobi, Kenya, pp. 1-30.

Konwar, G.R. (1988) Annual Report, CRIDA (1987-88), pp. 80-81.

Konwar, G.R. and Modh, Osman and Singh, R.P. (1988) Fodder and fuel wood production systems for sandy loam soils of semi arid Hyderabad region, In : *Promotion of fodder and fuel-wood trees.* Hedge N.G., Relwani L.L. and Kelkar V.D. Eds., B.A.I.F., Dev. Res. found ; Pune, pp. 78-79.

Mann, H.S. and Muthana, K.D. (1984) *Arid zone forestry,* ICAR, CAZRI, Monograph No-23 Jodhpur.

Mishra, R.R., Kaushal, R.C., Dyal, S.K.N. and Prabhu Shankar (1975) *Ann. Progress Report,* C.S.W.C.R.T.I., Dehra Dun, pp. 1-70.

Muhammad, S. (1984) Agroforestry practices, In : *Social Forestry, Training-cum-Workshop Proceeding,* Muhammad, S., Eds., B.A.U., pp. 45-55.

Noordwijk, M.V. and Dommergues, Y.R. (1990) Agroforestry and soil fertility. Root nodulation, The twelfth hypothesis, *Agroforestry Today,* Vol. 2(2), pp. 9-10.

Pathak, P.S. and Gupta, S.K. (1987) Soil amelioration role of *Leucaena* in the agroforestry system, *Proc. Int. workshop on Agroforestry for rural needs,* New Delhi.

Ramkrishna, Y.S. and Sastri, A.S. (1977) Microclimate under *Acacia tortilis* plantation, *Annual Progress Report,* CAZRI, Jodhpur, pp. 69-70.

Ramkrishna, Y.S. (1984) *Forest microclimate and its Importance in Agroforestry,* Shankarnarayan, K.A. Eds., ICAR CAZRI, Jodhpur, India, pp. 9-12.

Rao, M.R. and James H. Roger (1990) Agroforestry field experiments discovery the hard facts, Part 2, *Agronomic consideration,* Vol. 2(2), p. 11.

Reynolds, L ; Atta-Krah, A.N. and Francis, P.A. (1988) *Alley Farming with Livestock Guidelines,* Humid Zone Research Site, International live stock Centre for Africa, Nigeria, pp. 1-30.

Shannon, D.A., Kabaluapa, K.N. and Ngoyi, M.L. (1990) IITA and National legume programme of Zaire, AFNETA, Vol, 2, No. 1, pp. 8-9.

Singh, A. and Singh, M. (1980) Effect of various stages of shifting cultivation on soil erosion from steep hill slopes, *Indian Forester,* Vol. 16(2), pp. 116-121.

Singh, P. and Singh, S. (1987) Pest and Pathogen management in Agroforestry system, In : *Agroforestry for Rural Needs,* Khosla and Khurana, eds., Vol. 1 IUFRO workshop proceeding, pp. 153-177.

Verma, U.K. (1984) Soil conservation and its importance, In *Dry Land Farming and Water Management,* Mohsin, M.A. and Prasad, R.B., Eds., workshop proceedings, B.A.U. Ranchi, Bihar, pp. 1-205.

Young, A. (1989) Hypothesis for soil agroforestry research, *Agroforestry Today,* Vol. I, No. I, pp. 13-16.

Zonn, S.V. (1986) *Tropical and Subtropical Soil Science,* Mir Publishers, Moscow, pp. 1-422.

4

# Tree Management in Agroforestry System

The Management of trees in agroforestry system includes all operations that are required to get desired product from the unit area. The desired product to be produced depends on the policy and the objectives of management. Decision on policy needs an assessment of the requirements and needs of the participants in relation to the available physical resources, biophysical causes, labour and capital etc. After an assessment of available resources, extension workers/scientists/planners should guide the participants, how available resources can best be exploited to get desired product (design, silvicultural and agronomic management etc.). In agroforestry systems, management of tree species depends upon the choice of species, biophysical causes of land, labour and the capital. It is apparent that different tree species and even the same species in different agroclimatic conditions grow at differential rate. Besides the above, management of trees in agroforestry systems requires an understanding of the various factors that influence the growth of crops along with particular tree species. Knowledge of above factors enables to choose appropriate silvicultural and agronomic management to get desired produce on a sustainable basis from year to year. Thus, to manage trees in the agroforestry systems, farmers should have an idea about

(*i*) the choice of species and provenance, to get desired produce,
(*ii*) proper manipulation of canopy and stands in order to facilitate maximum penetration of sun rays,
(*iii*) nutrient recycling,
(*iv*) suitable harvesting methods, which permit the trees to regenerate through sprouting, and
(*v*) protection before harvesting and after harvesting.

Agroforestry systems aim regular continuous yields, from the same unit of land. Sustained yield of timber and fuel wood can be obtained in the following ways :

(*a*) If only one tree species is planted, all trees can be harvested at one time, *e.g.* Once every 8 to 10 years and then regenerated by coppicing, root suckers, etc.

(*b*) Intermediate yield from the trees may be harvested, if more than one species are planted, and farmers can also collect fire-wood annually, and fodder and green manure seasonally.

Farmers always want steady increase in yield from the same unit of land. In order to increase the yield trees must be managed to get desired produce. Thus, regularity and progressive increase in the yield can be possible by judicious manipulation as per objectives of management.

The management techniques enable a sustained yield of wood, firewood, fodder, etc. from trees over a long period of time which can be considerably greater than its-volume if it is simply allowed to grow and then felled (Foley and Bernard, 1984). Thus appropriate management of trees help optimal use of limited land by raising more economic valued trees and crops, sustained production of desired qualities and quantities of timber, Fuelwood etc., coupled with various environmental benefits. In this way this system contributes regular income to the participants.

The silvicultural manipulation depends upon the following condition :–

* Silvicultural requirements of the species in agroforestry system for sustained yield and renewal of tree crop in a given set of conditions.
* Characters of crops grown in the interspace.
* Demand of tree produce such as green manure/fodder, fuelwood, timber, etc.

Coppicing, pollarding, lopping, pruning and thinning are specialized technique for achieving the aims of agroforestry system.

**Stool Coppice - Management**

This is the method of producing coppice shoots growing from a stool or a living stump. In coppicing, whole tree is cut-off close to ground level. Shoots sprout from the residual stump technically referred to as the stool. Removal of the trees leading shoots results in hormonal suppression of lateral growth, which triggers growth of dormant buds on the side of the trunk. As stool diameter increases, the number of dominant bud shoots may increase 30 to 60 per cent faster than the number of callus shoots. Coppice growth is generally much more vigorous than normal seedlings growth and is characterised by both an upsurge of vertical development and retardation of lateral branching (Powlick, 1989).

Most broad leaved species when felled have the power to produce coppice shoots from the stump. The coppicing power, however, varies from species to species. The size of the trees is also important as the power of coppicing generally declines as the trees grows older. Shoots from the big trees tend to be weak and short

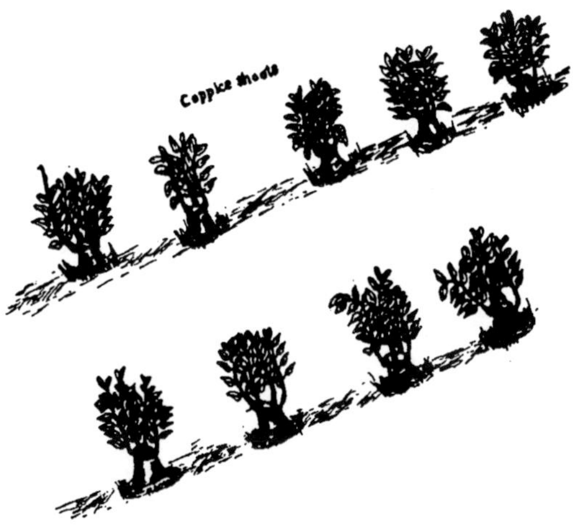


Fig. 14 : Coppicing

lived. Coppice shoots mostly depend on the root system of the parent tree, but some time they develop their own roots as the old roots rot away (Sharma, 1959). Management of forest by coppice system is very common in Indian forestry. *Eucalyptus globulus* in Tamil Nadu, tropical dry deciduous red sander (*Pterocarpus santalinus*) forest of Andhra Pradesh, sal forest of Gorakhpur, Kheri division of Uttar Pradesh, dry peninsular sal in parts of Madhya Pradesh, Bihar, Orissa etc., are managed under simple coppice, or coppice of two rotation system or the shelter wood coppice system. In agroforestry system, management of trees by coppice has been studied by many workers (Khanna, 1953; Basappa, 1986 and Bhimaya, *et al*; 1965).

Coppicing ability of *Prosopis juliflora* has been studied by Sugur (1988). Cutting *Prosopis* at 90 days interval yielded on an average 1.97 t/ha fuel wood for a corrected stand density of about 4500 plants/ha. Cutting at six months interval yielded 4.94 t/ha and a MAI of 9.88 t/ha. The increased total production could be due to bigger size and high density branch materials. The MAI slightly declined for the cutting at 270 days interval (8.73 t/ha/year). This could be covering up of trees due to faster growth and subsequent competition for nutrients and moisture (Sugur, 1988).

The coppicing power of babul varies greatly. Throughout the greater part of its habitat it coppices poorly, certain important exceptions have, however, been recorded where babul in Anantpur, Bellary, Chingleput, and Guntur districts in Southern Indian and under favourable conditions in ravine areas of Doab districts in U.P. has been regularly worked under coppice system for some years. In these localities, a large proportion of stools (90 to 120 cm in girth) produce shoots but when the stools are submerged in water for 3 to 4 months, they die. Coppice can be relied on to a fair extent in these parts provided the trees felled are not of large size. The precise reason why babul coppices in some localities and not in others is not satisfactorily understood. However, babul usually pollards well. In *Cassia fistula* the rate of growth of coppice-shoots is fairly rapid in the early years, but it slows down later. Measurement of one year old coppice shoots in Bhandara (Maharashtra) showed an average height of 1.9 m as against 2.2 m, for teak. Measurement recorded in nine years old coppice coupe at Bullawala (Dehra Dun), showed a mean girth of 19.6 cm and a mean height of 3.6 m of cassia as compared to 22.4 cm and 4.9 m respectively for sal (Annon, 1983). *Hardwickia binata* pollards well but its coppicing power varies a great deal. Its coppicing should in any case not be done at height lower than about 30 cm.

*Objectives of stool coppicing*

(*a*) To harvest fuelwood, poles and small sized timber within a short period.

(*b*) Due to adverse climatic and edaphic conditions large sized timber cannot be produced.

(*c*) Farmers need to invest less and get quick return on sustainable basis.

*Advantages*

* Regeneration from coppice shoot is more certain than regeneration by seeds.
* Growth of coppice shoots is faster than the seedlings. Mean annual increment of the coppice shoot is much higher than the shoot established by natural regeneration or by planting.
* Need to invest less and farmer gets higher income.

*Disadvantages*

* In agroforestry system, we prefer FGMPTs. These requires more nutrient for vigorous growth.
* After some years stool dies and cannot coppice. The gap has to be filled by the pot plants or by sowing seeds.
* Coppice shoot is often broken by wind from the place of attachment with stools.

*Points to keep in mind during coppicing*

* Trees should be coppiced only when attains good health and shade effect is too much on the inter-crops.
* Trees should be coppiced when it attains 3 to 5 meter height.
* Trees should not coppiced in dry period. In dry fallow period shade of trees help in moisture conservation.

* Coppicing should be done one week before the onset of the rains. This will prevent fungus attack and availability of moisture facilitates the growth of new shoots.
* Frequency of subsequent coppicing after the first cut should be decided after observing the health of the trees, crops likely to be grown in the field and use of foliage of the trees. For example, if farmers intend to use tree foliage as green manure, the coppicing should be done before ploughing the plot.
* The optimum cutting height for most of the coppicing trees should be kept in between 30 to 50 cm above the ground level. When cutting height is kept below 30 cm, poor shoot growth, is the result in most of the tree species.
* Initial cut must be angled, to permit rain water to run-off from the cut surface easily.
* In new coppiced tree many new shoots appears. If farmers want to collect fuelwood, then maximum six shoots may be retained. If farmers are interested in fodder then more shoots can be permitted to grow. If desired product is poles, only 2-3 straight shoots can be left.
* In case of poor shoot growth, mature trees should be harvested and replanted by new seedlings.

**Pollarding**

Pollarding is a process in which crown of the trees are harvested at least 2 meter above the ground level, *i.e.* beyond the reach of cattles in order to get flush of new shoots. In agroforestry systems mulberry and arjun (*Terminelia arjuna*) are often planted. Mulberry as well as arjun trees are managed by pollarding. Farmers generally start rearing silkworm after 3rd year. After completion of rearing, trees are pollarded to get flush of new leaves for rearing of next silk crops. One family can sustain on the rearing of mulberry silk worm on 1000 trees during the year. It has been assumed that one family will get work in making silk cloth out of the raw silk produced by 10 families (Tiwari 1983). In Chotanagpur plateau of Bihar farmers use pollarding as a management technique for *Teminalia arjuna*, tasar silk worm host plant under agroforestry system. At Ranchi it was observed that the pollarding of *Leucaena leucocephala* around crop fields in two rows provide highest annual forage yield (Sen, 1984). Live fences and border plantation of *Gliricidia sepium* are pollarded for the production of fodder, green manure and fuelwood (Wiersum and Dirjosoemarto, 1987). In Costa Rica production of *Erythrina* was highest when pollarded thrice a year for fodder (Russo and Budowski, 1986). Perennial pigeonpea (*Cajanus cajan*) are planted in alley cropping system and hedge is pollarded after maturity of grain. Leaves are used as fodder or as green manure and stems are used as fuelwood or fencing materials.

**Fig. 15 :** Pollarding

## Pruning

Pruning is cutting of branches or even young stem. This operation helps in minimizing shade effect on intercrops and maintain the quality of timber, fruits etc. Many workers Dijkman (1950), Das and Dalvi (1981), Osman (1981), Pathak, *et al.* (1981),

Fig. 16 : Pruning

Pathak and Patil (1982), Dugma, *et al.* (1988) have reported effect of height and intensity of pruning on the biomass production.

The pruning of the fruit bearing plant of *Emblica officinalis* is done after the cessation of the fruiting each year, while pruning of dead, diseases, broken, crossing branches are done regularly and suckers appearing from root stock are removed (Bajpai & Shukla, 1985). In Chotanagpur plateau of Bihar, eastern districts of Madhya Pradesh, parts of West Bengal and Uttar Pradesh, farmers manages lac host plant in their farm land and crops are grown in the interspace. Lac host plants are managed by pruning. According to Kumar and Mishra (1989) availability of succulent shoots of proper age is essential for the successful growing colonisation, vigorous growth and satisfactory survival of the lac insect. Pruning when properly done well before the onset of new flush of the host-plant, induces large number of suitable shoots and thus ensures efficient utilisation of insect input. Medium pruning for removing only the lac bearing or bearable portions along with the dead, dried or diseased portions is recommended. Heavy pruning results in stunted growth and reduction in the canopy size of the tree. Pruning time coincides with the harvesting time in the case of *kusum* and therefore, careful harvesting serves as pruning also and no separate operation is needed. In case of *palas* no separate pruning is required when *baisakhi* crop is harvested but for raising *katki* crop pruning is recommended during February - March.

The avacado (*Persea americana*) trees are pruned sparingly, mainly by heading back the central shoot in upright growing varieties, such as pollock, to develop a spreading habit (Muthukrishnan and Khader 1985). Yamdagni (1985) has pointed out that pruning the Ber trees is highly desirable, not only during its early years to build a strong framework but also in later years to obtain a profitable crop. The ber fruit is borne in the axil of leaves on the young, growing shoots of the current year. Hence a regular annual pruning is necessary to induce a good healthy growth which will provide maximum fruit bearing area on the tree. The best time of pruning is the hot, dry season when it sheds leaves and goes to rest. *Leucaena* hedgerows thrive well even with repeated five prunings annually for six consecutive years (Kang *et al.*, 1985). A field trials on effects of pruning intensities carried out on an oxic paleustalf in the humid zone of south western Nigeria with

*Leucaena leucocephala*, *Gliricidia sepium* and *Sesbania grandiflora* alley cropped with maize and cowpea show that *Leucaena, Gliricidia* and *Sesbania* are capable of tolerating multiple pruning (*Leucaena* twenty pruning combinations consisting of five pruning heights (25, 50, 75, 100 and 150 cm) and four pruning frequencies (monthly, bi-tri and six-monthly, *Gliricidia* and *Sesbania* hedgerows to nine pruning combinations consisting of three pruning heights (25, 50, and 100 cm) and three pruning intensities (monthly, tri and six monthly) (Duguma *et al.*, 1988).

In above woody species, biomass, drywood and nitrogen yield from the hedgerow pruning increased with decreasing pruning frequency and increasing pruning height. Biomass, drywood and nitrogen yields were in the following order *Leucaena* > *Gliricidia* > *Sesbania* (Duguma *et al.*, 1988). The multiple prunings had no ill effect on *Leucaena* plants. Pruning frequency had a greater effect than pruning height on survival of *Gliricidia* and *Sesbania* plants. When monthly pruned, about 25 per cent of the *Gliricidia* and all of the *Sesbania* plants died within six months. Even with lower pruning frequency *Sesbania* plants showed lower survival rates than *Gliricidia* or *Leucaena leucocepahala.*

The various pruning intensities of all the hedgerow species had more pronounced effects on the grain yield of the alley cropped cowpea than on maize grain yield. Higher maize and cowpea yields were obtained with increasing pruning frequency or decreasing pruning height (Duguma *et al.*, 1988).

***Points to keep in mind during pruning***

* The cut must be angled, to permit the rain water to run off above the cut surface. This reduce the incidence of fungal attack during rainy season.
* The branches should be cut 4-5 cm above the main stumps.
* Pruning must be started from the below first and then gradually proceed to top branches.
* The foliage obtained after pruning must be incorporated immediately into the soil, in order to get maximum nutrient release from the green manure. Pruned branches can be left in the field, until the leaves fall off. The dried

leaves will fall off within two weeks, Branches may be used as fuel wood. This requires less labour.

* Pruning must be done well in advance before sowing or planting the intercrops.

**Thinning**

Thinning comprises of felling or removal of a immature stands, so as to improve the growth and form of the trees that remain without making a permanent opening of the canopy (Negi, 1988). Thinning is an important silvicultural operation in traditional forestry. In agroforestry system thinning operation is carried out with the following objectives :–

* To provide suitable light condition to the crop.
* To eliminate the excess root competition between the trees and crops.
* To get earlier returns. If two or more tree species are planted, species having shorter rotation period is harvested earlier by thinning process. This also favours growth of selected timber/fruit tree species. In this way thinning helps farmers in getting early return before harvesting of main species. FGMPTs are planted along with fruit trees or timber tree species. FGMPTs are harvested early so that fruit trees or main timber producing tree species get more nutrient. Crops productivity also increases after thinning operation. Small size poles and fuelwood provide additional returns to the farmers.
* To bring down the number of coppice shoots to two per stool.
* To remove the dead, diseased and dying trees.

**Research Needs**

Many tree species are recommended for agroforestry system in different agroclimatic conditions. To improve the production potential in agroforestry systems research should be conducted in these aspects:

1. To study the coppicing power of indigenous and exotic multipurpose trees in different agroclimatic zones.
2. To study the optimum cutting height, best season, etc. for coppice operation.
3. To study proper age, height, season to get maximum foliage after pollarding.
4. To study the performance of green manure, leaf mulch in different season.

## REFERENCES

Annon, (1983). Silviculture of Indian trees (Vol.IV), Joshi, H.B. Loganey, R.N., Panaik, L.K., Chaudhari, D.C. and Chauhan R.J., (Eds.) *Controller of Publication,* Delhi, pp. 59, 197 & 214.

Bajpai, P.N. and Shukla, H.S. (1985). AONALA. In *Fruits of Indian Tropical and Sub-tropical,* Bose, T.K., Ed. Naya Prakash Publication, Calcutta, pp. 591-601.

Basappa, B. (1986). Coppicing of silver oak (*Grevillea robusta*), *My Forest*, Vol. 22(1), pp. 1-2.

Bhimaya, C.P., Kaul. R.N. and Ganguli, B.N. (1965). Studies on pre-sproutal stumps of *Prosopis juliflora, Annals of Arid Zone*, Jodhpur, Vo. 4(1) (C.R.)

Das, R.B. and Dalvi, G.S. (1981). Effect of interval and intensity of cutting of *Leucaena leucocephala, Leucaena Research Report,* Vol.2, pp.21-22.

Dijkman, M.J. (1950). *Leucaena* - a promising soil erosion control plant, *Econ. Bot.*,49, pp. 337-349.

Duguma, B., Kang, B.T. and Okali, D.V.V. (1988). Effect of pruning intensities of three woody leguminous species grown in alley cropping with maize and cowpea on an alfisol, *Agroforestry System,* Vol. 6, pp. 19-35.

Foley, G. and Barnard, G. (1984). Farm and Community forestry. *Tech. Rep.*, 3. Earthscan, London.

Kang, B.T.; Grimme, H. and Lawson, T.L. (1985). Alley cropping a sequential cropped maize and cowpea and *Leucaena* on sandy soil in southern Nigeria, *Plant and Soil,* 85, pp. 267-277.

Kang, B.T. and Duguma, B. (1985). Nitrogen management in alley cropping system. In : *Nitrogen management in farming system in humid and sub-humid tropics* Kang. B.T. and Vander Heide, J., Eds. Inst. for soil fertility (IB) Haren, Netherland, pp. 269-284 (C.R.).

Khanna, L.S. (1983). Khair coppice, *Indian Forester,* Vol. 79 (3), Dehra Dun.

Kumar, S. and Mishra, S.C. (1989). Lac-A lesser known cash crop, In: Grow more food, *Souvenir,* Ramkrishna Mission Ashram, Ranchi.

Muthukrishnan, C.R. and Abdul Khader, J.B.N. (1985) AVOCADO, In: *Fruits of India-tropical and sub-tropical,* Bose, T.K., ed. Naya Prakash Calcutta, pp. 465-478.

Negi, S.S. (1988). *Elements of General Silviculture,* Int. Book Distributor, Dehra Dun, pp. 1-316.

Osman, A.M. (1981). Effect of cutting intervals on relative dry matter production of four cultivars of *Leucaena, Leucaena Research Report.* Vol. 2, pp. 33-35.

Osmaston, L.S. (1909). Silvicultural notes on *Hardwickia binata, Indian Forester*, 35(6 & 7), Dehra Dun (C.R).

Pathak, P.S. and Patil, B.D. (1982). *Leucaena* research at the Indian Grass land Fodder Research Institute. In : *Leucaena Research in the Asian Pacific Region. I D R C,* Ottawa, Canada, pp. 83-88.

Pathak, P.S; Gupta, S.K. and Deb Roy, R. (1981). Production of aerial biomass in *Leucaena leucocephala, Indian Forester*, Vol.107 (7), p. 416.

Powlick. T. (1989) Coppice with care, *Agroforestry Today,* Vol. I No. 3, pp. 15-18

Rao, D.L.N. and Gill, H.S, (1985). Evaluation of perennial *sesbania* species for firewood production in Alkali Soil. In *Promotion of Fodder and Fuelwood Trees.* N.G. Hedge, Rewani, L.L. and Kelkar, V.D. (eds.), B.A.I.F. Dev Research Found, Pune, pp. 1-213.

Russo, R.O. and Budowski, G. (1986). Effect of pollarding frequency on biomass of *Erythrina poeppigiana* as a coffee shade tree, *Agroforestry System,* 4(2), (C.R).

Sen, P.P. (1984). Potential of planting *Leucaena leucocephala* on the *marginal land* at Titta village, Ranchi, *Annual Extension Report,* Faculty of Forestry, B.A.U. Ranchi, pp. 1-8.

Sharma, Y.M.L. (1959). *Elementary lessons in Forestry,* For Agricultural extension worker in India, F.R.I., Dehra Dun.

Sugur, G.V. (1988). Coppicing ability of natural stand of *Prosopis juliflora, My Forest,* Vol. 24(2). pp. 129-132.

Tiwari, K.M. (1983). *Social Forestry for Rural Development,* International Book Distributor, Dehra Dun, pp. 1-115.

Wiersum, F. and Dirjosoemarto, S. (1987). Past and Current Research with *Gliricidia* in Asia. In: *Gliricidia sepium Management and Improvement.* N.F.T.A., Hawai.

Yamdagni, R. (1985). BER. In: *Fruits of India Tropical and Sub-tropical,* Bose, T.K., (Ed.), Naya Prakash, Calcutta, pp. 528-536.

5

# *Crop Management in Agroforestry System*

To increase productivity on sustained basis, first step is inventory of soil and water resources. Second step is identification of biophysical causes of land and finally we decide cropping system which can provide greater stability and sustainability.

In any cropping system management depends upon biophysical causes of land, choice of species, labour and capital. Like pure crop management, agroforestry system also requires an understanding of the various factor that influence the growth of crops along with particular tree species. Knowledge of above helps in the management of crops and tree species to get desired produce on sustained basis with minimum deterioration of land.

To manage tree species, farmers should have an idea about the choice of species and provenance, proper manipulation techniques of stand in order to facilitate maximum penetration of sun rays, nutrient recycling, time of harvesting, etc.. Coppicing pollarding, lopping, pruning, thinning, etc. are accepted techniques of tree management. In agroforestry system knowledge of appropriate management technique of crops (grown in the interspace of trees) and tree species helps in optimal use of land resource. In agroforestry system, crop management depends upon :

* Characteristics of crop/vegetable/fruit/grass.
* Nutrients requirement.
* Water, Light and other requirements.
* In tree farming system or in forest tract suitable site, suitable soil, laying out, tillage, manuring has less significance where as in agroforestry system these plays important role to get desired produce.

**Suitable site/soil**

In agroforestry system, site plays an important role in selection of tree species as well as crop/vegetable/grass, etc.. Site indicates characteristic feature, limitation and treatment required. If in any watershed area we have land having slope 5 to 45°; land slope above than 25-30° should not be used for agriculture practice. These areas are only suitable for tree species. Same time land having slope below 5° are normally used for agricultural practice. Slope 5 to 10° or 10 to 20° may be considered suitable for agroforestry practice. If we feel that land is suitable for agroforestry practice, again we have to identify tree species and crops on the basis of species suitability, site capability, etc. If farmers decide to adopt silvihorticulture system, he has to borne in mind that the land should be high, well drained and spacing between crops should be such that penetration of light is not hampered. The management of silvihorticulture system will differ from other types of agroforestry system. A good supply of water is the chief consideration for agrisilviculture system. This is not so essential for silvipastoral system.

In plant growth, the soil plays important roles, namely (*i*) it serves as the anchorage, (*ii*) it provides a suitable environment for the growth of the plant, and (*iii*) it is a store house of plant food. Below the top 15 to 30 cm of surface soil lies subsoil. It represents more nearly the original soil since it is unattached by cultivation or manuring of the surface soil. It is much poorers in plant nutrient than the surface soil, and there is no advantage, but distinct disadvantage in bringing it to the surface. In such a condition, the subsoil may be improved by growing deep rooting crops which open the subsoil, increase its capacity to store organic matter and effect a certain transference of plant—food to the surface (Mitra Bahadur, 1981). In agroforestry system deep rooted tree species plays important role. Due to nutrient pumping action, trees carry nutrients from the lower most strata and make it available to the agricultural crops. In this system top 45 to 60 cm of soil should be well dug and cleaned and manure should be used in proper time as practised in pure cropping system.

**Tillage**

Quantity, quality and frequency of economic output from any piece of land are conditioned in a large measure by the nature and characteristics of the soil in the upper layer, suitable depth of which takes about hundreds years or more for its natural formation. The loss of the valuable topsoils results in a deterioration of the quality of land so much so that its use pattern has to undergo a complete change for the poorer (Sinha, 1984). The proper tillage is essential for obtaining suitable increase in the productivity of land.

In the word of Mitra Bahadur the word 'tillage' is applied to the different method of working the soil in order to secure the conditions needed for healthy growth of crops. By proper stirring up the surface soil which is intermixed with the fresh air, many chemical change takes place in that part of the soil and much of the plant—food is released. In fact, an adequate tillage is necessary to enable a soil to function properly. Generally speaking, the chief objective which tillage aims to secure are :—

1. To develop uniformity of soil conditions suitable for sowing of seeds and setting of plants.
2. To hasten and augment physical and chemical action in the soil which are conducive to plant growth.
3. To aid in the saving of soil moisture and to provide necessary action to retain the same.
4. To destroy and prevent the growth of undesirable weeds or other vegetation, and to destroy insects with their nests and other parasites which are harmful to crops.
5. To remove stubbles or other residues of the previous crops or incorporate these into the soil.

Great care should, therefore, be attached to tillage while preparing the soil in all stages and even while the crops are growing. It should always be remembered that the plot should be continuously hoed to keep the plot clean of weeds and free of insects and to aerate the soil. Proper soil working will minimise the application of manure and water.

It is generally observed that a hard pan is formed at the same depth due to continuous tillage. Sometimes due to translocation of finner soil particles and their deposition at certain depth below the soil surface a hard impervious layer is formed. Percolation of water in such soil is poor. Deep tillage helps to break such formation which ultimately increases the infiltration rates, and reduce the total volume of run-off and erosion of soil (Sinha, 1984).

**Fertilizer and Manure**

With greater availability of better seeds (high yielding varieties), irrigation facilities, plant protection measure and the supply of manures, can't produce desired higher yeilds. Use of chemical fertilizers is the key to achieve the high crop production. Among macro-nutrients, only three primary nutrients (N, P & K) are utilized by the crops in large quantities. These fertilizers are available in straight, compound or complex and mixed forms. Nitrogen containing fertilizers are called nitrogenous fertilizers; similarly phosphorous and potassium containing fertilizers are called phosphatic and potassic fertilizers, respectively.

Considering the high cost of fertilizers, their effective use demands utmost attention. Various methods of their application are based or soil condition and nature of plant growth.

Manures are organic substances rich in plant nutrients. These also provide food to soil micro organisms. Manures not only add most of the essential nutrients to the soil but also improve the physio-chemical properties of the soil. The organic manures are classified as bulky organic manures (FYM, Compost, etc.), green manure, concentrated organic manures (Oil cakes & Bone meal) and biofertilizer such as Azola, Blue green algae, Rhizobium and Azotobactor culture (Sacheti, 1985).

Knowledge about the judicious use of fertilizer and manure is much essential specially in farming System like agroforestry to get desired produce on sustained basis from an unit area. In agroforestry system we should be much careful in identifying deficiency symptoms. After identification of deficiency, cheapest source of nutrient fertilizer and manure should be used after estimating quantities of fertilizer/manure needed as per requirement of crops. We should always select cheapest fertilizer *viz.*, Urea is the cheapest source of nitrogen, rock phosphate is the cheapest source of phosphorous but its use is restricted only in acidic soil. Thus, generally superphosphate is recommended.

In agroforestry system, if use of fertilizer is essential, we should also be careful about the time and method of application. As we know nitrogen is essential throughout the growing period of crop, loss of applied N fertilizer to the soil takes place due to solubility in water, leaching and volatilisation. Thus, we should always recommend use of nitrogenous fertilizer in split dose in the coated or granular form or along with FYMs. In case of phosphatic and potassic fertilizer full dose is used at the time of basal dressing or sowing time. The various methods of fertilizer application *viz.* drill application, folliar application, broadcasting and placement methods are recommend as per need and growth pattern of crops and trees species in agroforestry system.

In agroforestry system, if trees are managed with the objective to use its green foliage to improve the soil fertility, is known as green manuring. In this system generally fast growing nitrogen fixing tree species are selected which produce abundant quantities of green foliage. Green foliage are harvested at periodic interval and incorporated into the soil. If foliage is used to increase fertility status of soil referred as green manure tree. The nutrients are released from the foliage. Nutrients are more readially available to the crops. Besides addition of organic matter, it also supply available nutrients to the next crop, improves moisture retention capacity and physio-chemical properties of soil.

The available results suggest that agroforestry cropping system alone maintain soil productivity, but use of moderate quantity of fertilizer may be the best solution to increase productivity. It is advisible to incorporate chopped green foliage, at 20 cm deep in the soil, one month before the succeeding crop. This facilitates proper decomposition of green manure.

The complete decomposition can be determined by estimating C.N. ratio of the soil. If C.N. ratio is about 10-15 : 1, organic residues are present in decomposed form.

In agroforestry system, root nodule bacteria (Rhizobium) of NFTs fix nitrogen. Legumes under favourable condition can fix from 40 to 194 kg/ha/yr of elemental nitrogen from the atmosphere (Sacheti, 1985). If requirement of nitrogen is more than this, problem can be solved by using Azotobacter, free living blue-green algae, *Azolla pinnata*, etc. Farm Yard Manure is an ideal and cheapest source in a village. It is well decomposed form of dung and urine of cattle. It contains N, $P_2O_5$ & $K_2O$ 0.5-1.5, 0.4-0.8 and 0.5-0.9 per cent respectively. It also contains micronutrients. After application of FYM to the soil, the nutrients are released slowly and are available to the plant over a long period of time.

In addition to improving the soil fertility status, when applied in large quantities, well decomposed FYM also improves the structures of the soil by improving the water holding capacity of light, soils and better aeration in poorely drained soils (Sacheti, 1985). The use of fertilizers, FYM, green manure, etc. help in luxuriant growth of crops and trees. In agroforestry system regular monitoring and identification of deficiency of nutrients in crops and trees and judicious application of green manure /FYM/ fertilizer as per need is essential to improve the fertility status of soil.

**Choice of crops and crop rotations**

In agroforestry system careful selection of crop species plays important role. If the soil have low moisture retention capacity, it is essential for most of the Kharif crops that they complete their life cycle within the safe moisture availability limits. It monsoon span in the area varies from 80-115 days it would seem advisable to fix the varietal maturity at around 85-110 days, depending on the monsoon span, for most of the crops. There are some notable exceptions like cotton and redgram, otherwise varieties are available to fit the required span. As a corollary it follows that in the kharif season, drought resistance capacity is relegated to a secondary role. No doubt it is important in view of the short and long drought spells often encountered during the season, but the capacity of the crops and varieties to attain a high production level within the safe period of around 100 days gains ascendency. It also follows that these varieties must possess proper resistance to diseases and pests whose incidence is high due to favourable moisture and temperature conditions.

In the rabi season, however, drought resistance crop and varieties is essential, since the crops have to grow entirely or for most of the period on a receding soil moisture regime. Hence, the crop should be either early enough to complete the grain formation before the moisture stress becomes limiting or should be late enough to be able to capitalised on the winter rains (Prasad, 1984).

One crop repeatedly grown on the same unit and tends, therefore, to cause an earlier exhaustion of the nutrients from the soil, while the demand for plant nutrients is also confined to that section of the soil in which the crop develops its roots. Also frequent repeatation of a crop on the same land tends to encourage the development of disease and insect pests special to it, which infest and ultimately destroy it. The alternation of crops of varying root characters and development, thus enables the resources of the soil to be more judiciously utilized, and thus prevents its exhaustion from the same layer of surface soil (Mitra Bahadur, 1981). Sinha (1984) has stated that a well balanced crop rotation keeps the land almost continuously covered, improves the physio-chemical properties of soil, reduces the bad effect of monocropping. He has suggested that clean tilled crops year after year depletes the soil nutrients in the most unbalanced manner. Therefore, fertility building crops should be incorporated in the general cropping pattern of the region.

In agroforestry system, mixed cropping *i.e.* growing of mixture of legume with other crop instead of growing only one clean tilled crop may be more beneficial. Intensive alley cropping plays important role in reducing the velocity of running water and it reduces soil and water loss. Its efficiency can be increased many fold by adopting strip cropping system. The tree hedge and alternate strips of erosion permitting (clean tilled row crops) and erosion resisting crops reduces water and soil loss and increase productivity of land. This system has great potential specially in hill farming system.

**Weed**

Undesirable plants growing in the specific season in the crop field and compete with crops nutrients, soil moisture, light and space. Removal of these undesirable plants help in boosting crop growth and yield. Its control is only possible, if we have knowledge of weeds habitat, growth behaviour and crop association. The most common weeds which are frequently encountered in the uplands, and medium lands during kharif, rabi season are as follows :—

**Kharif Weeds**

*Dactyloctenium aegypticum, Paspalidium flavicum, Setaria glanea, Eragrostis piloses, Triden procumbens, Digitaria longifolia*, etc.

**Rabi Weeds**

*Cynodon rotundus, Cynodon dactylon, Melitotus alba, Solanum nigrum, Senebirra pinnafufida, Ageratum conyzoides, Vicia hirsuta, Vicia sativa, Medicago denticulata, Gynoptalium indicum*, etc.

Rao (1984) has stated that it has been observed that there is no specific crop weed association, rather there is some association of edaphic conditions like moisture situation and texture of the soil with the type of weed flora apart from temperature which determines by and large the weed flora. This is exemplified by weeds of submerged, semi submerged and medium uplands and typical uplands whose weed flora distinctly differ from each other. He has also suggested that first and most efficient method of weed control without involving any expenditure is not to allow the weeds to set seeds by decapitating the influorescence of the flowering short (before the seeds are set) all around the cultivator's field as a routine practice. The second important method in dryland conditions is year round tillage. Repeated ploughing hoeing is able to physically destroy the weed seeds that come into contact with the implement apart from bringing into surface the weed seeds which will get weathered by sun and rain and get destroyed. The weeds which germinate, during the process of subsequent ploughing automatically gets destroyed. In addition, the repeated ploughing with every rain helps in opening up of the soil for more percolation and intake of rain water apart from breaking the capillaries which will prevent evaporation of valuable soil moisture.

Another management practice for weed control is identification of varities of crops which have early growth vigour. Brown Gora BR-23-19 of Chotanagpur area which has a very fast early growth rate does not suffer so much from weed competition as compared to other rice varieties which have slow early growth.

Where holdings are larger, chemical weed control can be effectively practiced by the peasants. The peasants should always used recommended dose of herbicides for different crops.

Above methods of management, minimise the menance of weeds in the agroforestry system.

**Insect Pests and Diseases**

Insect pests and pathogen causes extensive damage to trees and crops. The pest and pathogen population fluctuates depending on the nature of sylvan and agro-ecosystem. The management of pest and pathogen population in agroforestry system is different from pure agricultural crop or trees. Details about the pest and pathogen problems are discussed separately in chapter 15.

## Tips For Cereals, Pulses, Oil Seed and Vegetables Farming

| S. No. | Crops | Improved variety | Land | Seed rate | Sowing time | Cultivation | Fertilizer | Harvesting period |
|---|---|---|---|---|---|---|---|---|
| 1 | 2 | 3 | 4 | 5 | 6 | 7 | 8 | 9 |
| **(A) CERREAL** | | | | | | | | |
| 1. | Dhan(Paddy)<br>*Oryza sativa* | (*a*)BR 8 | Low land | 40kg/ha in transplan-ting. | June | After one month of sowing trans-planting is done | *Dwarft variety:* S.S.P., potash & Nitrogen @ 40, 20 & 80 kg/ha<br>*Tall variety :* S.S.P., Potash & Nitrogen @ 20, 20 & 40 kg/ha (Split dose) | 150-160 days |
| | | T 148 | • | • | • | • | • | • |
| | | Pankaj | • | • | May | • | • | • |
| | | Mansuri | medium land | 80 kg/ha broadcast-ing & 40 kg/ ha in trans-planting. | June | • | • | 130-150 days |
| | | IR 20 | • | • | • | • | • | • |
| | | Seeta | • | • | • | • | • | • |
| | | IR 8 | • | • | • | • | • | • |
| | | BR 34 | • | • | • | • | • | • |

(*Contd.*)

| 1 | 2 | 3 | 4 | 5 | 6 | 7 | 8 | 9 |
|---|---|---|---|---|---|---|---|---|
| | | Cauveri | Upland | 80 kg/ha broadcast-ing & 40 kg/ha in trans-planting. | May-June | After one month of sowing trans-planting is done | S.S.P., Potash & Nitrogen @ 20, 20 & 40 kg/ha (Split dose) | 100-115 days |
| | | Kiran | " | " | " | " | " | " |
| | | HYV Bala | " | " | " | " | " | " |
| | | *(b) Summer Paddy* | | | | | | |
| | | Pusa-33 | " | 50 kg/ha | March | " | " | 110-115 days |
| | | Pusa 2-21 | " | " | " | " | " | " |
| | | C.R. 44-46 | " | " | " | " | " | " |
| 2. | Gehu (Wheat) *Triticum spp.* | (a) Rainfed Condition | Medium Land | 100 kg/ha | October-November | Line sowing behind the plough and row spacing at 20 cm. | NPK @ 100, 50 and 25 kg/ha (Split dose) FYM @ 100-150q/ha | Crop harvest-ed in the morning |
| | | C 306 | " | " | " | | | |
| | | *(b) Irrigated* Janak | " | " | November | Line sowing row spacing 20-24 cm. Four to five time irrigation. | " | " |
| | | K 7414 | " | " | " | " | " | " |
| | | HP 1102 | " | " | " | " | " | " |
| | | Sonalika | " | " | " | " | " | " |
| | | (HD 1553) | " | " | " | " | " | " |

*(Contd.)*

| 1 | 2 | 3 | 4 | 5 | 6 | 7 | 8 | 9 |
|---|---|---|---|---|---|---|---|---|
| 3. | Jau (Barley) *Hordeum vulgare* | (*a*) Irrigated Ratna | Upland | 70-80 kg/ha. | November | Line sowing two to three irrigation at one month interval | NPK @ 40; 20 and 20 kg/ha (Split dose) | crop harvest-ed in the morming |
| | | BR 32 | " | " | " | " | " | |
| | | Joyti | " | " | " | " | " | " |
| | | DL 36 | " | " | " | " | " | " |
| | | (*b*) *Rainfed* | *Medium Land* | | | | | |
| | | Ratna | " | 100kg/ha | " | Line sowing | N&P 20 and 15 kg/ha (Split dose) | |
| | | BR 31 | " | " | " | " | | " |
| | | K 125 | " | " | " | " | " | " |
| 4. | Jai (Oats) *Avenus sativa* | H.R. 374 | Well-drained upland soil | 10 kg/ha | June | Two to three time ploughing and planting is essential | NPK @ 20, 20 & 20 (Split dose) | 80-115 days (Period varies in different varieties. |
| | | B.R. 2 | " | " | " | " | " | " |
| | | A. 404 | " | " | " | " | " | " |
| | | P.R. 202 | " | " | " | " | " | " |
| 5. | Marua (Finger Millet) *Eleusine coracana* | " | " | " | | " | " | " |

(*Contd.*)

| 1 | 2 | 3 | 4 | 5 | 6 | 7 | 8 | 9 |
|---|---|---|---|---|---|---|---|---|
| 6. | Makka (Maize) *Zea mays* | Suwan-1 | Upland | 18-20kg/ha | May to June | Sowing at 75 x 25 cm spacing (When moisture is availlable in the soil | NPK @ 30, 10 & 40 kg/ha. Again after 40 days Nitrogen @ 40 kg/ha. | 100-110 days |
| | | Ganga-5 | " | " | " | " | " | " |
| | | Vijay | " | " | " | " | " | " |
| | | Ganga safed-2 | " | " | " | " | " | " |
| (B) | **PULSES** | | well-drained soil | | | | | |
| *(a)* | **Rabi** | | | | | | | |
| 1. | Gram *Cicer arictinum* | B.R. 77 | | 70-75 kg/ha | October to November | Spacing 30 cm x 10 cm | N&K @ 15& 40 kg/ha | 130-140 days |
| | | Pantji-114 | " | " | " | " | " | 135-145 days |
| | | B.R. 78 | " | " | " | " | " | 140-150 days |
| 2. | Masoor (Lentil) | P.L. 8 | " | 30 kg/ha. | November | spacing 22 x 10 cm | " | 120-130 days |
| | | B.R. 25 | " | " | " | " | " | 115-125 days |
| 3. | Matar (Pea) *Pisum sativum* | Swarnrekha | well-drained soil | 70-75 kg/ha | October to November | Spacing 30 x 10 cm | N&P @ 15 & 40 kg/ha | 130-140 days for vegetables 100 days well filled pods, Changing colour light green should be plucked. |
| | | B.R. 12 | | " | " | " | " | " |
| | | Arkel | | " | October | " | " | " |

*(Contd.)*

| 1 | 2 | 3 | 4 | 5 | 6 | 7 | 8 | 9 |
|---|---|---|---|---|---|---|---|---|
| | (*b*) **Kharif** | | | | | | | |
| | (1) Moong | PS - 16 | well drained soil | 20 kg/ha | June to July | Spacing 30 x 10 cm | N&P @ 15 & 40 kg/ha | 60-70 days |
| | | Sunaina | " | | " | " | " | 60-65 days |
| | 2. Arhar (Cajanus cajan) | Laxmi | " | 20 kg/ha | June to July | Spacing 30 x 10 cm | " | 220-230 days |
| | | T 21 | " | " | " | spacing 25 x 10 cm | " | 150 days |
| | | BR 183 | " | " | " | " | " | " |
| 3. | Kulthi | Madhu | " | 20 kg/ha | August | Spacing 10 x 10 cm | " | 110 days |
| **(C) OIL SEEDS** | | | | | | | | |
| 1. | Tori (Mustard) *Brassica* spp. | Bhawani | " | 10 kg/ha | September to October | Distance between rows-30 cm. | NPK @ 25 25&10 kg/ha. | 100-120 days |
| | | BR 23 | " | " | " | " | " | " |
| 2. | Rai | BR 40 | " | " | " | " | " | " |
| | | Varuna | " | " | " | " | " | " |
| 3. | Til *Sesamum indicum* | Krishna | Upland | 10 kg/ha | June to July | | NPK @ 25, 40&20kg/ha. | 90-100 days |
| | | M3-1 (Black) | " | " | " | " | " | " |
| | | TC -25 | " | " | " | " | " | " |
| 4. | Ground nut *Arachis hypogaea* | Early A.K. 12-24 | Fertile sandy loom | 75-80 kg/ha shelled kernals | June to July | " | " | 100-105 days |

(*Contd.*)

| 1 | 2 | 3 | 4 | 5 | 6 | 7 | 8 | 9 |
|---|---|---|---|---|---|---|---|---|
| | | Early : G.G.2 | Fertile sandy loom | 75-80 kg/ha shelled kernals | June to July | Distance between rows-30 cm | NPK @ 25, 25&10 kg/ha | 100-105 days |
| | | Mid early : B.G. -1 | " | 100-110 kg/ha | June | " | " | 110-120 days |
| | | B.G. -2 | " | " | " | " | " | " |
| | | B.G. -3 | " | " | " | " | " | " |
| 5. | Soya bean | Alankar | " | 70-80 kg/ha | June to July | Spacing 30 x 5 cm | NPK @ 20, 80 & 40 kg/ha | 105-110 days |
| 6. | Nizer(Sarguja) (*Guizotia abyssinica*) | Oatacamund | upland | 6 to 7kg/ha | August to September | spacing 30 x 15cm | NPK @ 20, 20 & 10 kg/ha | 100-110 days |
| | | N-15 | " | " | " | " | " | " |
| 7. | Kusum(Safflower) | A 300 | " | 20 kg/ha | October | Row spacing 35-40 cm | NP&K @ 22, 20&10 kg/ha | 165-170 days |
| 8. | Tisi (Linseed) | Sweta | " | " | " | " | N&P @ 22 & 20 kg/ha | 165-175 days |
| | | Subhra | " | " | " | " | " | " |
| | | G 397 | " | " | " | " | " | " |
| **(D) VEGETABLEING** | | | | | | | | |
| 1. | Potato | Late : Kufri sinduri | Fertile sandy loam | 20-30 q/ha | October to November | Spacing 50 x 20 cm | Urea, S.S.P & muriate of potash @ 2.5 6 and 1.5q/ha respectively (Split dose). | 120 days or more |
| | | Patna white | " | " | " | " | " | " |
| | | Kufri white | " | " | " | " | " | " |

*(Contd.)*

| 1 | 2 | 3 | 4 | 5 | 6 | 7 | 8 | 9 |
|---|---|---|---|---|---|---|---|---|
| | | Early : | | | | | | |
| | | Great scot | Fertile sandy loam | 20-30 q/ha | October to November | Spacing 50 x 20 cm | Urea, S.S.P & muriate of potash @ 2.5 6 and 1.5q/ha respectively (Split dose). | 80-80 days |
| | | Kufri kuber | | " | | | " | " |
| | | Kufri chand- | " | " | " | " | " | " |
| | | dramukhi | " | " | " | " | " | " |
| | | Medium : | | | | | | |
| | | Kufri Joyti | " | " | " | " | " | 110-120 days |
| | | Kufri chamatkar | " | " | " | " | " | " |
| | | Kufri Badshah | " | " | " | " | " | " |
| | | Red round | " | " | " | " | " | " |
| 2. | Sweet Potato | Pusa suffaid | Sandy loam prefers slight to mode-rate aci-dic soil | 40 thou-sand/ ha | Cutting plan-ted during Sept. to October | 90 cm apart | Two part of bone meal, four parts of ground nut cake and one part of potassium sulphate @ 650 kg/ha | 110-120 days |
| | | F.13.4004 | | " | | " | " | |
| | | Pusa Lal : F.A.17 | | | | | | |
| | | Red | " | " | " | " | " | " |
| | | Gold husk | " | " | " | " | " | " |
| | | Pusa sunehri | " | " | " | " | " | " |
| | | B- 5941 | " | " | " | " | " | " |

(Contd.)

| 1 | 2 | 3 | 4 | 5 | 6 | 7 | 8 | 9 |
|---|---|---|---|---|---|---|---|---|
| 3. | Arum | Kalla kachu | Grows well on marshy land. | • | March to June | spacing 60 x 30 cm | FYM @ 60-80q/ha Super phosphate, muriate of potash and ammonium sulphate @ 130. 120 75 kg/ha Two top dressing with ammonium sulphate @ 75 kg/ha. | August to November |
| 4. | Brinjal | Pusa purple cluster | Loam and sandy loam | 500gm/ha | Crops:March to May-Nursery sowing: Mid June<br><br>Crops:June to Aug.Nursery sowing April Crops:Sept. to Jan. Nusery sowing-June. | 60 x 50 cm or 75 x 45 cm | Urea, S.S.P. & murate of potash @ 250, 400&100kg/ha or FYM 300 kg/ha. | After 12 weeks |

| 1 | 2 | 3 | 4 | 5 | 6 | 7 | 8 | 9 |
|---|---|---|---|---|---|---|---|---|
| | | Annamalai | Loam and sandy loam | 500gm/ha | Crops:June to Aug.Nursery sowing April Crops:Sept. to Jan. Nusery sowing-June. | 60 x 50 cm or 75 x 45 cm | Urea, S.S.P. & murate of potash @ 250, 400&100kg/ha or FYM 300 kg/ha. | After 12 weeks |
| | | Banaras giant | " | " | " | " | " | " |
| | | Manjri gota | " | 200gm/ha | " | " | " | " |
| | | Muktakeshi | " | " | " | " | " | " |
| | | Suphal | " | " | " | " | " | " |
| | | Vijay | " | " | " | " | " | " |
| | | Pusa hybrid-6 | " | " | " | " | " | " |
| 5. | Tomato | Tupali | Grows on variety of soils prefers light acidic or kaline soil | 150gm/ha | Very early variety: Nursery sowing: June-July Early variety. Nursery sowing: July-Aug. Late variety : Nursery sowing : Sept-Oct. | 60 x 60 cm | Urea, S.S.P & murate of potash @ 200, 250 & 100 kg/ha | After 3-4 months |
| | | Siouxe | " | " | " | " | " | " |
| | | Pusa hybrid-1 | " | " | " | " | " | " |
| | | Marglobe | " | " | " | " | " | " |
| | | Karnatak | " | " | " | " | " | " |
| | | Large red | " | " | " | " | " | " |

(Contd.)

| 1 | 2 | 3 | 4 | 5 | 6 | 7 | 8 | 9 |
|---|---|---|---|---|---|---|---|---|
| | | Pusa Ruby | Grows on variety of soils prefers light acidic or kaline soil | 150gm/ha | Very early variety: Nursery sowing: June-July Early variety. Nursery sowing: July-Aug. Late variety : Nursery sowing : Sept-Oct. | 60 x 60 cm | Urea, S.S.P & murate of potash @ 200, 250 & 100 kg/ha | After 3-4 months |
| | | Oxheart | " | " | " | " | " | " |
| | | S-120 | " | " | " | " | " | " |
| | | Pusa early-dwarf | " | " | " | " | " | " |
| | | Suttons early-scarlet | " | " | " | " | " | " |
| | | Sabour Prabha | " | " | " | " | " | " |
| | | Pusa Ruby | " | " | " | " | " | " |
| 6. | Cauliflower | Summer crop: Earli Kuari | Fertile sandy loam, | | | | | |
| | | | slight acidic to light alkaline | 500-600 gm/ha | Nursery sowing: Feb-March | 60 x 45 cm | Urea,S.S.P. & murate of potash @ 300, 500 & 100 kg/ha | 60 days after transplanting |
| | | Early | | | | | | |
| | | Pusa katki | | 600-700 gm/ha | Nursery sowing: June-July | 60 x 45 cm | | 56-60 days after transplanting |

(Contd.)

| 1 | 2 | 3 | 4 | 5 | 6 | 7 | 8 | 9 |
|---|---|---|---|---|---|---|---|---|
| | | Pusa Diwali | Fertile sandy loam, slight acidic to light alkaline | 600-700 gm/ha | Nursery sowing: June-July | 60 x 45 cm | Urea,S.S.P. & murate of potash @ 300, 500 & 100 kg/ha | 56-60 days after transplanting |
| | | 235-S | " | | " | " | " | " |
| | | Pusa Himjoyti | " | | " | " | " | " |
| | | Early patna | " | | " | " | " | " |
| | | Medium | | | | | | |
| | | Pusa katki | " | " | Nursery sowing: July to August. | 50 x 45 cm | " | More than 60 days after transplanting |
| | | 235-S | " | " | " | " | " | " |
| | | 74-6-C | " | " | " | " | " | " |
| | | 94-6-C | " | " | " | " | " | " |
| | | Pusa Dipali | " | " | " | " | " | " |
| | | Giant Snowball | " | " | " | " | " | " |
| | | Late | | | | | | |
| | | Indian snowball-16 | | 310 gm/ha | Nursery sowing : Sept to October | | NPK & @ 20, 125, 150 kg/ha | 60 days |
| | | Pusa snowball | " | " | " | " | " | " |
| | | Maghi | " | " | " | " | " | " |
| | | K-11 | " | " | " | " | " | " |
| | | Late Banaras | " | " | " | " | " | " |
| | | Dania (Kalimpong) | " | " | " | " | " | " |

(*Contd.*)

| 1 | 2 | 3 | 4 | 5 | 6 | 7 | 8 | 9 |
|---|---|---|---|---|---|---|---|---|
| 7. | Cabbage | Early : | | | | | | |
| | | Golden acre | Heavy to lighter soils | 500-600 gm/ha | September to October | Spacing 50 x 45 cm | N,P, & K @ 200, 125 & 130 kg/ha | After 3 to 4 months. |
| | | Kalimpong | " | " | " | " | " | " |
| | | Drum head savoy | " | " | " | " | " | " |
| | | Suttons pride of India | " | " | " | " | " | " |
| | | Sutton earliest | " | " | " | " | " | " |
| | | Larger bood red | " | " | " | " | " | " |
| | | Pusa drum head | " | " | " | " | " | " |
| | | Hybrid | | | | | | |
| | | Ganga | " | 310 gm/ha | " | " | " | " |
| | | Jumana | " | " | " | " | " | " |
| | | Kaveri | " | " | " | " | " | " |
| | | Cabbage No-8 | " | " | " | " | " | " |
| 8. | Onion | N -53 | Variety of soil, prefer deep friab-le loam soil | 10 kg/ha | September to November | Spacing 15 x 10 cm | Urea, S.S.P & muriate of potash @ 200, 300 & 100 kg/ha | Three to five months after transplanting |
| | Pusa red | | " | | " | | " | |
| | Pusa Ratnar | | " | | " | | " | |
| | Earti Grani | | " | | " | | " | |
| | Bellary of South India | " | " | " | " | " | " | |
| | Patnai Silver | " | " | " | " | " | " | |
| | Patnai Rid | " | " | " | " | " | " | |

(*Contd.*)

| 1 | 2 | 3 | 4 | 5 | 6 | 7 | 8 | 9 |
|---|---|---|---|---|---|---|---|---|
| 9. | Garlic | Fawari Gaddi | Well drained loam soil | 200 kg/bulbils/ha. | September to October | Spacing 15 x 75cm. | FYM 20 tonns/ha., side dressing of 125 kg ammonium sulphate after one month. | After 135-150 days |
| | | Rajalle gaddi | " | " | " | " | " | " |
| 10. | Carrot | Pusa Keshar | Light fertile soil | 4 kg/ha | November to December | 10 cm apart | FYM @ 250q/ha, Urea, S.S.P. Muriate of potash @ 2, 2.5 and 1 q/ha. | After 70 days |
| 11. | Radish | Pusa Deshi | " | 10 kg/ha | August to December | " | " | " |
| | | Pusa Chetki | " | " | " | " | " | " |
| | | Pusa Himani | " | " | " | " | " | " |
| | | Japani White | " | " | " | " | " | " |
| | | Jaunpuri Giant | " | " | " | " | " | " |
| | | Chinese White | " | " | " | " | " | " |
| | | Crimson Breakfast | " | " | | | " | |
| 12. | Chillies | For Vegetables : | | | | | | |
| | | Yellow-wonder | Well drained Light sandy loam, warm | 1.5 kg/ha | Spring crop : Nursery sowing : Nov. | Spacing 50 x 30cm | FYM @ 250 q/ha, Urea, SSP and muriate of | After 60-70 days |

*(Contd.)*

| 1 | 2 | 3 | 4 | 5 | 6 | 7 | 8 | 9 |
|---|---|---|---|---|---|---|---|---|
| | | | and humid climate | | Winter crop : Nursery sowing : July to Aug. Early crop : Nursery sowing : May to June. | spacing 50 x 30cm | potash @ 250 300 & 100 kg/ha. | After 60-70 days |
| | | Bharat | " | 1.5 kg/ha | " | " | " | " |
| | | Chinese giant | " | " | " | " | " | " |
| | | California wonder | " | " | " | " | " | " |
| | | Patna Red. | " | " | " | " | " | " |
| | | For spices | | | | | | |
| | | Jwala | " | " | " | " | " | " |
| | | N.P.-46 A | " | " | " | " | " | " |
| | | Sabour Arun | " | " | " | " | " | " |
| | | Puri-Red | " | " | " | " | " | " |
| 13. | French Beans (Kidney bean) | Bush crop S - 9 | Well drained sandy loam | 70-80 kg/ha | Jan to February & Oct-Nov. | Spacing 50 x 25cm | FYM @ 200 q/ha, Urea SSP & muriate of potash 150, 500 150 kg/ha. | Pods are harvested before maturity (50 to 80 days) |
| | | Contender | " | " | " | " | " | " |
| | | Pusa parbati | " | " | " | " | " | |
| | | Pole crop | | | | | | |
| | | B. Priya | " | 30 kg/ha | Jan. to Feb. May to June & Sept. to Oct. | Spacing 70 x 70 cm | | " |
| | Kentucky wonder | " | " | " | " | " | | |

(Contd.)

| 1 | 2 | 3 | 4 | 5 | 6 | 7 | 8 | 9 |
|---|---|---|---|---|---|---|---|---|
| 14. | Sem (Dolichos/hyacinth bean) | | | | | | FYM @ 200 q/ha, Urea SSP & muriate of potash 150, 500 150 kg/ha. | Pods are harvested before maturity (50 to 80 days) |
| | | Kanchan | " | 25 kg/ha | June to July | 2 x 2m | " | " |
| | | Dwarf sem | " | " | " | " | " | " |
| | | Velvet white flat | " | " | " | " | " | " |
| | | Red flat | " | " | " | " | " | " |
| | | Baghmaka | " | " | " | " | " | " |
| | | Altapati | " | " | " | " | " | " |
| 15. | Gwarfali(Guar) | Pusa Mousami | Well drained sandy loam | 15-16 kg/ha | June to July | Spacing 50 x 25cm | | Pods are harvested before harvestry (50-70 days) |
| | | Pusa sada bahar | " | " | " | " | " | " |
| 16. | Bodi(Cowpea) | Pusa do-fashli | " | 25 kg/ha | | Spacing 2 x 1.5 m | " | |
| | | Pusa Barsati | " | " | " | " | " | " |
| | | Yard long | " | " | " | " | " | " |
| 17. | Bhindi (Lady's finger) | Pusa sawani | All kinds of soil provided well drained and well manured | 10 kg/ha | Feb. to March to June to July | 45 x 25 cm or 60 x 25 cm | " | Six to seven weeks after transplanting |

(*Contd.*)

| 1 | 2 | 3 | 4 | 5 | 6 | 7 | 8 | 9 |
|---|---|---|---|---|---|---|---|---|
| | | | All kinds of soil provided well drained and well manured | 10 kg/ha | Feb. to March to June to July | 45 x 25 cm or 60 x 25 cm | FYM @ 200 q/ha, Urea SSP & mu-riate of pot-ash 150, 500 150 kg/ha. | Six to seven weeks after transplanting |
| | | Paivani kranti | " | " | " | " | " | |
| | | Pusa Makamali | " | " | " | " | " | |
| | | Lucknow Dwarf | " | " | " | " | " | |
| | | Long green | " | " | " | " | " | " |
| | | Velvet green | " | " | " | " | " | " |
| 18. | Parwai or Patal (Pointed gourd) | " | Sandy loam | Root cuttings are planted | Feb. to March | 1m apart | FYM @ 150q/ha | Three to four months |
| 19. | Khira (Cucumber) | Balam khira | Clay loam sandy loam soils. | 3 kg/ha | Summer crop : December to January Main crop : June to July | 1.5-2 m apart | FYM @ 150-200 q/ha Urea,SSP & M.P. @ 1.5, 2.5 & 1q/ha | After 90 days |
| | | Japanese long | | | " | " | | " |
| | | Poona Khira | | | " | " | | " |
| | | Pusa sanyong | | | " | " | | " |
| | | Darjeeling type | | | " | " | | " |

(Contd.)

| 1 | 2 | 3 | 4 | 5 | 6 | 7 | 8 | 9 |
|---|---|---|---|---|---|---|---|---|
| 20. | Bottle gourd (Kadoo) | Pusa Manjiri | Clay loam sandy loam soils. | 5-6 kg/ha | Summer crop : December to January Main crop : June to July | 2 to 3m apart | FYM @ 250 q/ha Urea, SSP & M.P. @ 2.5, 3.5 & 1 q/ha. | Fruit ready for picking after 80-100 days of sowing |
| | | Pusa Meghdoot | " | " | " | " | " | " |
| | | Dholi safed | " | " | " | " | " | " |
| 21. | Pumpkin (Kumhara) | Arka suryamukhi | | " | " | " | | " |
| | | Red pumpkin | | " | " | " | | " |
| | | Globe perfection | | " | " | | | " |
| | | Co- 1 | | " | " | | | " |
| 22. | Ridge gourd (Jhigni) | Satputia | | 5kg/ha | | 1.5 to 2 m apart | | After 80 days |
| | | Pusa Nasadar | " | " | " | " | | |
| 23. | Sponge gourd (Nenua) | Long white | | | " | 2 to 2.5 apart | | After 80 days |
| | | Long green | " | " | | " | | |
| | | Pusa chikni | " | " | | " | | |
| 24. | Bitter gourd (Karela) | Arka Harit | " | 5-6 kg/ha | | 1.5-1.5 m apart | FYM @ 200 q/ha, Urea, S.S.P M.P. @ 1.5, 2.5 & 1 q/ha | After 80 days |

(*Contd.*)

| 1 | 2 | 3 | 4 | 5 | 6 | 7 | 8 | 9 |
|---|---|---|---|---|---|---|---|---|
| 25. | Palak sag | All green | Light loam | 20-25 kg/ha | September to November | 15 cm apart | S.S.P. Urea @ 100 & 40 kg. After every cutting top dress with urea is essential. | After 40 days |
| | | Pusa Joyti | | " | | | " | " |
| | | Vergenia Savoy | | " | | | " | " |

## REFERENCES

Mitra Bahadur, Rai D.N. (1981). *Grow more vegetables and eat more vegetables*, Revised and enlarged by Bose, P.K. and Roy, B.B., Calcutta University Publication, pp. 1-53.

Prasad, R.B. (1984). Technology for dryland agriculture on watershed basis, In : *Proceeding of Dryland farming and watershed management*, Mohsin, M.A. and Prasad, R.B., eds., B.A.U., Ranchi, pp. 72-81.

Rao, M.S.S. (1984). Integrated weed management in dryland agriculture, *paper presented in workshop-cum-Discussion seminar for S.M.S. of W. Bengal, Orissa, Bihar, and Uttar Pradesh*, B.A.U., Ranchi, pp. 1-12.

Sacheti, A.K. (1985). *Fertilizers and manures : Instructional-cum-practical manual*, N.C.E.R.T., New Delhi, pp. 1-87.

Sinha, A.K. (1984). Approved practices of soil conservation for watershed management, In: *Proceeding of Dryland farming and Watershed management*, Mohsin, M.A. and Prasad, R.B., eds., B.A.U., Ranchi, pp. 84-102.

# 6

# *Multipurpose Trees and Shrubs for Agroforestry System*

In some tropical countries deliberate cultivation of trees for fire wood is already financially profitable. The rising costs of other fuels is likely to make this increasingly attractive. Trees grow well in the humid tropics. The heat and humidity ensure high rates of photosynthesis and short harvest cycle, thereby making these areas suitables for supporting appreciable amount of wood energy on sustainable basis (Myers, 1980). In arid zones, any fuel wood becomes a revenue producing product, while in other part of country the sale of wood for fuel is of secondary importance. Trees are the cheapest renewable sources of energy, fodder, green manure and timber and when these products are incorporated in farm land (not only) soil productivity is increased but also present dependence on high cost inputs is reduced. Farm forestry which is synonymous with organic farming will not only help to prevent further desertification which is taking place today but also assist in producing more food grains, milk and meat for the people. Bhaskar and Swami Rao (1985). The natural forests can hardly supply our daily requirements of firewood, fodder and timber and in this context farm agroforestry remains the only answer to meet our daily requirements during 21st century.

In the past few decades to get fuel, fodder, timber etc., potential of multipurpose tree species in agroforestry systems has been well recognised. A large number of multipurpose trees and shrubs are deliberately retained or incorporated on farms in the subsistence farming system on the steep slopes and around the fields. The contribution of these trees is the production of fodder and firewood and their protective function in reducing the erosion hazards and thereby making crop production possible in those steep slopes where profitable cropping would otherwise be extremely difficult (Fonzen and Oberholzer, 1984). Indian farmers are integrating multipurpose trees in agroforestry or farmforestry from time immemorial. Farmers in Tamil Nadu integrate numerous species of multipurpose trees and shrubs on their farm lands in close association with agricultural crops and/or live stock. The dominant among them are *Borassus flabellifer*, *Tamarindus indica*, *Ceiba pentandra*, *Acacia leucocephloea* and *Acacia nilotica*. These woody perennials are better able to cope with poor site conditions. Their increasing integration on farmlands represents a strategy to minimise risk of crop failure (Jambulingam and Fernandes, 1986).

Traditionally managed farmforestry requires latest scientific intervention in order to get sustained yield from the same unit of land. This is possible through selection of suitable multipurpose tree species for particular agroclimatic zone. Thus, collection, storage and distribution of multipurpose tree germ plasm are necessities of the present days. Multiple use of lands and multipurpose trees should be the pillars of strength for our rural economy.

**Multipurpose Trees**

Multipurpose trees can be used for more than one purpose. It provides food, fodder, fuelwood, timber, minor tree products and shelter material to subsistence farmers. All plants (herb, shrub and tree) can be use for more than one purpose. In agroforestry system multipurpose trees are those woody perennials which are managed by farmers deliberately in order to get more than one intended products besides positive contribution to soil fertility and sustainable crop production (Jha, 1991).

When a particular tree species is grown in the forest area, to get timber, may not be referred as multipurpose tree species. If it is grown in farm land to get intended products (food, fodder, fuel wood, shelter material) would be placed in multipurpose category.

Our knowledge on the long term effects of the multipurpose trees incorporated in the agroforestry system is rather inadequate. We do not have good answer between trees and field crops in an environment where moisture regime and soil fertility are limited and longer gestation period after which farmers expect to get returns. Intensive research is needed for both these.

Above problem can be tackled by judicious screening of fast growing, short rotationed woody species. Since the last decade, many shrubs are getting importance in agroforestry cropping

Fig. 17 : Bamboos (Monocotyledons) and ideal multipurpose species

Fig. 18 : Bamboos House

system. The most commonly used agroforestry multipurpose woody shrubs are *Cajanus cajan*, *Sesbania* spp. etc. Thus, in agroforestry system tree species as well as woody shrubs (usually less than six metres in height and branches from the base) together are referred as multipurpose tree species (Ong and Daniel, 1990).

Van Carlowitz and Burley (1984) defined MFTs as trees grown deliberately or kept and managed for preferably more than one intended use, usually economically motivated major products and or services in any multipurpose land use system, especially agroforestry system.

According to Burley (1985) the prefix "multipurpose" appears to be used most commonly when a species is deliberately grown at one site which is capable to produce more than one product or

**Fig. 19 :** Flowers and Vegetables grown under multipurpose tree species (Neem) in Rarha Village, Ranchi. (*Courtsey* : Faculty Forestry, B.A.U. Ranchi).

benefit. These include, timber, fuelwood, medicines, tooth picks, human food and animal fodder (including flower for bees and leaves for silkworm), and service attributes such as shade, shelter, soil conservation and improved soil fertility.

Within the category of fast growing multipurpose tree species, the nitrogen fixing trees contribute greatly to the productivity and sustainability of many agricultural and forestry ecosystems. Plants cannot grow without nitrogen, and many tropical soils have low supplies of this nutrient. Although nitrogen is abundant in the earth's atmosphere, this nitrogen is not available for plant growth. NFTs convert atmospheric nitrogen into a form available to plants and Atmospheric nitrogen is "fixed" by either *Rhizobium* bacteria or an actinomycetes *Frankia* in nodules present on the roots of NFTs. "Fixed" nitrogen is available for NFTs growth in exchange for a supply of energy to the microorganisms. This buit-in "Fertilizer factory" concentrates nitrogen in the leaves of NFTs, making them valuable for animal feed and fertilizer (Glover, 1989). Halliday and Nako (1982) prepared the list of more than 100 species of woody perennials (legumes and non legumes) thought to have the potential for nitrogen fixation. Patil and Pathak (1977) have listed promising MPTs and shrubs for different agro ecological zones in India. There are number of MPTs in our country, but most of them are not fit as per definition of multipurpose trees.

**Objective of planting FGMPTs (Fast growing multipurpose tree species) in agroforestry system :**

(*a*) To gain maximum degree of self adequateness.

(*b*) To make use of the land to the best advantage.

(*c*) To increase the productivity of farmland by increasing the nutrient status of the soil (green leaf manure, nitrogen fixation by root nodules, root decomposition), conserving water and soil, by mitigating the dessicating effects of hot wind and by providing fuelwood, fodder, timber and minor forest product to the farmers.

(*d*) To improve the condition of cattle by providing nutritious leaf fodder, during lean period.

(*e*) To supplement the farmers income.

**Characteristic of ideal MPTs**

While selecting MPTs for agroforestry system, only such species should be introduced which show proper growth, with ability to increase productivity of site. A good multipurpose tree species should have following attributes:

1. Production of firewood, timber, green manure, fodder on sustainable basis.
2. Fast growing and ability to produce a flush of fresh shoots when cut from near the ground level.
3. Minimum competition should be between crops and multipurpose tree species. The problems like sunlight penetration etc. can be tackled by judicious canopy management.
4. Tree species having nutritious and palatable leaves are preferred, to supplement leaf fodder during lean period.
5. MPTs having ability to fix nitrogen in addition to other productive benefits should be preferred.

**The multiple benefits of MPTs**

(*a*) *Protective role*

(*i*) *Supply of nutrients* :-The principal role of FGMPTs is supply of nutrients to the intercrops. Nair (1984) has shown that one of the advantages that is most commonly attributed to agroforestry is the possibility for improvement of fertility status of agricultural lands through additional amounts of nitrogen that could be added to the system by the tree legume component. Although legumes are not the only group of woody species that have a potential role in agroforestry, the family leguminosae offers by far the greatest choice of woody species for agroforestry in terms of economic use as well as ecological adaptibility. *Gliricidia* prunings have a high nutrient content, containing 3-4.5 per cent nitrogen, 0.2-0.3 per cent phosphorus , 1.5-3.5 per cent potassium, 1.4 per cent calcium and 0.4 -0.6 per cent magnesium. *Gliricidia* produces between 300 and 500 kg.(dry weight) of prunings per hectare as each cutting depending on site condition, tree spacing and frequency of cutting contributing a substantial amount of nutrients (Glover, 1989). Tree roots through nutrient pumping action carry minerals from the below soil surface and make it available to the agricultural crops. Parennial pigeon pea produces more litter than oher types. In an experiment conducted at ICRISAT, short and medium term genotypes of perennial pigeonpea produced less than 1.0 tonne of litter (ICRISAT, 1986), and a four-year pigeonpea fallow under study at IITA yielded 12.8 tonnes per hectare of air dry matter, excluding woody parts (IITA, 1981). Superior nitrogen fixation has also been observed in slow-maturing types. Pigeon pea's beneficial effect on soil fertility is reflected in higher yields obtained from succeeding crops. For example, maize planted after a medium duration pigeonpea crop showed a 32 per cent increase in dry-matter yield and a 57 per cent increase in grain yield compared with control (Kumar Rao, *et al.*, 1981).

**Fig. 20 :** Bamboos foliage as mulch to decrease the velocity of run-off and conserve moisture and soil.

(*ii*) *Conserving soil and water*

In agroforestry systems, foliage of multipurpose tree species are used either as mulch or green manure. This acts as sponge and decrease the velocity of run-off and absorp more water and conserve soil. If tree species are not introduced along with crops, high intensity rains causes a considerable amount of soil to be eroded away. Presence of tree cover on the farm land reduce the intensity of rain fall on the soil and thus minimizes splash and sheet erosion. When MPTs are planted along the contours, the root hold the soil together and reduce the soil loss of a hill side.

(*iii*) *Weed reduction*

The presence of MPTs prevents weed growth in the field by its shade and wind break effect. Wind speed are reduced greatly by the tree cover. Which also prevents dispersal of weed seeds. Budelman (1988) has reported result of two trials comparing the leaf mulch of *Leucaena leucocephala, Flemingia macrophylla* and *Gliricidia sepium* for their effectiveness to reduce weed growth. Of the three mulch materials only that of *Flemingia macrophylla* shows promise in retarding weed development. Budelman also suggested that other factors can be mobilized as well that improve effect and/or durability of mulches.

(B) *Productive role*

Selection of right MPTs and application of sound silvicultural techniques in agroforestry systems help in the sustained production of fuel wood, fodder, building material, silk, and other minor tree products. Introduction of suitable MPTs supplement the farmers income and make use of land to the best advantage. Agroforestry farming on co-operative basis can help in the development of the potential of an industry by providing raw material on sustainable basis. The Western India Match Company (WIMCO) has been promoting adoption of agrisilviculture practice by planting poplars in large scale along crops. Indian farmers can play important role in suppling pulp wood for paper industry, minor forest product for minor tree product based industries. This is only possible when FGMPTs would be introduced in large scale on a co-operative basis.

(C) *Socio-Economic*

Increased employment potential is one of the main goals of scientific agroforestry practices. The agricultural labour will get employment during lean period. Jha (1988) has claimed that the introduction of FGMPTs in agroforestry system will definitely solve the requirements of the rural people and will solve the problem of unemployment.

Fig. 21 : Populus an ideal multipurpose trees : ready for felling (Courtsey : Parmar).

**Woody Biomass Production Potential of MPTs**

The growing demand of fuel wood, timber can be fulfilled by growing FGMPTs along with crops, vegetables, etc. In general, the multipurpose trees produce good quality of fuel wood. Within short period farmers can earn more from fuelwood. Swaminathan and Ravindran (1989) have studied the suitability of four tree species for use in dry zone silvipasture system and to determine relative forage productivity. Among the four species tried Subabool was found the best, attaining a mean height of 4.08 m and a girth of 14.68 cm (2½ years age), a highly significant growth compared, to the others. There was no significant difference in the growth among Sissoo, Hardwickia and Albizia. Thus the data reveal that while there was no significant differences in growth among the local species, Subabool was performed significantly better than the others, indicating a better growth rate and suitability to the site.

A comparison of the biomass production expressed in different parts of the plant species tried indicates that *Dalbergia sissoo* has the highest ratio of forage (twigs and leaves) to the total biomass produced, *i.e.* 20.30 per cent followed by *Hardwickia* (18.77%), *Leucaena* (16.68%) and *Albizia* (10.79%). But in terms of the unit production of forage subabool is found having the highest forage productivity, producing 4.036 t/ha followed by *Albizia* (0.588), *Hardwickia* (0.48) and *Dalbergia* (0.358). Deb Roy (1988) has observed that under silvipastoral system at the age of 9 years *Albizia lebbeck* produce a mean bole biomass of 77.86 kg which varies from 58.26 to 95.39 kg/tree. In case of *Albizia procera* trees of the same age recorded higher mean bole biomass of 87.54 kg. Bole biomass varies from 50.12 to 147.35 kg/tree (Deb Roy 1987). Pathak (1987) reported 74.3 t/ha/yr biomass productivity of *Leucaena Leucocephala* at 5 years rotation.

**Potential as Fodder**

A few multipurpose trees have been tested under agroforestry system in details to know their ability to produce palatable fodder and period of fodder availability. Under silvipastoral system selection of tree species could be either for one purpose or for dual purpose *i.e.* fuel-cum-fodder. Singh, *et al.*, (1988) have noted that production potential of silvipastoral system at Pantnagar. Six treatments combination consisting of three forage bushes (subabool, sesbania and desmenthes) and two grasses guinea grass + stylosanthes and nandigrass + stylosanthes were tested. Maximum green forage yield was recorded from sesbania + nandigrass + stylosanthes 1808 q/ha followed by 1757 q/ha, 1658 q/ha green forage in sesbania + guinea grass stylosanthes and desmenthes + nandigrass + stylosanthes respectively. According to Jha (1990), when *Leucaena leucocephala* was planted along with napier grass, plant height and girth under silvipastoral system showed increase in height and girth 14.9 and 17.5 per cent, respectively over control. In control plot, *leucaena leucocephala* produced 41.5, 431.5 and 448.5 q/ha fresh weight and 96.4, 101.5 and 105.5 q/ha dry weight in the year 1985, 1986 and 1987 respectively. Napier grass produced 130.0, 121.5 and 114.5 q/ha fresh weight and 28.1, 26.7 and 24.0 q/ha dry weight in the year 1985, 1986 and 1987 respectively. Under silvipastoral system (*Leucaena* + Napier) produced 574.1, 588.2, 622.6 q/ha fresh weight and 129.3, 133.7 and 137.5 q/ha dry weight in the year 1985, 1986 and 1987. He has concluded that under silvipastoral system there is increase in height, girth of trees as well as total forage and fodder production is more in comparison to control plot.

**Genetic improvement of FGMPTs**

In India research in breeding of FGMPTs is in rudimentary stage. We have not studied genetics of most of the fast growing multipurpose tree species. In India breeding (selection) so far done in conventional forestry is oriented to production of timber not the requirements of agroforestry. If agroforestry is to occupy the place that it deserves—more attention is required to be given to the genetic aspects both in respect of trees as well as agricultural crops. The production of genetically pure seed of most of the long rotation timber producing species requires more than 50 years time. In agroforestry system, we introduce FGMPTs having short rotation period. Tree scientist may get genetically pure seeds within a decade provided a co-ordinated project on tree improvement approved by the Government of India in different agroclimatic zone (Shah, 1988).

Khosla (1988), has stated that the genetic improvement of tree depends on the objectives and management factors influencing the production variable. The optimum selected genotype for current management policy is not automatically the optimum for a changed mamagement policy. For meaningful improvement, therefore, it is essential to understand the model of tree management as partitioned into management, environmental and genetic factors:

1. Management factors, artificial environment super imposed on natural conditions.
2. Environmental factors: Natural conditions as represented by climatic, topological and pedological complex.
3. Genetic factors : species, provenance and individual genotype, only chromosomal inheritance.

In order to utilize most effectively natural variation of economic traits, the genetic and environmental components must be essential. The best estimate that can be made of the actual genotype of the tree is reduce the environmental components of the equation to a minimum. The less the environmental effect, the closer the phenotype approaches the genotype.

**Role of MPTs in Village Economy**

Patil (1987) has studied the role of agroforestry in relation to socio-economic conditions of the farmers of Dhule district in Maharashtra. Nine villages were sampled and from these villages 100 cultivators were interviewed with the help of a questionnaire.

Relatively higher proportion of farmers was in the range of 31 to 60 years age group *i.e.* 67 per cent, while younger age group of 18 to 30 year constituted only 14 per cent. 58 per cent Scheduled tribes and 41 per cent of non-B.C category adopted agroforestry. The percentage of illiterate cultivators was 23 per cent, while 77 per cent cultivators was found educated. 71 per cent respondents had 4 to 5 and above 5 members in their families. Only 29 per cent families had 2 to 3 members. About 56 per cent farmers belonged to marginal, small and medium land holding groups, and 44 per cent had enough land holdings. Out of 336.20 ha total land holding of 100 farmers, only 28.80 ha land was irrigated., *i.e.* only 8.93 per cent., this factor directly affected the survival of planted tree species. Farmers also did not make efforts for application of fertilizers and irrigation from other sources. Annual fuel wood requirement of the farmers ranged from 10 to 30 quintal/year. This shows the importance of fuelwood, in the village economy, hence emphasis was to grow more fuelwood species. The average number of cattle/family was 3.89 Per capital fodder requirement was 8 kg per day.

The farmer's choice of species were *Eucalyptus teriticornis, Leucaena leucocephala, Mangifera indica, Tamarindus indica, Zizyphus mauritiana, Syzygium cumini, Acacia nilotica, Tectona grandis, Hardwickia bianata*, and other tree species like Mahua, etc. The farmers introduced tree species as per their choice in agroforestry system.

It is apparent from Table 6.1 that, percentage of survival of all most all species was very low *i.e.* only 29.1 per cent. This is due to poor rains received by the area during last two years. Causality replacement was not done by many of the respondents.

**Table 6.1 :** Estimation of outplanting and seedling survival percentage

| Sl. No. | MPTs | No.of farmers participated in planting | Total No. of seedlings planted | Total No, survived | Survival per cent |
|---|---|---|---|---|---|
| 1. | *Eucalyptus* hybrid | 96 | 1,24,293 | 36018 | 29.1 |
| 2. | *Leucaena leucocephala* | 32 | 4,800 | 1272 | 26.5 |
| 3. | *Mangifera indica* | 16 | 220 | 66 | 30.0 |
| 4. | *Tamarindus indica* | 13 | 270 | 85 | 31.5 |
| 5. | *Zizyphus mauritiana* | 17 | 500 | 325 | 65.0 |

(*Contd.*)

| Sl. No. | MPTs | No.of farmers participated in planting | Total No. of seedlings planted | Total No, survived | Survival per cent |
|---|---|---|---|---|---|
| 6. | *Syzium cumini* | 9 | 180 | 37 | 19.7 |
| 7. | *Acacia nilotica* | 7 | 715 | 250 | 42.5 |
| 8. | *Tectona grandis* | 4 | 500 | 146 | 29.2 |
| 9. | *Hardwickia bianata* | 14 | 1270 | 432 | 33.8 |
| 10. | Other species like Mahua etc., | 13 | 150 | 42 | 26.8 |
| | Total | | 1,32,898 | 38711 | 29.1 |

**Source :** Patil (1987).

Fertilizers were not applied due to lack of irrigation facilities. However, farmers took special care and provided irrigation to *Zizyphus mauritiana* which showed 65 per cent survival. Farmers choice of the species has been given in Table 6.2.

**Table 6.2 :** Distribution of respondents according to choice of species planted.

| Sl.No. | MPTs | No. of Farmers responded Yes. |
|---|---|---|
| 1. | *Eucalyptus* hybrid | 57 |
| 2. | *Leucaena leucocephala* | 22 |
| 3. | *Mangifera indica* | 29 |
| 4. | *Tamarindus indica* | 2 |
| 5. | *Ziziphus mauritiana* | 14 |
| 6. | *Syzigium cumuni* | 8 |
| 7. | *Acacia nilotica* | 13 |
| 8. | *Tectona grandis* | 19 |
| 9. | *Hardwickia bianata* | 11 |
| 10. | Other forest species | 6 |
| 11. | Total no. of farmers who did not get species of their choice | 23 |

**Source :** Patil (1987).

It is clear from Table 6.2 that in the surveyed village, maximum number of respondents preferred to plant *Eucalyptus* spp. followed by mango, subabool and teak. Good response to eucalyptus and subabool was due to fast growing character of these two species and more planting material was also made available by forest department. However, 23 respondents were not able to get the species of their choice for planting. Regarding considering eucalyptus as a MPT Davidson (1988), has pointed out that multitude of uses of *Eucalyptus* spp., ranging from timber, poles, posts, oils, medicines, honey, pulp and paper, fuelwood, charcoal, reconstituted wood products etc. makes it eligible for inclusion in MPT list, its incapacity to fix nitrogen and providing fodder notwithstands. The disadvantage of not being able to fix nitrogen in case of eucalyptus can be overcome by planting a leguminous ground cover of understoery. Choice of a fodder species for this helps to overcome the other disadvantage.

Out of the 100 respondents, only 32 have harvested subabool for green fodder and 17 have harvested fruits from *Ziziphus mauritiana* plantation. The income generated is given in Table 6.3.

**Table 6.3 :** Yield obtained from tree species

| Sl. No. | MPTs | Yield kg/ha | Rate in Rs. | Income in Rs. |
|---|---|---|---|---|
| 1. | *Leucaena leucocephala* | 6325 kg green fodder | 50.00 quintal | 3162.50 |
| 2. | *Ziziphus mauritiana* | 3332 kg fruits | 1.50/kg | 4998.00 |

**Source :** Patil (1987).

Remaining farmers did not get any income from other tree species, because growth of plants is very slow. However, all respondents were getting adequate yields from agriculture crops.

The yield of agricultural crops sown along with multipurpose tree species is presented in Table 6.4.

Though good yields were generated from high value crops like cotton and ground nut. Maximum land area was under Jowar (Sorghum) and Bajra (Pearl/millet) in Dhule district. It is clear that by incorporating MPTs in farming system, farmers are able to get good income from crops as well as from tree species. Further, no adverse effect on the productivity of crops during initial growth period of MPTs was observed.

**Table 6.4 :** Distribution of respondents (farmers) according to yield obtained from agriculture crops

| No.of Farmers | Crop | Yield/ha (in quin.) | Rate/Quital (in Rs.) | Income (in Rs.) | Cost/ha (in Rs.) | Profit/ha (in Rs.) |
|---|---|---|---|---|---|---|
| 58 | Jowar | 22.6 | 130 | 3588 | 1875 | 1713 |
| 39 | Bajra | 15.0 | 170 | 2550 | 1190 | 1340 |
| 26 | Cotton | 9.0 | 516 | 4644 | 1550 | 3094 |
| 17 | Groundnut | 9.0 | 550 | 4950 | 2680 | 2270 |
| 14 | Pulses | 7.0 | 450 | 3150 | 1140 | 2010 |
| 13 | Wheat | 15.0 | 250 | 3750 | 1450 | 2300 |
| 9 | Gram | 8.0 | 450 | 3600 | 1170 | 2430 |
| 15 | Other | 6.0 | 400 | 2400 | 1200 | 1200 |

**Source :** Patil (1987).

Respondents were interviewed to know the effect of trees on the yield of agricultural crops. They were in the opinion that there is some effect of tree growth on yield of the agril. crops. However, at this stage they are not able to assess the exact effect. They were also in the opinion that the reduction in yield of agricultural crop will be recovered by the trees in terms of getting fodder, fuelwood, timber, etc. Farmers were following the agroforestry practice since 1982-83 but they are facing difficulties mentioned below :

* Lack of finance.
* Lack of technical guidance regarding planting, after care and management of MPTs.
* Lack of irrigation facilities, and fear of higher water consumption of by eucalyptus.
* Lack of supply of seedlings as per choice.
* Lack of self confidence towards the tree planting along with crops.

Patil has suggested that more financial assistance should be given for tree planting to economically poor section. If planting of MPTs under agroforestry programmes is seriously practised on a large scale for 20 years, poverty of rural people can be eradicated.

**Future Research Thrusts**

(1) Establishment of seed orchard of MPTs to produce improved seeds.

(2) Study on seed treatments, nursery techniques and field management to produce desired product in different climatic and edaphic condition.

(3) Evaluation of large number of MPTs, through comparative trials of different seed source (both native and exotic) in each agroclimatic zone.

(4) Evaluation of leguminous and non-leguminous MPTs and assessment of their nitrogen fixation ability.

(5) Assessment of multiple characters of MPTs to know their multiple use in each agroclimatic zone.

(6) To study vegetative propagation method, appropriate to different MPTs following aspects needs to be investigated (Gupta and Bagchi, 1986) :

(*a*) Suitable age of plant from which maximum cuttings may be made.

(*b*) Suitable measurements of branches from which cuttings may be made.

(*c*) Suitable season for rooting of cuttings to be found out in different localities. Accent may be on rooting of cuttings in field condition.

(*d*) Length and diameter of the twig to be taken for cutting for maximum success.

(*e*) Genotypical difference in rooting ability from different trees.

(*f*) Nutritional requirements for higher rooting and survival.

(*g*) Induction of rooting by hormones, etc.

(*h*) Suitable season for transplanting of rooted cuttings.

(*i*) Cost of raising 1000 cuttings.

(*j*) Viability after field transplantation.

(*k*) Cost of raising plantation through cuttings.

(*l*) Viability loss due to storage.

(*m*) Suitability for tissue culture technique and cost benefit ratio.

(7) Identification of major MPTs which may be grown under different set of factors (climatic, topography, soil, geology, etc.)

(8) Evaluation of potential of MPTs under different agroforestry system in On-station and On-farm trials.

(9) Training of different levels of functionaries.

(10) Transfer the technology to the farmers field with the full involvement of farmers.

(11) Short term and long term credits to the farmers to introduce multipurpose tree species in their farm land.

(12) Creation of marketing infrastructure for the products of MPTs.

**MPTs for specific edaphic situation**

***(1) Waterlogged areas***

*Dalbergia sissoo, Eucalyptus* spp., *Butea monosperma, Albizia procera, Pongamia pinnata, Syzygium cumuni, Terminalia arjuna*, etc.

***(2) Coastal Sandy***

*Anacardium occidentale, Prosopis* spp., *Casuarina equisetifolia*, etc.

***(3) Alluvial soils***

*Leucaena leucocephala, Populus deltoides, Acacia nilotica, Azadirachta indica, Dalbergia sissoo, Albizia lebbeck, Bombax ceiba, Gmelina arborea*, etc.

***(4) Calcareous soils***

*Tamarix aphylla, Acacia nilotica, Azadirachta indica, Melia azedirach, Pongamia pinnata, Eucalyptus camaldulensis*, etc.

***(5) Acidic soils***

*Albizia lebbeck, Albizia procera, Gmelina arborea, Melia azedirach, Pongamia pinnata, Dalbergia sissoo, Acacia auriculiformis, Tectona grandis, Eucalyptus tereticornis, Terminalia arjuna, Azedirachta indica* etc.

***(6) Deep sandy soils***

*Acacia tortilis* (Var. *raddiana*), *Acacia senegal, Prosopis juliflora, Prosopis cineraria, Hardwickia bianata, Cassia siamea, Azadirachta indica, Albizia lebbeck, Tamarix articulata, Colophospermum mopane*, etc.

**Top Feed MPTs of Temperate and Tropical Region of India**

| Family | Species | Palatability | Period of availability |
|---|---|---|---|
| 1 | 2 | 3 | 4 |
| **(A) TEMPERATE REGION :** | | | |
| (*a*) MELIACEAE | (*i*) *Toona ciliata* | Fair | February-December |
| | (*ii*) *Toona serrata* | Fair | February-November |
| (*b*) SAPINDACEAE | (*i*) *Acer caesium* | Moderate | May-February |
| (*c*) ROSACEAE | (*i*) *Prunus cerasoides* | Fair | July-April |
| (*e*) MORINGACEAE | (*i*) *Moringa oleifera* | Moderate | April-February |
| (*f*) URTICACEAE | (*i*) *Morus serrata* | Moderate | April-December |
| (*g*) SALICACEAE | (*i*) *Populus ciliata* | Good | March-October |
| | (*ii*) *Salix acmophylla* | Good | May-December |
| | (*iii*) *Salix daphnoides* | Good | April-December |
| | (*iv*) *Salix tetrasperma* | Good | April-December |
| (*h*) CUPULIFERAE | (*i*) *Quercus dilatata* | Good | Throughout year |
| | (*ii*) *Quercus incana* | Moderate | May-February |
| **(B) TROPICAL REGION :** | | | |
| (*a*) SIMAROUBACEAE | (*i*) *Ailanthus excelsa* | Good | April-February |

(*Contd.*)

| 1 | 2 | 3 | 4 |
|---|---|---|---|
| (*b*) MELIACEAE | (*i*) *Azadirachta indica* | Moderate | Throughout year |
| | (*ii*) *Melia azedarach* | Moderate | March-February |
| (*c*) PAPILIONACEAE | (*i*) *Dalbergia sissoo* | Moderate | February-December |
| | (*ii*) *Sesbania grandiflora* | Good | June-November |
| | (*iii*) *Sesbania sesban* | Good | Throughout year |
| (*d*) CAESALPINIACEAE | (*i*) *Bauhinia purpurea* | Moderate | April-November |
| | (*ii*) *Bauhinia variegata* | Fair | April-November |
| | (*iii*) *Hardwickia bianata* | Moderate | May-January |
| (*e*) MIMOSACEAE | (*i*) *Acacia nilotica* | Moderate | April-February |
| | (*ii*) *Acacia senegal* | Moderate | April-January |
| | (*iii*) *Acacia tortilis* | Moderate | May-February |
| | (*iv*) *Albizia lebbeck* | Moderate | April-December |
| | (*v*) *Albizia procera* | Moderate | July-March |
| | (*vi*) *Prosopis cineraria* | Good | Throughout year |
| | (*vii*) *Prosopis juliflora* | Fair | Throughout year |
| | (*viii*) *Leucaena leucocephala* | Good | Throughout year |
| (*f*) COMBRETACEAE | (*i*) *Anogeissus pendula* | Moderate | February-December |
| (*g*) URTICACEAE | (*i*) *Artocarpus heterophyllus* | Good | Throughout year |
| (*h*) EUPHORBIACEAE | (*i*) *Emblica officinalis* | Moderate | April-February |
| (*i*) GRAMINEAE | (*i*) *Arundinaria racemosa* | Good | Throughout year |
| | (*ii*) *Bambusa arundinacea* | Good | Throughout year |
| | (*iii*) *Dendrocalamus strictus* | Good | Throughout year |
| (*j*) RHAMNACEAE | (*i*) *Zizyphus nummularia* | Good | Throughout year |

*(7) Shallow sandy loam soils*

*Acacia tortilis, Prosopis juliflora, Ailanthus excelsa, Dichrostachys glomerata, Holoptelia integrifolia*, etc.

*(8) Saline sodic soils*

*Acacia nilotica, Tamarix articulata, Terminalia arjuna Salvodora oleoides, Sesbania sesban.*

**IMPORTANCE MPTs FOR DIFFERENT AGROCLIMATIC ZONE:**

*(a) Temperate Zone*

*Acacia mearnsii, Acer ceasium, Aesculus indica, Alnus nepalensis, Celtis australis, Eucalyptus globulus, Fraxinus floribunda, Olea cuspidata, Populus* spp., *Quercus leucotrichophora, Q. dilatata, Q. semecarpifolia, Robina pseudoacacia.*

*(b) Moist temperate zone*

*Acer campbellii, Acer ceasium, Alnus nitida, Robinia pseudoacacia, Salix babylonica, Salix daphnoides.*

*(c) Dry sub-tropical zone*

*Acacia modesta, Albizia lebbeck, Albizia procera, Annogeissus latifolia, Azadirachta indica, Bauhinia purpurea, Dalbergia sissoo, Dendrocalamus strictus, Eucalyptus camaldulensis, Ficus religiosa, Melia azedarach*

*(d) Moist sub-tropical zone*

*Alnus nitida, Anogeissus latifolia, Bauhinia purpurea, Emblica officinalis, Grewia oppositifolia, Grevillea robusta, Grewia elastica, Morus alba, Toona ciliata.*

*(e) Dry tropical zone*

*Acacia auriculiformis, Acacia catechu, Acacia nilotica, Acacia senegal, Acacia tortilis, Ailanthus excelsa, Albizia lebbeck, Azadirachta indica, Cassia siamea, Dendrocalamus strictus, Emblica officinalis, Eucalyptus tereticornis, Leucaena leucocephala, Mangifera indica, Melia azedarach, Prosopis chilensis, Prosopis cineraria, Prosopis julliflora, Tamarindus indica, Tectona grandis, Ziziphus mauritiana.*

***(f) Moist tropical zone***

*Acacia auriculiformis, Acacia catechu, Acacia nilotica, Ailanthus excelsa, Albizia lebbeck, Albizia procera, Anogeissus latifolia, Artocarpus chaplasha, Artocarpus heterophyllus, Azadirachta indica, Bambusa* species, *Bombax ceiba, Cassia siamea, Dalbergia sissoo, Dendrocalamus strictus, Emblica officinalis, Eucalyptus tereticornis, Gmelina arborea, Leucaena leucocephala, Mangifera indica, Moringa oleifera, Morus alba, Syzygium cumini, Tectona grandis, Ziziphus mauritiana.*

***(g) Wet-tropical zone***

*Albizia odoratissima, Albizia procera, Anogeissus latifolia, Artocarpus heterophyllus, Emblica officinalis.*

## REFERENCES

Bhaskar, V. and Swami Rao, N. (1985). Place of farm forestry in the production of firewood, fodder and timber in the 21st century some strategies, *My forest*, Vol. 21(4) pp. 265-267.

Budelman, A. (1988). The performance of the leaf mulch of *Leucaena leucocephala, Flemingia macrophylla* and *Gliricidia sepium* in weed control, *Agroforestry system*, 6, pp. 137-145.

Burley, J.(1985). Global needs and problems of the collection, storage and distribution of multipurpose tree germ plasm, *ICRAF*. Publication, pp. 1-179.

Davidson, J. (1985). Setting aside the idea that Eucalyptus are alwase bad, *Working paper No. 10*, F.A.O., U.N.D.P. Project BGD/79/017.

Deb Roy, R. (1987). Growth and aerial biomass production of *Albizia lebbeck* under silvipastoral system. Paper presented at the *International workshop on Agroforestry* for rural needs, New Delhi (C.R.).

Deb Roy, R. (1988). Biomass production potential of *Albizia lebbeck* under silvipastoral system. Int. Book. Distributor, DehraDun, Cited from *Promotion of fodder and fuelwood trees*, Hegde, N.G., Relwani, L.L. and Kelkar, V.D. eds. BAIF, Pune, pp. 71.

Fonzen, F.P. and Oberholzer, E. (1984). Use of multipurpose trees in hill farming system in western Napal, *Agroforestry system*. 2, pp.187-197.

Glover, N. (1989). *Gliricidia production and use*, NFTA, Hawaii, pp. 1-42.

Gupta, P.K. and Bagchi, S.K. (1986). Methods and needs for vegetative multiplication of fuelwood species, *Myforest*, Vol. 22(1), pp. 113-119.

Halliday, J. and Naka, P. (1982). The symbiotic affinities of woody species under consideration of nitrogen fixing tree, *NFTAL Project*, Univ. of Hawaii, Honolulu, Hawaii.

International Crop Research Institute for the semi-arid tropics (1986). *Annual Report 1985*, Patancheru, India, p. 380.

International Institute of Tropical Agriculture (1981). *Annual report 1980*, Ibadan Nigeria. p. 186.

Jambulingam, R. and Fernandes, E.C.M. (1986). Multipurpose trees and shrubs on farmland in Tamil Nadu State (India), *Agroforestry system*, 4, pp. 17-32.

Jha, L.K. (1988). Forestry schemes to uplift the farmers of Chotanagpur plateau, *B.A.U. Publication*, Ranchi, pp. 1-26.

Jha, L.K. (1990). Role of MPTs in rehabilitation of *Jhumland, Workshop*, Social Forestry support office, Shillong.

Jha, L.K. (1990). Forage production under silvipastoral system. Paper presented in Seminar on *Environment assessment and management through Social Forestry in tribal region*, Mandelian Society of India and B.A.U., Ranchi.

Khosla, P.K. (1988). Genetic improvement of agroforestry tree *Commemorative lecture* In: *Social Forestry for rural development*, Toky, O.P. (ed.), Department of Forestry, Hissar, pp. 120-126.

Kumar Rao, J.V. D.K., Dart, P.J., Matsumoto, T. and Day, J.M. (1981). Nitrogen fixation of pigeon pea, *In proceeding of the International Workshop on pigeonpea*, Patancheru, India; ICRISAT, pp. 190-199.

Myers, N. (1980). Conversion of tropical moist forests, *National Academy of Sciences*, Washington, D.C., pp. 1-205

Nair, P.K.R. (1984). Soil productivity aspects of agroforestry, *ICRAF Science and Practice of Agroforestry-I*, Narobi, Kenya, pp. 1-86.

Ong, C.K. and Daniel, J.N. (1980). Traditional crop sparks new interest as a multipurpose tree, *Agroforestry Today*, Vol. (2), pp. 4-8.

Pathak, P.S. (1987). *Leucaena leucoephala* based production system for wasteland in social forestry. In: *Social Forestry for Rural Development* (Khosla, P.K. and Kohli R.K. (eds.) pp. 86-95.

Pathak, P.S. (1993). Multipurpose trees and shrubs for agroforestry system. In: Agroforestry-Indian scenario, Jha, L.K. and Sen-Sarma P.K., (eds.), Ashish Publishing House New Delhi, pp 31-60.

Patil, B.D. and Pathak, P.S. (1977). Energy plantation and silvipastoral system for rurual areas, *Invention Intelligence*, V-12, pp. 79-87.

Patil, D.R. (1987). *Thesis*, Diploma in Social Forestry, S.F.S. College, Dehra Dun, U.P.

Shah, S.A. (1988). *Forestry for people*, India Council of Agricultural Research, New Delhi, pp.1-147.

Singh, R.P., Bang, O.P.S and Anand Prakash (1988). AICRP on Agroforestry, *Annual Report*, G.B. Pant University of Agricultural and Technology, Pantnagar, pp. 24-25.

Swaminathan, M.H. and Ravindran, D.S. (1989). Evaluation of tree species for use in dry zone silvipastoral system, *Myforest*, Vol. 20(4), pp. 373-376.

Van Carlowitz, P.G. and Burley, J. (1984). Multipurpose tree germplasm, *ICRAF Publication*, Nairobi Kenya.

7

# *Forage and Fodder Production through Agroforestry*

National Commission on Agriculture in its report (1976) emphasized to protect, conserve and develop forest, and at the same time to create more and more forest to improve the socio-economic condition of the rural community. It has been stressed that taking up the programme of raising trees in village community land, wasteland, farmers agricultural land, marginal land and degraded forest close to the habitation could meet the forest and forage requirements. While undertaking above programme, fodder species and fuelwood species become the priority choice.

According to Deb Roy (1989) the present livestock population is over 474 million comprising more than 50% of the world's buffalo, 20% of cattle besides 103 million goat and about 55.5 million sheep. But the livestock products such as milk, meat, wool, etc. are too meagre to meet our fast increasing human population of more than 800 million which is estimated to exceed 1000 million by the end of centuries. A committee appointed by Government of India in the year 1985 estimated that availability of green fodder was only 224 million tonnes as against the requirement of 612 million tonnes to feed the present livestock population. This clearly indicates that green fodder resources of the country fall short of the requirements. In India, only 4 per cent of cultivated land grows fodder. According to Khoshoo (1986) by 2000 A.D. India would require additional land area of 10 million hectare for fodder only. Singh (1988) however, estimated that in order to ensure high productivity and economic return from livestock, the fodder requirement by 2000 A.D. could be around 1233 million tonnes on dry matter basis.

In the chapter search for sustainable development [ICRAF Strategy to the 2000 (1990)] it has been emphasized that pasture degradation is the major problem of the semi-arid and arid zones, where livestock represents an important, often the principal, means of livelihood and economic base. Over 60% of the world's productive dry lands are believed to have suffered moderate to severe degradation (Bene *et al.*, 1977). Outside the semi-arid zone, and especially in densely populated highland areas, there may also be severe degradation of areas allotted to grazing amidst predominantly arable farm land. The underlying causes of pasture degradation are increasing human population and growth in livestock numbers beyond the carrying capacity of pasture lands. Indeed, grazing areas are actually shrinking in many places, due to expansion of agriculture. In addition the fact that grazing areas are often exploited on a communal basis may create special problems. There may be little incentive for individual livestock owners to improve pasture limiting animal numbers or introducing improved grasses, legumes or trees.

Livestock constitutes an important component of the economy in the village. The economic return from such poorly fed livestock is very small, while the damage caused by such a large livestock population to the vegetative cover and environment is very serious. Low productivity of livestock is mainly due to poor nourishment (Jha and Sah, 1990). Jha also pointed out that even after survey at village level it was difficult to estimate the quantity of fodder required for each cattle, but villagers of South Bihar informed that about a head load grass (20 to 25kg) is required for 5 cattles *i.e.* on an average of 5 kg per cattle per day.

In the present context to fulfil the minimum requirements of forage and fodder, we should give more emphasis to improve the productivity of grasses, fodders in marginal and sub-marginal wastelands (Chart No. 1). These lands are not economical for agricultural crop production due to various biophysical causes. This position has to be changed by introduction of good palatable varities of fodder and forage and scientific management practices. Silvipastoral practices can ensure high yields of good quality of fodder, besides maintaining the productivity of such sites. These trees can tap the lower layers of the soil, which will otherwise not be tapped by grasses Silvipastoral practices will thus ensure better utilization of site (Jha, 1988). If all the pasture land are tackled, the yield will be multiplied every year and full requirements of grass and fodder are met with. Grass cover will also conserve soil and moisture and improve fertility status of soil.

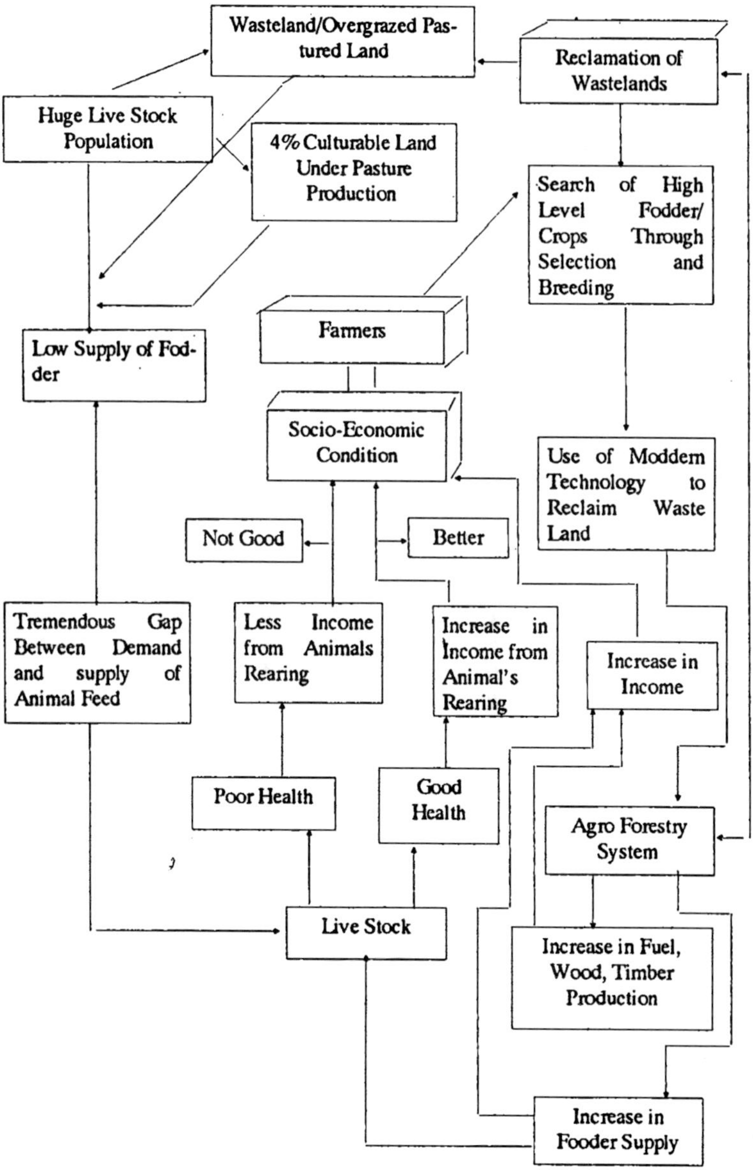


**Chart No. 1 :** A schematic diagram showing fodder and forage production in wastelands through agroforestry system and improvement of socio-economic condition of farmers.

**Area Under Wasteland**

Lal (1989) in the State of Forest Report mentions that the new assessment gives the figure of actual forest cover at 64.01 million hectare. The previous assessment based on the visual interpretation of 1981-83 space data had given the figure of forest cover in the country 64.20 million hectare. Thus, there has been a reduction of 0.19 million hectare, of forest cover during a period of 4 years. The annual rate of loss worked out 47,500 hectare.

It is apparent that 1989 study estimates the actual forest cover in the country at 64.01 million hectare against the official recorded area of 75.1 million ha. The percentage of the actual forest cover to the Geographical area works out at 19.47 as against the figure of 22.8 which is infact, the percentage of the recorded forest area to the geographical area.

According to NWDB report (1986) the extent of wastelands in our country is around to 158.06 million hectares (Table 7.1).

With the increase in human population more and more area of forest was cleared. When people got more civilized and started living in settlements they cleared as much of the forest as possible to acquire land ownership. With further advancement in civilisation more land was needed for agricultural produce, for improving the standard of living and gradually vast area was converted into waste lands. Now it is difficult to provide adequate quantity of food to cattles. Tiwari (1983) suggested that Dairy Industry had a vast potential for employing the landless people. In all suitable areas, planting of fodder trees like *Sesbania, Moringa* , etc. can be taken up in a big way to provide leaf fodder. Once leaf fodder is available, it should not be difficult to provide milch cattle to the landless families through schemes of Intensive Rural Development (IRD) etc.. and make them economically sustainable on a permanent basis. It will be necessary to earmark specific areas where fodder trees have been planted so that user may have interest in preserving such planted trees and utilising the leaves on a scientific basis so that the trees are not destroyed as is happening to day.

Above problems *i.e* improvement in forest cover as per our National Forest Policy and increase in fodder and forage productivity can be only solved if degraded forest lands and under utilized wastelands are developed and reclaimed by fuel and fodder species. In order to achieve the objective our the then Prime Minister

**Table 7.1** :Categories of land under different types of wasteland in India (NWDB, 1986).

| Sl. No. | Category | Area in million hectares |
|---|---|---|
| 1. | Water eroded | 73.60 |
| 2. | Degraded forest | 40.00 |
| 3. | Riverine | 2.73 |
| 4. | Ravines and gullies | 3.97 |
| 5. | Shifting cultivation | 4.36 |
| 6. | Sand dunes | 7.00 |
| 7. | Water logged | 6.00 |
| 8. | Saline/alkaline wastelands | 7.50 |
| 9. | Wind eroded | 12.90 |
| | Total : | 158.06 |

(Late) Sri Rajiv Gandhi had emphasized the need of bringing five million hectares of land every year under fuelwood and fodder plantations. With the thrust on increase in grain productivity, there is minimum possibility for increasing the fodder/forage production from the present agricultural land. We can achieve our target by reclaiming the wastelands under forage and fodder or by introducing two or three tire system of agroforestry in farmers land. However, our main target area is wastelands. Rehabilitation of wastelands for the rehabilitation of weaker section of society should be our motto. (Chart No. 2)

**Need for Defining Wasteland for Fodder and Forage Production**

In my opinion, it is essential to define wastelands that can be reclaimed under fuel/fodder plantation under agroforestry systems to increase the productivity of animal feed and the fertility status of soil.

Various experts tried to define areas that can be considered as wasteland. Patel (1985) has suggested that "The land where no greenery can be sustained is termed as wasteland". Cultivated land is turned into the marginal and wasteland due to water logging and salinity. According to S.P.W.D., any land which is not producing green biomass consistent with the status of soil and water must be treated as wasteland. SPWD has suggested that areas under wasteland can be classified with regard to ownership *i.e.* (*i*) Forest

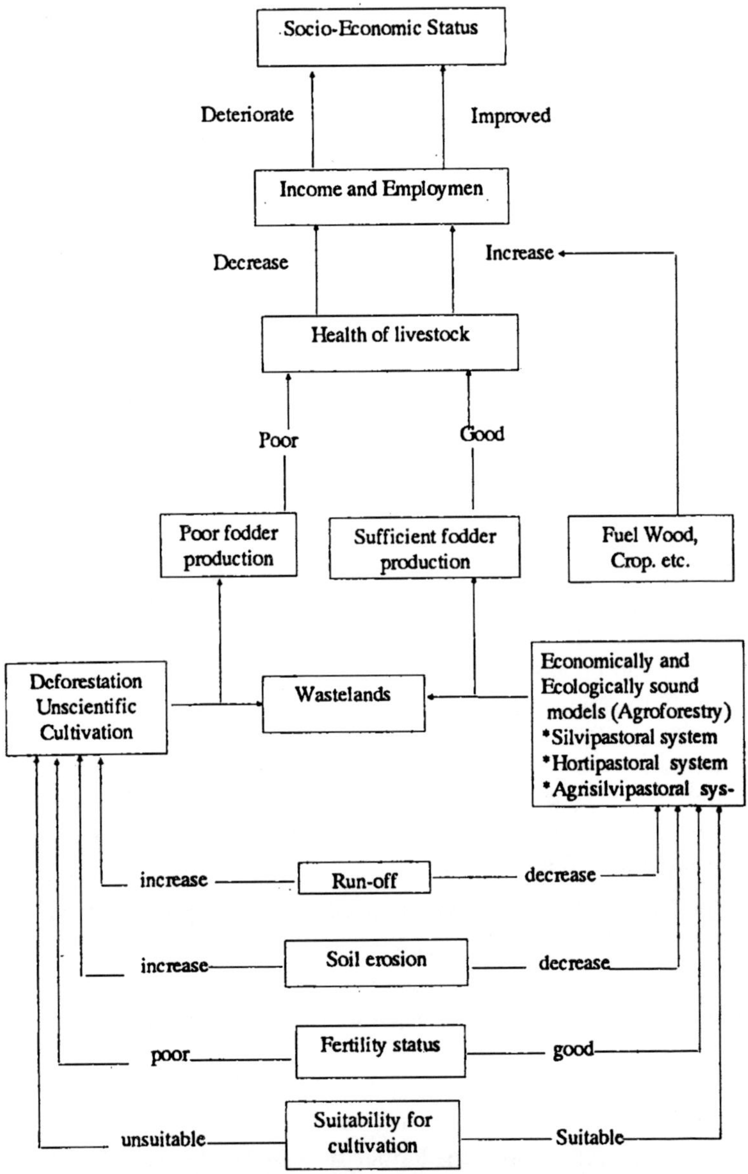


**Chart No. 2 :** A schematic diagram showing agroforestry as an economically and ecologically sound models to increase the productivity (fodder, fuel, etc.) in wasteland.

land-a considerable extent of degraded forest may be classified as wasteland; (*ii*) farm lands— consists degraded portions both inside the fields as well as on inter cultivation bunds etc. (*iii*) public land— (*a*) owned by government departments such as railways, canals, public works, education institutes, etc., (*b*) village community land includes village common land, gram sabha land, grass lands, gochar lands, lands under water tanks, etc.

Malhotra and Sharma (1988) opined that wasteland could be defined as (*a*) all vacant land lying unused due to negligence or having become unproductive and non-profitable due to factors like grazing, erosion, removal of vegetation and developed toxicity at root zone etc.; (*b*) areas where estimated cost of inputs for raising agricultural crop was more than net estimated profits.

Jha (1991) has defined wasteland as "Those land which may not be put under proper use due to various biophysical problems and practice of agriculture will be non-profitable without desired treatment as per site condition". He has suggested that to improve the productivity, suitable designs/models for each agroclimatic zone are required to be developed. He has also stated that the fertility status of the wastelands can be improved by introducing silvipasture system.

On 24.4.86, a meeting was held at Vigyan Bhavan, New Delhi to finalise definition of wasteland. After day long deliberations the following conclusions were unanimously arrived at, as presented by Director, SRAC, Uttar Pradesh. Wasteland was defined as "That land which is degraded and is presently lying unutilised (except as current fallow) due to different constraints".

Wastelands would consist primarily of (*i*) Culturable wasteland and (*ii*) Unculturable wasteland.

**"Culturable wastelands" would comprise**

1. Gullied and/or ravinous land.
2. Undulating upland.
3. Surface waterlogged land and marsh.
4. Salt affected land.
5. Shifting cultivation area.
6. Degraded forest land.
7. Degraded non-forest plantation.

8. Sandy area.
9. Mining and industrial wasteland.
10. Strip lands.
11. Pasture and grazing land.

**"Unculturable wasteland" would comprise**

1. Barren rocky/stony wastes/sheet rock area.
2. Steep sloped area.
3. Snow covered and/or glacial area.

## AGROFORESTRY SYSTEM FOR FODDER PRODUCTION

Ecologically, the retrogressive trend in the vegetation dynamics from woodland habitats to rangelands, grasslands and ultimately the wastelands in the tropics and sub-tropics has been a process that had been operative over thousand of years. In an attempt to give a turn in the opposite direction towards *i.e* restoration of available resources, one of the possible and promising alternatives appears to be agroforestry systems of production. Several grain, vegetables and fruit crops are commonly intercropped at the same unit of land to meet the human needs. However, the other

**Fig. 22 :** Grasses grown along with *Dalbergia sissoo* at Kumharia village, Ranchi

Fig. 23 : Grasses grown along with *Melia azadirach* at Kumharia village Ranchi (Courtsey ; Faculty of Forestry, B.A.U. Ranchi).

components *i.e.* feed and fodder requirements for the animals are unfortunately and invariable lost sight of. Such anamolous situation creates imbalance in the mixed farming system causing additional pressure on the culturable land for fodder production (Singh, 1984).

The system of fodder-forage production or fodder, forage and food production aims as obtaining food, forage and fodder from the same patch of land and this can be achieved through following approach.

**Silvipastoral system**

Under silvipastoral system of land management improved pasture species are introduced along with tree species. Here selection of tree species could be either for timber alone or for dual purpose *i.e* fuel-cum-fodder. We can say that under this system grass or grass legumes mixture are grown along with woody perennial simultaneously on the same unit of land. Combining trees, grasses and legumes, also helps to conserve soil moisture and finally improve the nutrient status of soil. This system may be subgrouped into two :

(*a*) Silvipastoral system with controlled grazing for domestic animals.

(*b*) Silvipastoral system in which domesticated animals are not permitted for controlled grazing. Forage are utilised for stall feedings. Some workers are of opinion that this system comes under agrisilviculture (King 1979).

This system of land management is best for the wastelands even with marginal fertility. The trees can tap the nutrients from the deeper layers, which will not be tapped by forage species. This system may be introduced as a two or three tier system.

(*a*) *Two tier system*

In this system, multipurpose tree species are integrated with grass on the same piece of land. Tree species are selected as per choice of farmers. While selecting tree species, farmers should always consider biophysical aspects of the land. Perennial grasses can be grown by sowing or root slips transplanting after establishment of tree species or in the same year tree species can be introduced along with grass. In the initial years their is no problem of competition for solar light and grass is not affected by shade etc. After 3rd-4th years these problems can be tackled by appropriate canopy management. Tree foliage can be used as fodder, mulch or green manure.

(*b*) *Three tier system*

Under three-tier system, one more component, pasture legumes are introduced in above (two) tier system. In this system, perennial, short statured legumes are introduced. This occupy the lower most canopy due to its spreading habit. In this system each component is managed in such a way that they remain complimentary to each others. Singh (1984) has stated that the advantages of growing grass and legumes among the fodder trees are that during the wet season or the period in which these forage species remain green, animals can look to the ground for green foliage but during the lean period animals can look up for the tree foliage as top feed for sustenance. Legumes also provide complete ground cover as well as high nutritive fodder. Complete ground cover intercepts the precipitation, decreases the run-off and increases water holding capacity of soil. The shade effect of tree is reduced by proper lopping management which are used as leaf fodder during lean period. This may also be used as green manure or

Fig. 24 : Silvi-pastoral : Grasses grown along with *Acacia auriculiformis* at Tilta village, Ranchi, (*Courtsey* : Faculty of Forestry, B.A.U., Ranchi).

mulch. Ultimately tree foliage helps in improving the fertility status of soil.

### Hortipastoral System

In this system fruit trees are raised and managed in the farmland along with grass species. This practice can be adopted even in old orchards or where trees are planted at larger spacing. Here, also farmers can increase the productivity of the site by introducing three-tier system *i.e.* pasture legumes besides the grass species. The perennial legume will occupy the lower most tier. Thus, the farmers will achieve dual benefit *i.e.* fruits for human consumption and fodder for animal consumption. Fruit tree species and grasses are selected keeping in mind biophysical attributes of the soil and available market for disposal of excess fruits. Grass species like Ginger grass (*Cymbopogon flexuosus*) may also be exploited for extraction of valuable cosmetic oils. From the roots of vetiver grass scented mats are prepared by the farmers. These scented mats are used in offices during summer season. Farmers can also sow grass species like *Erianthus munga*, *Erianthus ravennae*, etc. for thatching of roofs.

**Agrisilvipastoral system**

This is a more complex system. This system is the result of an union of silvipastoral and agrisilviculture system. Under this system, land is managed to get food crops and cash crops, fuelwood and farmers also utilise the same piece of land for animal rearing.

**Silviagripastoral and fisheries**

As discussed in the earlier chapter "Types of agroforestry system", this is the most complex system of farming, although we do not have relevant data related to this system and therefore, requires indepth study. In some states of India, trees are planted on the terraced land around the pond. If fruit trees are planted at closer spacing *e.g.* Nariyal (*Cocus nucifera*) supari (*Areca catechu*) etc., they cause shade on the ground surface, some shade tolerant crops/vegetables/spices/grasses are grown under tree cover. If tree species are planted at wider spacing then other cash crops/food crops/grasses are grown by the farmers. The ponds are used for Makhana (*Euryale ferox*) cultivation and fish breeding. The dried leaves of tree species, crops and makhana leaves etc., enrich the pond nutrients. The system is characterised by long path of material energy flow within an Ecosystem.

**Fodder and foliage production under alley cropping**

Under this system crops/grasses are grown in the interspace between two hedgerows which are established by trees or woody shrubs. If tree species are established in the hedgerows, space between the two tree species or woody shrubs are utilised by grass species as fillers. The cut material of hedgerows are utilised as green manure (chopped foliage mixed in the soil) or used as mulch (spread over the soil surface). The green manure adds considerable quantity of organic matter to the soil in addition to nitrogen. For this foliage of leguminous trees/shrubs is incorporated under wet weather conditions into the soil. When foliage is chopped, this should be incorporated in the soil at a depth of 20-25 cm. Sufficient time should be allowed for proper decomposition of green manure before sowing the cash crop/food crop/grasses. It is important to mention that the nutrient release from the green manure depends on the age of the tree and climate of a particular zone. During fallow periods, tree species (hedgerow) are allowed to grow to produce firewood. If farmers can afford to use Farm Yard Manure, the foliage of nitrogen fixing tree species/shrubs can be better used as animal feed. Alley cropping system also improves soil structure. Foliage when used as mulch reduces soil temperature and controls soil erosion. Alley cropping is best suited for hill cultivation. As the vegetative hedgerows create effective barriers, silt gets deposited behind hedgerows and natural permanent terraces are formed. Further, the bacteria living in the root nodules of leguminous trees fixes 2/3rd of the total atmospheric nitrogen.

**Choice of grass species under silvipastoral system**

The selection of species/varieties/strains would vary as per the agro-climatic condition and biophysical attributes of particular zone. Besides this, cropping pattern, design and socio-economic conditions of farmers also greatly influences success of particular species of grass.

On the basis of life duration, grasses are classified as follows :-

**(a) Short duration grass**

If farmers desire to grow short duration grasses, they have to be made aware about the productive and nutritive grass best suited in a specific edaphic and climatic condition. The grass species should have the ability to sustain more than one cut and to self seeding. It should produce maximum fodder yield within short period. In this connection, Singh (1984) has reported that among the short duration grasses, the well adapted as well as highly productive and nutritive grasses for the plateau region is Dinanath grass (*Pennisetum pedicellatum*). It is seeded in June with shallow tillage @ 15kg/ha. Green fodder yields upto 800 q/ha. About 20 per cent dry matter can be expected, if adequate compost and chemical fertilizers have been used at the time of showing. Jha and Sen (1988) while reporting the results of lab to land extension project at Ranchi have reported that Dinanath yielded up to 350 q/ha under silvipastoral system in wasteland, where the farmers were not able to grow even a single blade of grass.

**(b) Perennial grass**

Perennial grass plays very important role specially in hills. I would like to give more emphasis here that large scale farming of perennial grass under silvipastoral system on wastelands in plains as well as in hill slopes will help in bridging the gap of demand and supply of fodder. The decision for the choice of species depends upon villagers. But extension workers should explain about the best suitable perennial grasses for a particular zone in open as well as under shade condition. To convince the farmers, demonstration of on-station models should be considered as most important factor. Farmers should know the best grasses in open land and grass under shade condition. Singh (1984) has stated that in rainfed condition thin napier (*Andropogon gayanus*) and signal grass (*Brachiaria brizantha*) can be grown either through seeds or cuttings during June-July. In four or five cuttings, farmers can harvest green fodder up to 400-500 q/ha with 20 per cent of dry matter and 6.8 per cent crude protein. Guinea grass (*Panicum maximum*), Rhodes grass (*Chleris gavana*) and Nandi grass (*Setaria ancepts*) can be sown in June-July with herbage yield potentials up to 500 q/ha, 10-20 per cent dry matter and 6-8 per cent crude protein.

### Legumes

As discussed earlier, introduction of third component *i.e.* legume under agroforestry (three-tier) system, raises the productivity of site. Like grass species, legumes are also seasonal and perennial. Among the seasonal legume species most common are stylo (*Stylosanthes humilies*) and Horse gram (*Dolichos biflorus* Cv. Leichhardt). Their herbage yield potential is up to 300 q/ha. Among the perennial legume. *Stylosanthes quianesis* Cv. *schofield.*, *Stylosanthes scabra* Cv. *Seca, Stylosanthes hamata* Cv. Verano, *Macroptilim atropurpureum* Cv. Siratro are suitable for rainfed farming.

Desirable characteristics of forage species :

1. High nutritive value, palatable and easily digestible.
2. Suitable for making good quality hay or silage
3. Easy and quick regenerative power.
4. Good resistance against adverse climatic and edaphic factors *e.g.* acidic soil, water logging etc.
5. High nodulating and carbohydrate reserve abilities in forage legumes for subsequent regeneration following frequent cutting and compatibility with associated crops in mixed/intercropping system (Lal, 1990).

### Pasture Production under Agroforestry System

Extensive research work on various components of agroforestry to increase the productivity of fodder and forage on wastelands have been carried on in several International and National Research Institutes and Universities.

**Pasture productivity**

The International Livestock Centre for Africa (ILCA) has carried out extensive research on hedgerows fodder production system in Nigeria. These systems are helpful for small and medium scale livestock producers. *Gliricidia* is planted in hedgerows 4 metres apart with 4 rows of pasture grass (*Panicum* spp.) planted between the hedgerows. This system produces 20 tonnes of dry matter/ha/yr of mixed grass gliricidia fodder which can feed approximately 3 heads of cattle. An alternative design, which is equally productive and has a higher proportion of protein, has also been tested. In this, hedgerows are spaced 2.5 m apart with 2 rows of grass between the hedgerows. This design requires a higher level of management because the trees must not be allowed to shade the grass and preventing overgrazing.

The farm "La pacifica" located in Cost Rica uses gliricidia as a protein supplement during the dry season for beef and dairy cattle production. Gliricidia fodder is harvested from live fences which occupy a 12 m wide strip around the farm. Trees are harvested on a 3-5 year rotation. Twenty trees occupying 480 $m^2$ (40 x 12 m) produce 300 kg of fresh gliricidia fodder (75 kg dry fodder). The fresh fodder is chopped and mixed with King grass (*Pennisetum purpurium*) and (*Pennisetum typhoides*). The gliricidia-king grass mixture consists of 30% gliricidia and 70% king grass, on a fresh weight basis. Cattles are fed 2 kg/day of fresh gliricidia-king grass as a grazing supplement. A live fence around one hectare on this farm provides enough gliricidia for approximately 2 heads of cattle, using a 5-year rotation with 150 days of supplemental feed. Trees managed on a 3-5 year rotation also produce fire wood and fence post for use and sale (Nancy Glover, 1989).

Shankar *et al.*, (1976) have studied the influence of *Prosopis cineraria* on the range grasses, and showed that the dry grass yield under khejri was 2.3 tonnes/ha as against 1.66, 1.32, 0.85 and 0.75 tonnes/ha under *Tecomella undulata, Albizia lebbeck, Prosopis juliflora* and *Acacia senegal* respectively. Paroda *et al.* (1977), during silvipastoral studies at Jodhpur, tested compatibility of *Cenchrus ciliaris* with *Acacia tortilis, Colophospermum mopane,* and *Leucaena leucopcephala.* They found 98.5% survival of *A. tortilis,* followed by 68.1% of *Leucaena Leucocephala* and only 26.7% of *C. mopane.*

The dry forage yield of grass ranged between 10 to 16 q/ha. However Paroda and Muthana (1979) have found the best tree grass combination of *Azadirachta indica* + *Cenchrus* species and *A. tortilis* + *Cenchrus* species. Muthana and Shankaranarayan (1978) found no significant difference in forage production under different tree species in their early stage of growth.

Deb Roy *et al.,* (1990) studied on the establishment and forage production of *Cenchrus ciliaris* and *Cenchrus setigers* in association with *Acacia tortilis* and *Leucaena leucocephala. Acacia tortilis* recorded an establishment varying from 80-100 per cent *Leucaena leucocephala* exhibited more than 75 per cent establishment. In general the growth was better with wider spacing compared to closer one. Statistical analysis of pooled data revealed that the forage yields amongst the treatment were not significant. Forage yields were significantly affected by the amount of rainfall in various years. Maximum forage yield (average of various treatments) of 51.44 q/ha was recorded during 1975 followed by 33.69, 30.90, 30.18 and 24.28 q/ha during 1974, 1972, 1976 and 1973 respectively. They had also mentioned that forage production had some relation with rainfall. Higher the rainfall, higher was the forage production. The average forage production for five years comes to 34.23 q/ha. *Cenchrus ciliaris* in association with *Acacia tortilis* gave an average forage production of 35.0 q/ha while in association with *Leucaena leucocephala* it gave higher production (40.7 q/ha). In contrast, *Cenchrus setigerus* gave higher production (31.2 q/ha) in association with *Acacia tortilis* compared to *Leucaena leucocephala* (29.85 q/ha).

Mann and Muthana (1984) discussed about the silvipastoral system in the Arid zone. The objective was to study the establishment and growth of economic tree species, *viz., Holoptelia integrifolia* in association with grasses to find out the possibilities of suitable tree grass combination. The forage estimation revealed that 7 quintals of *Cenchrus ciliaris* and *Cenchrus setigers* per hectare per year was produced on an average without adversely affecting the growth in height at the tree species up to fifth year of plantation. From the sixth year, 50 per cent of the trees were lopped for fodder an average yield of 310 kg dry fodder per hectare was recorded.

Shankarnarayan (1984) conducted silvipastoral studies at CAZRI, Pali in 1970, involving treatments comprising of four tree species (*Acacia tortilis, Azadirachta indica, Albizia lebbeck* and

*Holoptelia integrifolia*), two weedings (Hand weeding and application of weedicides) and four grasses (*Cenchrus ciliaris, Costetigerus* species, *Dichanthium annulatum* and *Panicum antidotale*) and with no grass as control. Results showed that there was no significant difference in the dry grass yield under different tree species but the maximum forage yields of 36.1 and 14.4 q/ha were recorded in 1975 and 1976, respectively. The yield difference under weed control by manual labour or weedicides was not very marked. The mean dry forage yield of *Dichanthium annulatum* was 28.0 q/ha followed by *Cenchrus ciliaris* (25.1 q/ha) and *Panicum antidotale.*

Singh and Joshie (1985) have experimented to explore the possibility of utilising highly degraded riverbed lands (with 74% boulders) in Doon valley. The four top feed tree species *viz. Albizia lebbeck* (Siris), *Grewia optiva* (Bhimal), *Bauhinia purpurea* (Kachnar) and *Leucaena leucocephala* (subabul) at 4 m. espacement formed the main treatments, other treatments were two lopping intensities (*i.e.* 50 p.c and 75 p.c.) and two grasses *viz. Chrysopogon fulvus* (Gorda) and *Eulaliopsis binata* (Bhabar) at 1 m x 0.5 m espacement in between the tree species as subtreatments. From such lands 55.63 q/ha of dry Bhabar grass and 33.45 q/ha of gorda grass were harvested. Siris allowed maximum grass growth followed by subabul, kachnar and bhimal (in 1st four years). In the fourth year, subabul surpassed all the tree species in allowing more growth of Bhabar. Singh (1984) has observed that under silvipastoral system trees like *Leucaena leucocephala, Albizia lebbeck, Tephrosia,* etc., when planted in sufficiently spaced rows, combination of compatible grasses and legumes in between the rows can be sown in mixture as two or 3-tier system, instead of allowing the weeds to grow in the vacant space. The best combination for 3 tier system under the dry land condition of plateau region has been found to be growing a combination of sadabahar or signal grass with stylo as understorey of fodder trees, whereby total green fodder yield of 1000 t/ha/yr could be obtained.

According to Pathak (1984), very high production of forage (108.30 t/ha) could be obtained by intercropping *Leucaena leucocephala* (2 x 1 m) with guinea grass and berseem. Leucaena can be pollarded 4 times, to increase the production efficiency.

Singh (1988) have observed that silvipastoral technology is successful in areas with 1000 mm rainfall with a dry season for nine months. It has potential to produce 3.0-7.5 t/ha/yr fodder and 4.5-6.0 t/ha/yr fire wood. In multitierd system and additional grain yield of 0.6 t/ha/year can also be obtained. In such situation, it is possible to maintain one adult animal/ha during lean period. Sharma (1989), in hortipastoral studies in sandy rangelands of arid Rajasthan has observed that Buffel grass (*Cenchrus ciliaris*), Jujube (*Zizyphus mauritiana*) system yielded upto 12 q/ha/year of forage in normal rain fall year with about 200-250 plants/ha and this did not affect the forage yield.

Deb Roy (1986) reported higher annual forage production of 6.3 t/ha in *Albizia lebbeck-Cenchrus-Stylo* silvipastoral system compared to 4.78 t/ha/yr. in *Albizia procera-Cenchrus-Stylo* silvipastoral system. Dhukia (1989) reported that field trial on rabi forages (berseem, oats, bokla, metha) and rabi grain crops (Wheat, grain) was conducted under Shisham (*Dalbergia sissoo*) and Siris (*Albizia lebbeck*) plantations for a period of three years (1984-85 to 1986-87) on 2-4 years old plantation of H.A.U Hissar. Among forages berseem followed by Oats gave maximum yield during all the three years under both the plantations. Fodder yields decreased under 4 years old trees as compared to 3 years old plantation. The grain yield of wheat was higher under both the plantation. Yields of both wheat and gram were higher under shisham trees which declined with increasing age of the trees. The depressive effect of intercrops with the advancement of age on yields indicate a negative association between yield of intercrops and age of trees. Sharma, *et al.*, (1988) have reported higher grass production of 7.6. t/ha under *Robinia pseudocasia* compared to 5.2 t/ha under *Bauhinia variegata* and 3-5 t/ha in natural grassland.

Konwar *et al.*, (1988) reported that *Leucaena leucocephala* seedlings were planted in contour furrows spaced 7.5 m apart, at an inter-row spacing of 2m. The inter-row spaces were sown by *Cenchrus ciliaris* and *Stylosanthes hamata* each on 50 per cent land. The pasture was harvested every year, while leucaena was allowed to grow as a tree for pole and fuel wood. *Leucaena leucocephala* and *Stylosanthes hamata* system was more productive (41.5 q/ha) than *Leucaena leucocephala* and *Cenchrus ciliaris* (24.8 q/ha).

Singh and Prasad (1989) reported result of intercropping studies in subabool with annual cereals and legume. The observation revealed that narrow spacing (2 m x 1 m) of subabool performed better over the wider spacing (3 m x 1 m) in total green as well as dry matter yields. Teosinte, Jowar and maize as pure crop or teosinte and maize in association with cowpea sown as intercrops between subabool line (2 m x 1 m) gave on an average maximum green forage yield (23.6 to 26.2 t/ha).

Jha (1990) reported results of an on farm experiment at village Tilta (Ranchi). *Leucaena leucocephala* were planted along with Napier grass. Plant height and girth under silvipastoral system showed increased of 14.9 and 17.9 per cent respectively over control. In control plot, *leucaena leucocephala* produced 415.5, 431.5 and 448.5 q/ha fresh weight and 96.4, 101.5, and 105.5, q/ha dry weight in the year 1985, 1986 and 1987 respectively. Napier grass produced 130.0, 121.5 and 114.5 q/ha fresh weight and 28.1, 26.7 and 24.0 q/ha dry weight in the year 1985, 1986 and 1987 respectively. Under silvipastoral system (Leucaena + Napier) produced 574.1, 588.2, 622.6 q/ha fresh weight and 129.3, 133.7 and 137.5 q/ha dry weight in the year 1985, 1986 and 1987. Jha (1990) concluded that under silvipastoral system there was increase in height and girth of trees as well as total forage and fodder production relative to control plot.

Four tree species safeda, subbabul, kikar and khair were planted at different spacing. The grass showed good performance under *Acacia* species, because of wider spacing. Grasses Bhabar and Anjan yield maximum under *Acacia catechu i.e.* 98.8 q/ha/year/. CPI-59956 and CPI-59972 yielded maximum under *Acacia nilotica.* Overall maximum yield (450 q/ha) was given by CPI-59956 (425 q/ha) and poorest (81 q/ha) by Bhabar grass (Anon; 1988).

Grewal and Abrol (1987) have observed that in alkali soils fresh forage production of Karnal grass (*Diplachne fusca*) with *Acacia nilotica* was 7.7 t/ha in the first year and 16.5 t/ha in 2nd year.

Singh and Gill (1988) studied effect of management practices on mesquite (*Prosopis juliflora*), Karnal grass (*Leptochloa fusca*) agroforestry system on a sodic soil. Mesquite was planted at three spacing *i.e.* 2 x 2 m, 3 x 3 m, and 4 x 4 m, and the effect of lopping was observed. Trees were planted in trenches, 30 cm deep and 30 cm wide dug across the plots. These were filled with mixture of original soil and gypsum ( 3 kg plant$^{-1}$). Karnal grass was planted through root cutting in August, 1984 and lopping of mesquite was done in November, 1985 and December 1987. Lopping improved mesquite growth significantly. Green forage yield from Karnal grass in a growth period of 40 months was 62 t/ha$^{-1}$ under lopped and 49 t/ha$^{-1}$ under unlopped condition.

Singh, *et al.*, (1988) have reported production potential of silvipastoral system at Pantnagar. Six treatments combination consisting of three forage bushes (Subabool, Sesbania and Desmenthes) and two grasses Guinea grass + Stylosanthes and Nandi grass + Stylosanthes were tested. Maximum green forage yield was recorded from Sesbania + Nandi grass + Stylosanthes (1808 q/ha) followed by 1757 q/ha, 1658 q/ha green forage in sesbania + guinea grass + stylosanthes and Desmenthes + Nandi grass + Stylosanthes, respectively.

Arya (1989) studied mean dry grass production of 81.95 q/ha under popular plantation at 4 x 7 meters spacing as compared to 55.69 q/ha in 3 x 3 m spacing, where as under Eucalyptus trees planted at 3 x 3 m spacing yield was only 14.92 q/ha.

Mishra *et al.*, (1988) have studied a long term interaction trials of plum (*Prunus saliciana*) with *Populus deltoides* 'G-3' which were established at Kalaghat area, Solan during, 1984. The density of plum plant was kept at 278 plum tree/ha. The poplars were planted as gap filler among plum trees and on the boundaries. They reported that within 3-5 years age, poplar plants have been able to attain 7.81 to 8.81 cm average diameter and 5.82 to 6.90 m average height. The plum plants attained 2.81 to 3.60 m average height and 4.48 to 6.05 cm average diameter. The dry grass yield varied from 30.5 to 56 q/h/yr.

Singh (1986) investigated, the compatibility and productivity of grass under natural forest of Utis (*Alnus nepalensis*) and Chillowance (*Schima wallichii*) with tree density 28-48 per 100 sq.m (at Sikkim AICRP on agroforestry Centre). Four shade loving grasses amliso (*Thysanolaena agrostis,* Dhotesaro (*Setaria palmifolia*), Napier (*Pennisetum purpureum* and Love grass (*Eragrostis curvula*) were planted at 60 x 30 cm spacing. Love grass completely failed to grow. Amliso did not give any yield in 1984 but it produced consistently, higher yield of 30 and 50 q.ha$^{-1}$ dry matter in third and fourth year. Napier grass yielded about 31 q.ha$^{-1}$ dry matter in 1st year but the yield reduced significantly in the third (14 q.ha$^{-1}$) and the fourth year (24 q.ha$^{-1}$).Dhatesaro gave about 14 and 25 q.ha$^{-1}$ dry matter in 2nd and third year, respectively. Thus amliso produced more forage than napier and dhotesaro. Napier grew well only where tree densities were lower (20-22/100 sq.m.) but failed to give yield at higher tree densities. In contrast amliso and dhotesaro gave consistantly higher yield from third year onward under high tree densities of (25-48/100 sq.m.). In the same report he has also mentioned result of another experiment. A representative wasteland area was selected with 60 per cent slopes. The land was susceptible to land slides. The area was rocky with gravelly soils. Nevaro (*Ficus hookerii*) saplings were planted at a spacing of 4 x 3.5 m during monsson season of 1981. These were allowed to grow unlopped for three years and were lopped in fourth and fifth year. Ten grasses (Love grass, Para grass, Napier grass, Orchard grass, Amliso grass, Dhotesaro, Nandi, Setaria, Fodder triticate, Rye grass and tall fescue) and seven legumes (Sevri, Hedge lucerne, Veranostylo, Brazilian stylo, Townsville stylo, White clover and Scabra stylo) were introduced during monsoon in fifth year based on single tree plot treatment. Tree was in the centre of 2 x 1.75 sq.m. plots. Observation on growth of tree in fifth year showed variation in plant height between 3.5 to 5.6 metres. The collar girth of trees varied from 28 to 56 cm. The tree having larger collar girth were those possessing bifurcated and sometime trifurcated stem. The range of variation in single stemmed tree was 28 to 40 cm. Data on forage yield showed on an average 4.70 ± 1.06 kg green fodder yield per tree in the fourth year. This gave about 24.3 to 39.4 $q.ha^{-1}$ green fodder with dry matter yield of 7.0 to 11.4 $q.ha^{-1}$. Fodder yield from tree increased in fifth year by 81 per cent *i.e.* 4.70 to 8.54 kg. green matter per tree. This is largely due to more branching as a result of lopping in fourth year. The yield varied from 3.00 to 14.35 kg per tree reflecting variation in growth of trees. Consequently total green fodder yield varied from 21.42 to 102.50 $q.ha^{-1}$ with a dry matter yield of 6.21 to 29.73 $q.ha^{-1}$. Among the introduced grasses napier (*Pennisetum purpureum*), amliso (*Thysanolaena agrostis*), paragrass (*Brachiaria mutica*), dhotesara (*Setaria palmifolia*), Love grass (*Eragrostis curvula*) and nandi setaria (*Setaria sphacelata*) gave higher green forage yield. Amliso and napier gave about 22-23 $q.ha^{-1}$ dry matter yield. This was followed by dhotesara (11.37 $q.ha^{-1}$) love grass (9.80 $q.ha^{-1}$), para grass (8.30 $q.ha^{-1}$) and nandi setaria (7.70 $q.ha^{-1}$).

Thus, the total forage yields obtained under silvipasture system on steep slope were 204.4, 156.0, 130.2, 114.7, 99.2 and 89.8 $q.ha^{-1}$ green matter, having napier, amliso, love grass, dhotesara, nandi, setaria and paragrass, respectively. None of the legumes reached the growth that can be harvested but Brazilean stylo grew comparatively well. Growth of hedge lucerne and sevri was very poor. Cloves, towns-ville and Scabra stylo completely failed. This showed the possibility of growing napier, amliso and dhotesara successfully under nevaro in Sikkim.

**Influence over nutrient status of soil**

Silvipastoral system gradually improves the nutrient status of soil, due to continuous flow of nutrient from litter fall root biomass, which help in improving organic carbon, nitrogen, phosphorus contents of soil.

Singh *et al.*, (1979) studied change in soil fertility in two silvipastoral experiments, one initiated in 1969 and the other in 1973 at the IGFRI, Jhansi. Silvipastoral experiments (I) comprised of the trees *Leucaena* species ($T_1$) or *Acacia* species ($T_2$) in combination with natural grasses ($G_0$), seeding with *Cenchrus setigers* ($G_2$) was initiated in 1969. Another silvipastoral experiments (II) comprised of fodder-cum-fuel trees, *Hardwickia bianata* ($T_1$) or *Albizia amara* ($T_2$) was initiated in 1973. The result indicate that growing of *Cenchrus setigers* ($G_2$) over natural grass ($G_0$) caused 24, 6.7 and 50 per cent increase in organic-C, total -N and organic-S respectively. Thus, in five year *Cenchrus setigers* ($G_2$) grass seeded in tree space resulted in a turnover of 3.4 tonnes of organic matter, 67 kg each of total-N and organic-S per ha. Further, seeding with *Cenchrus ciliaris* ($G_1$) yielded higher organic carbon and total-N over those of natural grass ($G_0$) in silvipasture system which worked out to a turnover of one tonne of organic matter and 45 kg of total N per ha in five years. In silvipastoral experiment II, seeding with *Cenchrus ciliaris* ($G_1$) over natural grasses ($G_0$) caused a percentage increase of 6.5 and 2.4 in organic -C and total -N, respectively, which amounted to a yearly turnover of one tonne of organic matter and 22.4 kg of total -N per ha.

Jha (1990) studied silvipastoral system with napier grass along with the *Leucaena leucocephala* at Ranchi. The experiments showed increase of organic carbon, available-K and pH of soil. In the initial stage level of organic carbon, available-P and available-K and pH were recorded 0.4%, 29 kg/ha, 152 kg/ha and 5.8 respectively. After three years level of organic carbon, available-P, available-K pH were recorded 0.49 per cent, 46 kg/ha, 179 kg/ha and 6.5 respectively.

**Fig. 25 :** Perennial grass grown along with *Leucaena leucocephala* (Courtsey : Manish Kumar)

## Micrometrological parameters

Silvipasture system influence the micrometrological parameteres, which influences the productivity of site. Only few relevant literature are available in this field. Hazra and Patil (1986) studied micrometrological condition under silvipastoral system. The air temperature under the tree cover was found to be lower (36.8 to 37.8°C) than in non-tree open cover (38.4°C). The leaf temperature of grasses under different tree covers varied from 31.5°C to 32.2°C, whereas it was 33.8°C in open. Hazra and Patil have thus concluded that the biomass production is a physiological processes and is largely controlled by the micrometrological parameters particularly radiation availability and to some extent by air temperature difference and relative humidity.

Mann and Muthana (CAZRI, Monograph 1984, series-23) stated that net radiant energy recorded underneath *A. tortilis* tree cover during August revealed that it remained very poor from 14 to 30 and 15 to 24 per cent, respectively, of the energy received in the open area. Similar trend was also recorded during September. Poor germination and ultimate failure of *Cenchrus ciliaris* grass establishment underneath the tree cover in spite of abundant soil moisture availability during kharif season can be explained in terms of micrometrological conditions.

Hazra (1989) recorded that at Jhansi some of the perennial grasses like guinea grass, napier, bajra hybrid and setaria were most suitable for agroforestry system and the average reduction in yields were observed to be 13%, 25% and 26% respectively with 35% reduction in radiation availability as compared to open canopy of 7-8 year old *A. lebbeck*. Even with these three crops, there is a great variation at the varietal level with regards to their sensitivity to radiation. The variety hamil of guinea grass, IGFRI-6 of napier-bazra hybrid and norok of setaria were found to be least affected by radiation, curtailment and yield reductions were 9 per cent, 20 per cent and 13 per cent respectively as compared to open.

Thus it is apparent that under silvipastoral system sustainable productivity of site is possible by means of judicious canopy manipulation.

**Fodder Tree Species Suitable in Different Climatic Zone :**

| Sl. No. | Common name (species) and uses | Fruit/seed collection time | sowing time | Pre sowing treatment | Plantation techniques |
|---|---|---|---|---|---|
| 1 | 2 | 3 | 4 | 5 | 6 |
| **(A) Temparate Region** | | | | | |
| 1. | Silver wattle (*Acacia dealbata*) Fodder fuel and tannin. | June-July | October | Scarification Hotwater | Direct sowing entire transplanting |
| 2. | Black Wattle (*Acacia mearnsii*) Fodder, fuel and tannin. | November | September October | -do- | -do- |

(*Contd.*)

| 1 | 2 | 3 | 4 | 5 | 6 |
|---|---|---|---|---|---|
| 3. | Horse chestnut (*Aesculus indica*) Fodder, timber and fuel. | October | October | Stratification | Stump planting and entire transplanting. |
| 4. | Kharik (*Celtis australis*) Fodder, fuel, timber and medicinal. | October-November | October-November | -do- | Direct sowing. |
| 5. | Kimu (*Morus serrata*) Fodder, toys and furniture | March-April | May-June | -do- | Branch cutting and direct sowing. |
| 6. | Kau (*Olea ferruginea*) Toys, fodder, sticks and fatty oil. | September -October | September-November | Scarification | Direct sowing, root suckers and Branch cutting. |
| 7. | Padam (*Prunus cerasoides*) Fodder, fuel, timber and fruit. | May-June | July-August. | -do- | Branch cutting and entire transplanting. |
| 8. | Banj (*Quercus incana*) Fodder, agricultural implements, fuel, acorns are eaten by monkey squirrel and bears. | August-November | February-March | -do- | Branch cutting and entire transplanting. |
| 9. | Bis (*Salix alba*) Fodder, fuel and criket bats. | — | — | — | Branch cutting. |
| 10. | Majnu (*Salix babylonica*) Fodder, fuel and ornamental plants. | — | — | — | Branch cutting |
| 11. | Bashroi (*Salix daphnoides*) Fodder, fuel, Basket. | — | — | — | Branch cutting. |
| **(B) Sub-tropical Region** | | | | | |
| 1. | Phulai (*Acacia modesta*) Fodder, fuel, timber Gum and Tannin | October-December | February-March | Hot water Treatment | Branch cutting direct sowing and entire transplanting. |

*(Contd.)*

| 1 | 2 | 3 | 4 | 5 | 6 |
|---|---|---|---|---|---|
| 2. | Siris (*Albizia lebbeck*) Fodder, fuel, timber and medicinal | October-November | October | — | Direct sowing, Branch cutting & transplanting. |
| 3. | Siris (*Albizia chinensis*) Fodder, fuel and timber. | December-February | After seed collection | Scarification & hot water treatment | Stump planting, entire transplanting and direct sowing |
| 4. | Bakli (*Anogeissus latifolia*) Fodder and fuel. | April-May | -do- | Soaking in water | Entire planting. |
| 5. | Khairwal (*Bauhinia purpurea*) Fodder, fuel and Gum. | January-April | March to April | — | Stump planting entire transplanting and direct sowing. |
| 6. | Khirak (*Celtis australis*) Fodder, fuel, timber sports goods. | October-November | October-November | -do- | direct sowing. |
| 7. | Shisham (*Dalbergia sissoo*) Fodder, fuel and timber | December-March | March-April | — | Root and shoot cuttings and direct sowing |
| 8. | Aonla (*Emblica officinalis*) Fodder, fruit, timber, fuel and tannin | November-January | February-March | Hot water | Budded stump, branch cutting and entire transplanting. |
| 9. | Pipal (*Ficus religiosa*) Fodder, fuel and packing boxes. | — | — | — | Budded stump, and entire transplanting. |
| 10. | Bhimal (*Grewia oppositifolia*) Fodder, timber, fibre. | September-October | September-October | — | Entire transplanting. |
| 11. | Subabul (*Leucaena leucocephala*) Fodder, fuel and timber | July-October | Soon after collection | Soak in acid water | Direct sowing and entire transplanting. |
| 12. | Sahjan (*Moringa oleifera*) Fodder, ediable fruits and flowers etc. | April-June | Soon after collection | Soak in cold water | Branch cutting and direct sowing. |
| 13. | Tut (*Morus alba*) fodder, firewood and timber | April-June | May-June | — | Branch cutting, stump planting, direct sowing and entire transplanting. |

*(Contd.)*

| 1 | 2 | 3 | 4 | 5 | 6 |
|---|---|---|---|---|---|
| 14. | Bis (*Salix alba*) Fodder fuel and bats | — | — | — | Branch cutting. |
| 15. | Majnu (*Salix babylonica*) Fodder, fuel and ornamental plants. | — | — | — | Branch cutting. |
| 16. | Laureil (*Terminalia crenulata*) Fodder, timber and oxalic acid (obtained from bark) | March-April | Just after collection | Alternate dry and hot-water treatments | Direct sowing and entire transplanting. |
| **(C) Tropical Region** | | | | | |
| 1. | Khair (*Acacia catechu*) Fodder, fuel, timber and Khair. | February-March | March-April | Scarification and Hot water treatment. | Stump planting, direct sowing and entire transplanting. |
| 2. | Babul Pissi (*Acacia farnesiana*) Fodder, fuel, gum and tannin. | May-July | June-July | -do- | Direct sowing and entire transplanting. |
| 3. | Phulai (*Acacia modesta*) Fodder, fuel, timber and tannin. | October-December | February-March | -do- | Branch cutting, direct sowing and entire planting. |
| 4. | Babul (*Acacia nilotica*) Fodder, fuel and timber. | April-June | May-June | -do- | Direct sowing and entire planting. |
| 5. | Khor (*Acacia senegal*) Fodder, fuel and Gum | April-June | June-July | -do- | -do- |
| 6. | Israeli Babul (*Acacia tortilis*) Fodder and Fuel. | November-February | February-March | Hot water | Entire planting and direct sowing. |
| 7. | Lallei (*Albizia amara*) Fodder, timber, fuel and wood. | March-April | May-June | -do- | Direct sowing and entire transplanting. |
| 8. | Siris (*Albizia lebbeck*) Fodder, fuel and timber. | January-March | February-March | -do- | Direct sowing, entire transplanting and stump planting. |
| 9. | Chaplash (*Artocarpus chaplasha*) Fodder, timber and fuel. | June-July | July-August | -do- | Stump planting, direct sowing and entire planting. |
| 10. | Kathal (*Artocarpus heterophyllus*) Fodder, fruit and timber. | April-June | June July | -do- | Direct sowing and entire planting. |

*(Contd.)*

| 1 | 2 | 3 | 4 | 5 | 6 |
|---|---|---|---|---|---|
| 11. | Neem (*Azadirachta indica*) Fodder, fuelwood and medicine. | May-August | July-August | Hot water | Root sucker, direct sowing and entire planting. |
| 12. | Khairwal (*Bauhinia purpurea*) Fodder, fuel and Gum. | January-April | March-April | -do- | Stump planting, entire transplanting and direct sowing. |
| 13. | Shisham (*Dalbergia sissoo*) Fodder, fuelwood and timber. | December-. March | March-April | -do- | Root and shoot cutting, direct sowing and entire transplanting. |
| 14. | Aonla (*Emblica officinalis*) Fodder, fruit, fuel wood, timber and medicine | October-February | March-April | -do- | Budded stump, branch cutting and entire planting. |
| 15. | Pippal (*Ficus religiosa*) Fodder and timber | - | - | - | Branch cuttings. |
| 16. | Jarul (*Lagerstroemia speciosa*) Fodder, timber and Medicinal. | January-February | February-March | Soak in water | Direct sowing and entire transplanting. |
| 17. | Subabool (*Leucaena leucocephala*) Fodder, timber and fuel wood. | July-October | July | Soak in acid water. | Direct sowing and entire transplanting. |
| 18. | Sahjan (*Moringa oleifera*) Fodder, ediable fruits and flower. | April to June | July | Soak in water | Stump planting and direct sowing. |
| 19. | Jangli jalebi (*Pithecellobium dulce*) Fodder, fruit and timber. | April-May | April-May | -do- | Branch cuttings, entire planting and direct sowing. |
| 20. | Karanj (*Pongamia pinnata*) Fodder, fuel, timber and seed oil. | March-May | July-August | -do- | Stump planting direct sowing and entire planting. |
| 21. | Bilayati babul (*Prosopis juliflora*) Fodder, fuel and soil conservation. | May | June | Scarification, soak in water. | Direct sowing. |
| 22. | Jand (*Prosopis cineraria*) Fodder, fuel and soil conservation. | May-August | June-August | -do- | Direct sowing and entire transplanting. |
| 23. | August (*Sesbania grandiflora*) Fodder, vegetables and Medicine | July-October | July-October | - | Direct sowing and entire transplanting. |

(*Contd.*)

| 1 | 2 | 3 | 4 | 5 | 6 |
|---|---|---|---|---|---|
| 24. | Toon (*Toona ciliata*) Fodder and timber | April-June | Just after collection | Scarification, soak in water | Stump plantings direct sowing and entire transplantation. |
| 25. | Ber (*ziziphus mauritiana*) Fodder, fruit and fuel. | February | March | Soak in water | Grafting and entire planting. |

## Conservation of surplus Fodder for lean period

May-June and Nov.-Dec. are considered as lean period for green fodder. Thus, only solution is conservation of extra forages for lean period. Sometimes above process also helps in reducing the toxic constituents available in the original forages.

### Silage Preservation

Harvested forage respires till the availability of oxygen in it. After depletion of oxygen, anarobic condition prevails. Anarobic condition influences the growth of lactic acid forming bacteria. These bacteria converts water soluble carbohydrates into lactic acid and reduces the pH level to less than 4.5. Low pH level obstracts the growth of bacteria and enzyme action and ultimately preserves the silage. The moisture contents of the mass is reduced to approximately 70 per cent. This prevent the growth of undesirable bacteria like clostridia, which produce butyric acid. Good silage remains free from $NH_3$ and butyric acid smell. Percentage of lactic acid between 5-9% is considered good for good quality of silage.

Good quality of silage is obtained from natural grass. Grasses should be cut before flowering stage. In case dry matter is above 45 per cent, much attention is to be paid in pressing. Nutritional quality of grass silage (50% DM) can be further improved by mixing it with 1 per cent urea (on fresh weight basis, sprayed with 1:1 solution in water ) at the time of ensiling (Singh 1990).

### Hay making

In field condition hay making process takes 3-4 days. This period increase with the increase in humidity and ultimately inferior quality hay is produced. To make good quality of hay following steps should be followed (Singh, 1990).

**Fig. 26 :** Need of appropriate conservation of surplus forages to feed the animals adequately. (*Courtsey* : Pankaj Mishra, B.A.U, Ranchi).

(*a*) Early harvesting of crops.

(*b*) Chopping and drying.

(*c*) Drying on hay racks or tripod stands.

Heat drying reduces the respiration losses. Chemical additives such as propionic acid, formic acid, anhydrous ammonia reduces the mould growth during storage even at 20% moisture.

**Research needs under agroforestry to produce maximum animal feed**

To study the available genetic resources and genetic improvement of suitable tree and grass species for silvipastoral system.

Identification and evaluation of multipurpose tree species having deep root system, nitrogen fixing ability, coppicing ability and nutritious palatable leaves for each agroclimatic zone.

To study the best combination/combinations of trees and grass species under silvipastoral system for each agroclimatic zone.

To study the best silvipasture design for rehabilitation of hill slopes.

To establish the simplest design model under silvipastoral system (*e.g.* Alley cropping etc.), which may be followed by the farmers with minimum efforts.

To study nutrient recycling hydrological balance, cost benefit analysis, etc.

To conduct physiological studies related to rate of transpiration, photosynthesis activities, etc.

To study the effect of tree management (lopping, coppicing, pollarding, etc.) on the productivity of fodder, forage, timber, fuelwood under silvipastoral system.

To study the role of allelopathy, if any, productivity of silvipastoral system.

To study the role of mycorrhiza and nonsymbiotic nitrogen fixing bacteria in relation to the productivity of crops and trees under silvipastoral system.

More emphasis on-farm research trial

(*a*) To test on-station design in the different psycho-socio-economic conditions of farmers.

(*b*) Modification in on-station trial/design as per psycho-socio-economic and biophysical constraints of the farmers and instroduction of modified model by the farmers under supervision of scientist/extension worker. This will create desired socio-economic impacts on the farmer's life. These village laboratories will be a continuing features on long term basis.

**Developmental priorities**

Identification of leadership pattern for motivation.

Short term course to the beneficiaries on various aspects of silvipastoral system, supported by research findings based on on-station and on-farm research.

In the agroforestry system financial assistance is essential, otherwise farmers will not take interest in any type of plantation scheme, specially after care of saplings. Hence, short term and long

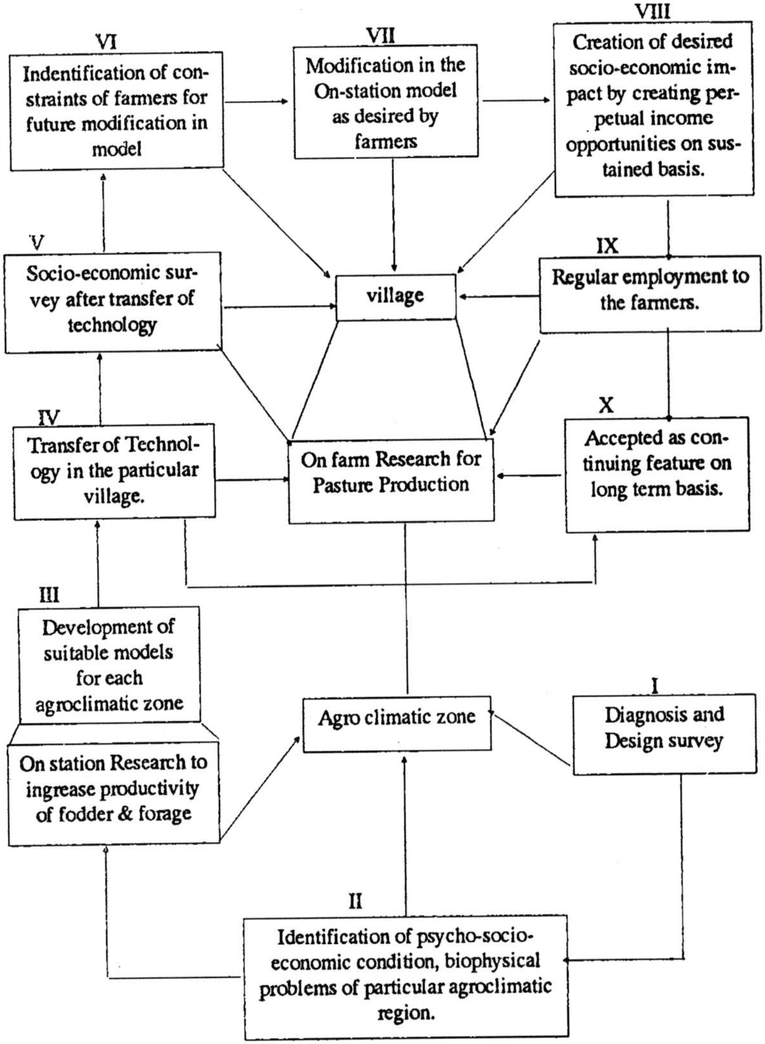


**Chart No. 3:** A schematic diagram showing on farm research in the farmers wasteland to create desired socio-economic impact.

term financial assistance and other types of incentive should be provided to the farmers, so that poor beneficieries can actively participate in above project.

Establishment of social/agroforestry Co-operative Society to co-ordinate the activities. Credit should be provided through Co-operative society only.

Permanent allotment of lands to the beneficieries wherever feasible.

Monitoring, to assess the achievement.

## REFERENCES


Annual report (1986). *AICRP project on Agroforestry*, ICAR, N. Delhi.

Annual Report (1988). *AICRP Project on Agroforestry*. Department of Forestry and natural resources, P.A.U., Ludhiana.

Arya, S.P. (1989). *Natural grassland production as influenced by tree species under silvipastoral system*, unpublished Data, Department of S.A.F report UH and F, Solan.

Bene, J.G, Beall, H.W. and Cote, A. (1977). Trees, food and people: Land management in the tropics Ottawa, *IRDC*, p. 52 (Cited from ICRAF, strategy to the year 2000 [1990] p.19).

Deb Roy, R. (1986). Studies on some aspect of renewable energy (fuel) and pasture production in an ecosystem of *Albizia procera* and *Albizia lebbeck* with grass and legume, *Ph.D. thesis*, Jiwaji University, Gwaliore. Cited from R. Deb Roy (1990) Silvipasture production system, NARP training programme, Jhansi.

Deb Roy, R. (1990). Silvipasture production system present status and furture prospects in India. *Ist NARP Training programme on agroforesty forage production and animal nutrition*, I.G.F.R.I., Jhansi, pp. 14-30.

Dhukia, R.S. (1989). *Agroforestry Newsletter*, Vol-I, No.I, pp. 3-4 National Centre for Agroforestry, ICAR, Jhansi.

Extract from the minutes of the meeting to finalise definition of wastelands and to discuss the methodology for creation of appropriate data base held on 24/4/86 at Vigyan Bhavan, New Delhi Wasteland Development for fuelwood and fodder production, F.R.I, Dehradun, pp. 14.

Grewal and Abrol I.P. (1987). Effect of rain water conservation in an alkali soil on trees and grass growth and soil amelioration, In: *Afforestation of salt affected soil, Int. sym.*, Vol.3, pp. 87-104.

Hazra, C.R. and Patil, B.D. (1986). Forage production under silvipastoral system; Light and temperature relation, *Indian J.Range. Mgmt.*, 7(1), pp. 33-36.

Hazra, C.R. (1989). Soil and climatological studies in relation to agroforestry research, *Ist NARP Training programme on Agroforestry, forage production and animal nutrition*. 26 Dec. 1989 to 25th Jan. 1990 I.C.F.R.I., Jhansi.

Jha, L.K. and Sen, P.P. (1988). *Lab to land Extension Report*, Department of Social Forestry, Faculty of Forestry, B.A.U., Ranchi, pp. 1-10.

Jha, L.K (1988). Forestry schemes to uplift the farmers of Chotanagpur plateau, *Tec. Bull.* B.A.U. publication, pp. 1-26.

Jha, L.K. (1990). Forage production under silvipastoral system, Paper presented in *Environment Assessment and Management through Social forestry in triabal region*, Mandelian Society of India and B.A.U. Ranchi.

Jha, L.K. and Sah R.B (1990). Impact of social forestry/agroforestry practies in the village economy (A case study), *Mendel*, 7(3-4), pp. 295-301.

Jha, L.K. (1991). Forage and fodder production through agroforestry, Lecture delivered in *Training Programme, Social Forestry support office W.L.D.B., Govt. of India.*

King, K.F.S. (1979) Key note address to the National Seminar on agroforesty India, delivered on May 16, 1979.

Konwar, G.R; Osman, M and Singh R.P. (1987). Fodder and fuelwood production system for sandy loam soils of semi-arid Hyderabdad region, In *Promaotion of fodder and fuel wood trees*, Hedge, N.G., Belwani. L.L and Kllkar, V.D. (1988) eds., B.A.I.F., Development Research foundation, Pune, India.

Lal, J.B. (1989). *State of Forest Report*, Govt. of India, Forest survey of India, (Ministry of Environment and Forest) 25, Subhash Road, Dehra Dun.

Lal, M. (1990). Agronomic aspects of promising fodder crops. *Ist Narp training programme on Agroforestry, forage production*, I.G.F.R.I, Jhansi.

Malhotra, P.P. and Sharma, K.K. (1988): Need for difining wasteland for afforestation purposes, *Wasteland Development for fuelwood and fodder production, Proceeding*, F.R.I. and Colleges, Dehra Dun, PP. 3-10.

Mann, H.S. and Muthana, K.D. (1984). Arid zone forestry ICAR, CAZRI, Jodhpur, *Monograph*, No. 23.

Muthana, K.D. and Shankaranarayan, K.A. (1978). Scope of silvipastoral management in arid region. *Souvenir on arid zone research and development*, CAZRI, Jodhpur.

Mishra, V.K. Sharma, S.K. and Verma, K.S. (1988). *AICRP on Agroforestry annual report*, ICAR: Department of silviculture and Agroforestry, Dr. Y.S. Parmar University Horticulture and Forestry, Solan.

Nancy Glover (1989). Gliricidia production and use. The *Nitrogen Fixing Tree Association* (NFTA), P.O. Box, 680, Waimanalo, Hawali, U.S.A.

National Commission on Agriculture (1974). The forestry sector, Part IX, Delhi.

Paroda, R.S. Muthana, K.D. and Sharma, S.K. Ed. (1977). Silvipastoral studies. CAZARI, *Ann.Rep*. pp.-71.

Paroda, R.S. and Muthana, K.D. (1979). Agroforestry practices in the arid zone, *Proc. Agroforestry seminar*, Jaiswal, P.L., eds, Impal, May, 16-18, New Delhi, ICAR pp. 69-79.

Patel, V (1985). Wasteland Development for more bio-mass, *National workshop on utilization of wasteland for Bio-energy*, April 27-29, Pune.

Pathak, R.S. (1984). Autecology of *Leucaena leucocephala* (Lam.) de. wit., IGFRI, *Ann. Rep*, 45 : 141-143.

Pathak, P.S (1988). Management of subabul for optimizing production, In: *Promotion of fodder and fuelwood trees*, Hedge, N.G.; Relwani, L.L. and Kelkar, V.D. (Eds.), B.A.I.F, Development Research foundation, Pune, India.

Shankar, V., Dodhich, N.K. and Sexena, S.K. (1976). Effect of Khejri (*Prosopis cineraria*) on the productivity of range grasses growing in its vicinity, *Forage. Res.*, Vol. 2, pp. 91-96.

Shankarnarayan, K.A. (1984). Silvipastoral system—A pragmatic approach to efficient integrated land management, pp. 133-142, In : *Agroforestry in arid and semi-arid zone*, Shankarnarayan, K.A., eds., CAZRI, Jodhpur, India.

Sharma, S.K.; Verma, K. S. and Mishra, V.K. (1988). Natural grass yield as influenced by various fodder tree species under pastoral, *Annual Research report*, 1988-89, pp. 83-84, Department of S.A.F., U.H. and F., Solan.

Sharma, S.K. (1989). *Agroforestry Newsletter*, Vol. I. No. I, p. 3 National centre for Agroforestry, ICAR, Jhansi.

Singh, A.P. (1984). Forage and pasture management technology for dryland. Agriculture In : *proceeding of workshop for S.M.S.* Mohsin, M.A. and Prasad, R.B., eds., Department of Extension, Ministry of Agriculture. Govt. of India and B.A.U., Ranchi, Bihar.

Singh, A.P. (1984). Grasses in Agroforestry In : *Proceeding of Social Forestry workshop*; Muhammed, S. eds., Birsa Agril. University, Ranchi, pp. 94-102.

Singh, A.P. and Prasad, N.K. (1989). *AICRP for Research on forage crops, Annual Report*, Main Centre, Ranchi, B.A.U., Ranchi.

Singh, A.P. (1990). *Conservation of fodder as silage and hay*, Lecture delivered to the participants of Ist NARP Training on Agroforestry, Forage production and animal nutrition at I.G.F.R.I., Jhansi.

Singh, D., Mannikar N.D. and Srivastava, N.C. (1979). Effect of silvipasture system on changes in soil fertility, *J. Indian. Soc. Soil. Sci.*, Vol. 27(4), pp. 486-488.

Singh, G. and Gill, H.S. (1988). *Effect of some management practices on Mesquite-Karnal grass agroforestry system on sodic soil*, Field visit of delegation to All India workshop on Agroforestry to the experiments on afforestation of sodic soils, Central Soil, Salinity. Research Institute, Karnal, pp. 12-13.

Singh, H. and Joshie, P (1985). Bouldery riverbed lands of Doon valley can profitably be put undergrasses and trees for meeting fuels, fodder and fibre requirements, *Annual report*, C.S. and W.C.R. and T., Inst; Dehradun, p.7.

Singh, K.A. (1986). *A.I.C.R.P. on Agroforestry project Report*, ICAR; Sikkim centre, Tadong, Gangtok.

Singh, P. (1988). Indian Rangelands. Status and improvement plenary address, *3rd Int. Rangeland Congress*, I.G.F.R.I. Jhansi, p. 40.

Singh, P. (1990). Forage and fuel production through agroforestry system in wasteland. Lecture delivered at the 1st NARP Training programme on Agroforestry, forage production and animal nutrition at IGFRI, Jhansi.

Singh, R.P., Bang, O.P.S. and Anand Prakash (1988). *AICRP on Agroforestry, Annual report*, G.P. Pant. University of Agriculture and technology, Pantnagar, India, pp. 24-25.

Strategy to the year 2000, ICRAF, KENYA Publication, pp. 19-20.

Tiwari, K.M. (1983). *Forestry for the rural poor, Indian forester*, Vol. 109, No. I, pp. 1-

8

# *Production Potential of an Alley Cropping in Agroforestry System*

Today alley farming system is recognised as a best techniques in stabilizing and improving farmland productivity in most of the country. Biophysical causes and socio-economic problems of farmers can be tackled by adopting alley cropping system in large scale. This system can play an important role in increasing production of wood, food, fuel, fodder and generate employment potential through out the year.

Infact, alley cropping is rarely practised by the Indian farmers. Keeping in mind, diversity in locality factors and socio-economic problems of farmers, many agroforestry centres were established under supervision of ICAR and ICFRI at their Regional Research Stations and by State Agricultural Universities. In my opinion, this system would be a best, socially accepted cropping system in near future in our country.

**Alley cropping (Intercropping)**

Alley intercropping, is also known as hedgerow cropping or hedgerows intercropping system. Alley intercropping is a farming system in which trees or shrubs are planted in rows, along contourline in slope land. Trees/shrubs alley are managed as hedgerows by periodical cuttings and prunings and crops/fodders, etc. are grown in alleys between shrubs/tree hedgerows. In general, alley cropping helps in minimization of soil loss, decrease surface run-off and increase soil nutrient status. Thus, this system helps in increasing the land use efficiency. Traditional farming system is devided into cropping and fallow period, while in alley cropping both periods are combined together and helps in improving the productivity and also protect land from degradation.

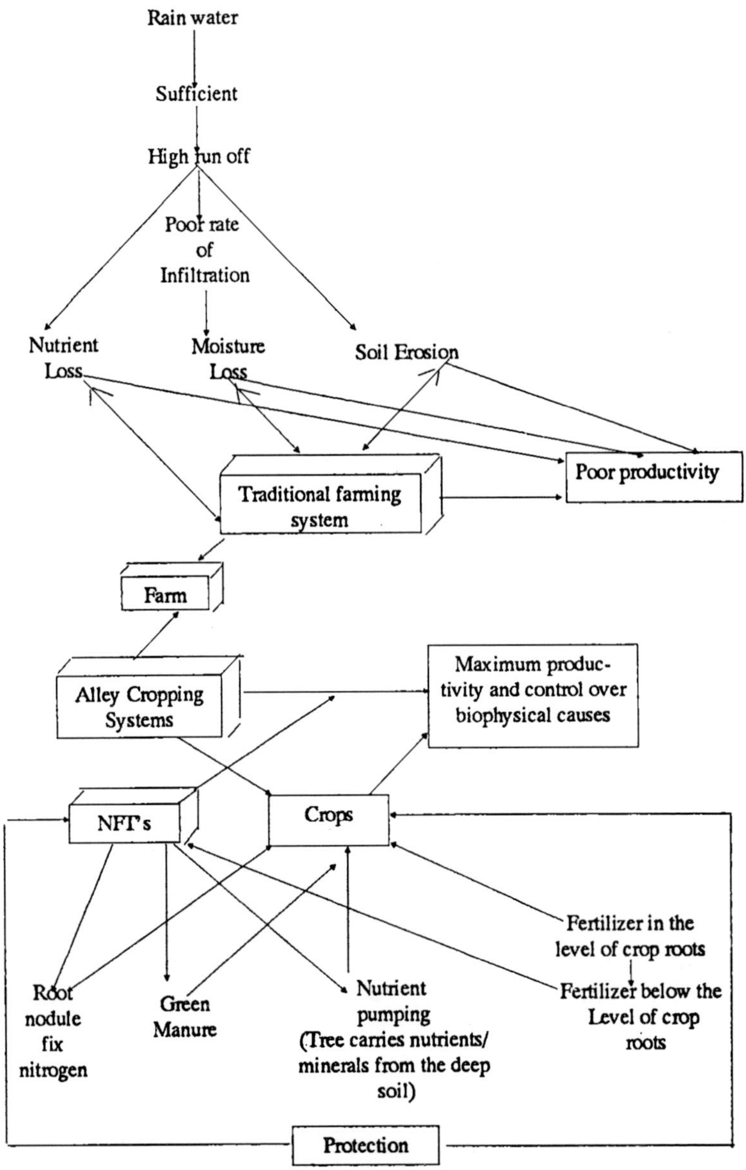


**Chart No. 4:** A schematic diagram showing alley farming system : as a potential solution to tackle the biophysical causes of Indian farm land.

**Fig. 27 :** Alley cropping : Agri-silvi-pastoral.

**Types of alley cropping systems**

*(A) Intensive hedgerows cropping*

When tree species are planted at close spacing within rows and managed to improve nutrient status of soil, soil and water conservation and regular supply of fuelwood, fodder and agricultural crops, vegetables, etc. are grown in between the 'alleys' formed by trees, is known as *intensive hedgerows cropping system.* Width between the two hedgerows depends upon the socio-economic requirements and biophysical causes of the land. If farmers can use tractor, spacing between rows should be nearly 5 M or more.

This system can be devided in to four sub types :

(*i*) Forage intensive alley cropping system

(*ii*) Green manure-cum-mulch alley cropping system.

(*iii*) Forage-cum-mulch alley cropping system.

(*iv*) Forage-cum-green manure alley cropping system.

(*i*) *Forage intensive alley cropping system*

In order to fulfil the requirements of fodder, suitable fodder trees/shrubs are planted to establish hedgerows and crops/vegetables, etc. are grown in the interspace between the alleys formed by the fodder trees/shrubs. This system is known as Forage alley cropping system. In this practice forage are obtained after lopping and pruning of hedgerows. Productivity of fodder as well as intercrops depends upon management of species, alley width, cutting height, pruning intensities, etc.

(*ii*) *Green manure-cum-mulch alley cropping system*

This systems has dual objective. The hedgerows are maintained to get nitrogen-rich green manure for maintaining soil fertility and mulch to improve the water retention capacity of soil. The hedgerows limit soil erosion.

In this system fertility status of soil is maintained in natural process, similar to traditional fallow period.

(*iii*) *Forage-cum-mulch alley cropping system*

In this cropping system, foliage of the hedgerows are used as mulch. Extension worker should suggest the farmers to use foliage as mulch, if he is confident that the foliage mulch will absorb little moisture and permit the rain water to percolate rapidly into the soil. This practice also improves soil structure. Foliage of the hedgerows are used as forage during lean period.

(*iv*) *Forage, mulch-cum-green manure alley cropping system*

In this system hedgerows are managed to use as green manure, mulch and forage. The foliage are harvested well in advance and after chopping incorporated in the soil, so that foliage can get sufficient time to decompose. Intercrops are sown or planted after complete decomposition of green manure. Sometimes 40-50 per cent foliage are utilised as green manure and rest foliage are harvested after establishment of crops, and spread over in the interspace as mulch. During off-season hedgerows foliage are used as forage.

***(B) Wide-alley intercropping***

Tree species are planted at wider spacing within rows (as per silvicultural requirements of the species). The trees are allowed to grow in natural condition. This system does not improve the nutrient status of soil like intensive hedgerows system. Beside this, tree causes shade effect on the intercrops and decrease the productivity of site. This problem can be solved by judicious management of tree canopy and tree species must be planted in East-West direction to reduce the shade effect. The tree branches can be pollarded to facilitate penetration of sun rays.

**Fig. 28 :** Alley cropping : Periodical pollarding/pruning to use foliage as green manure or mulch.

Wide alley hedgerows cropping becomes more useful, if N.F.T.'s hedge forming species/perennial grass species/vetiver grass/sisal are planted as fillars at closer spacing in the space between the trees within rows. The grass species/hedge forming nitrogen fixing trees species prevents soil erosion and also conserves moisture. The leaves of *Vetiveria zizanioides* etc. can be used as mulch vetiver roots contain strong aromatic oil, acts as a good repelents for most of the insect and rodent pests.

**Management of intensive alley farming system**

(*i*) ***Ist Year*:**

*Ist season*

* Land preparation and sowing or planting of crops, vegetables, etc.
* Plant the seedlings of tree species/shrubs within the crops in East-West orientation. Seedlings should be planted after onset of rains or as directed by the extension worker.
* In hilly terrain (sloping land) shrubs should be planted on one side of every three or six ridge.
* Tree species/shrubs should be planted across the slope on the contour line. In steep hill sides, contour lines should be determined accurately.
* Hoeing and weedings of shrubs/N.F.T.'s along with the crops is essential.

*2nd season*

* Select crops, having ability to improve the fertility status of soil.
* Fertilizer, farm yard manures may be needed in the initial year in marginal, sub-marginal land to boost the growth of the crops as well as hedge forming species.
* Fertilizer should be used only after consultation with extension worker.

*(ii)* ***2nd Year***

*First season*

* Crops should be grown on either side of 'alley' formed by the hedge forming trees species/shrubs.

* If, hedgerows are well established, than hedge foliage can be harvested 20-40 days before the succeeding crops and green foliage should be incorporated in the soil at the time of ploughing of field.

* Ensure proper and deep incorporation of green matter.

* Maximum time should be given to decompose the green matter. We can study complete decomposition by estimating C:N ratio of the soil. The ratio should be 10-15 :1. It indicates that organic residues are present in decomposed form. Here it is essential to mention that to facilitate decomposition of green manure, farmers should use super phosphate @ 100 kg/ha. Gypsum or super phosphate are commonly used as preservatives. They absorb ammonia and do not let it escape easily. The percentage of phosphorus in the manure increases and this is effectively utilised ; particularly in acid soil. Under this system, supply of nutrients are properly managed through proper crop management practice. *Leucaena leucocephala* contains N, P and K,1.5, 0.2 and 1.2 per cent respectively. *Gliricidia* contains, N, P and K, 4.15, 0.27 and 3.0 per cent respectively and *Moringa* contains N, P and K, 3.83, 0.26 and 2.22 respectively (Sacheti, 1985).

  During the non cropping season the hedgerows may be allowed to grow, to produce fuelwood. It also provides cover to the soil during dry season and helps in soil and water conservation and restrict weedgrowth. In the dry season foliage can be cut at regular intervals for animal feed.

*Fourth/Fifth/Sixth Year*

* N.F.T.'s/shrubs are harvested and replanted just after onset of rains or pollarded.

**Management of wide alley intercropping system**

As discussed earlier under wide alley intercropping system generally multipurpose nitrogen fixing tree species/fruit trees/timber trees are planted at wider spacing and leguminous

**Figs. 29&30 :** Alley cropping : Perennial pigeon pea grown in alleys, intercropped with wheat and mustard.

shrubs or other hedge forming species are planted within row, as fillars at close spacing in space between the tree species.

* In the first year the row to row distance may be kept less. Because in the initial year shade effect will be very less. Crops can be grown in the interspace.
* In the subsequent year, tree becomes larger and creates more shade effect over intercrops. This problems can be tackled by judicious canopy management.
* When tree becomes marketable or can be personally utilised by farmers ; hedgerows and trees within the rows are removed to widen space between two alleys. This allows greater availability of light to the intercrops.
* The harvested rows are planted by new seedlings.
* After few years, the trees are harvested in order to get timber/poles and again planted by new seedlings. This cycle may continue for long time.

**Alley cropping and formation of natural terrace in hilly terrain**

In this system to make barrier more effective, trees are spaced densely 5-6 cm. within rows. If nitrogen fixing tree species are planted at larger gap, than gap between trees will facilitate water flow and soil erosion, nutrient loss will be more. To tackle the problem, space between tree species, within rows are used for planting perennial grass/pigeonpea/Vetiver grass, etc. If farmers are willing to plant fruit trees or tree species having longer rotation period (timber species), spacing between tree species are kept more. In above condition, *Leucaena leucocephala*/vetiver grass/any suitable perennial grass/pigeonpea, etc. may be planted as fillars within rows. The hedge forming tree species/shrubs/grasses are harvested after proper establishment. After establishment of hedgerows, tree species are managed under coppice system. Tree species (hedge forming) are generally cut about 10-15 cm above the ground surface. Vetiver/perennial grasses cut down to ground level.

After establishment of hedgerows, soil is filtered out of the run-off and run-off oozes through the hedge formed on the contour. Silt will be deposited behind the plant and water flows on the slope, with slow speed, with very low quantity of silt mass. The deposited silt will form natural terrace, with in few years. Thus, in this system natural terrace is formed without extra labour.

## Tree management under alley cropping system

In order to maintained nutrient status of soil and reduce shade effect caused by tree/shrub species, (allowing maximum penetration of solar energy, etc.) careful tree management is essential.

In this system trees are managed by coppicing, pollarding and side prunings.

### Coppicing

This is the method of producing coppice shoots growing from a 'stool or a living stump. Powlick (1989) has stated that in coppicing whole tree is cut off close to ground level. Shoots sprout from the resulting stump-referred to technically as the stool. Removal of the trees, leading shoots results in hormonal suppression of lateral growth, following which dormant buds on the side of the trunk begin to grow. He has mentioned that as stool diameter increases, the number of dormant, bud shoots may increase 30 to 60 per cent faster than the number of callus shoots. Coppice growth is generally much more vigorous than normal seedlings growth and is characterized by both an upsurge of vertical development and retardation of lateral branching.

### Some points to keep in mind during coppicing

* Trees should be coppiced only when attains good health and shade effect is to much on the intercrops.
* Trees should be coppiced when it attains 3 to 5 metre height. Growth of the tree depends upon edaphic, climatic and other factors of the locality
* Coppicing should be done one week earlier to onset of the rains. This will prevent fungus attack, and availability of moisture facilitate the growth of new shoots.
* Frequency of subsequent coppicing after the first cut should be decided after observing health of the trees, crops likely to be grown in the field and use of foliage of the trees. For example, if farmers willing to use tree foliage as green manure, the coppicing should be done before ploughing the plot.
* The optimum cutting height for most of the coppicing trees should be kept in between 30 to 50 centimetres

above ground surface. If trees are cut below 30 centimetres causes fungal attack, and ultimately decay of the stump. If cutting height is kept below than 30 centimetres poor shoot growth observed in most of the tree species.

* Initial cut must be angled, to permit rain water to move above the cut surface easily.
* Tree should not coppice in dry period. In dry fallow period shade of tree species helps in moisture conservation.
* In new coppiced tree many new shoot appears. If farmer wants to collect fuelwood than maximum six shoots may be retained. If farmer is interested in fodder than more shoots can be permitted to grow. If desired products is poles, than 2-3 straight shoots can be left.
* In case of poor shoot growth, mature trees should be harvested and replaced by new seedlings.

**Good coppicers**

In alley cropping system trees having good coppicing power are planted. Based upon the coppicing power, tree species are classified into good coppicers, poor-coppicers and non-coppicers. Some examples of good coppicers are :

1. *Albizia* species
2. *Acacia* species
3. *Emblica officinalis*
4. *Syzygium cumini*
5. *Butea monosperma*
6. *Azadirachta indica*
7. *Tectona grandis*
8. *Leucaena leucocephala*
9. *Salix* species
10. *Eucalyptus* species
11. *Anogeissus* species
12. *Grevillea robusta*

**Pollarding**

Pollarding is a process in which crown of the trees are harvested at least 2-metre above the ground level, above the reach of cattles in order to get flush of new shoots. Farmers of Chotanagpur plateau of Bihar extensively use pollarding as a management technique for *Terminalia arjuna*, tasar host plants under agroforestry system. Some other species managed by pollarding are :-

1. *Salix* species
2. *Grewia oppositifolia*
3. *Grevillea robusta*
4. *Acacia albida*
5. *Hardwickia binata*

**Pruning**

Pruning is cutting of branches or even young stems. This operation helps in minimizing shade effect on intercrops and maintain the quality of timber.

**Some points to keep in mind during prunings**

* The cut must be angled, to permit the rain water to flow above the cut surface. This reduce the incidence of fungal attack in rainy season.
* The branches should be cut 4-5 cm. above the main stumps.
* Pruning must be started from the lower branch and than gradually approach top branch.
* The foliage obtained after pruning must be incorporated immediately into the soil, in order to get maximum nutrient release from the green manure. Pruned branches can be left in the field, until the leaves fall off. The dried leaves will fall off within two weeks. Branches may be used as fire wood. This requires less labour. Pruning must be done well in advance before sowing or planting the intercrops.

**Choice of species**

Selection of tree species and crops under alley cropping system depends upon following factors :

* Socio-economic requirements of farmers.
* Climatic factors had led to great influence over establishment of vegetation, in different zone.
* Topographic factors like configuration of the land, altitude, slope, aspect, etc. influence soil formation, nutrients and moisture level in soil. Thus, selection of tree species as well as crops must be done after

considering topographic, and edaphic factors of particular zone.

* Effect of a particular tree species on crops and vice-versa.
* Effect of tree species, crops on the site condition (*e.g.* allelopathy effects and its significance on production).

## FOOD, FODDER AND FUEL PRODUCTION UNDER ALLEY CROPPING SYSTEM

Extensive research work on various components of alley cropping system to increase the productivity of site have carried out in several International and National Research Institute and Agricultural Universities.

Kidd and Taogaga (1985) reported that an alley cropping trial using six nitrogen fixing species was established in 1982 at Nafanua, Upolu. The first taro *Colocasia esculenta* crop and a rattoon crop were harvested with poor results, as planting was done in dry season. Taro was again planted in late 1983 and early 1984 with favourable results achieved when harvested after seven months. Tree biomass of *Calliandra calothyrsus*, *Erythrina* species, *Gliricidia sepium*, *Leucaena leucocephala* (k-8), *Samanea saman, Sesbania grandiflora* were recorded 12.2, 3.0, 7.9, 10.7, 6.9 and 6.1 dry t/ha/yr respectively. The taro yield along with *Calliandra calothyrsus, Erythrina* species, *Gliricidia sepium, Leucaena leucocephala* (k-8), *Samanea saman, Sesbania grandiflora* and in control plot were recorded 3.10, 4.32, 5.42, 4.58, 3.44, 3.89 and 3.55 wet t/ha/cycle. Measurements were taken on 25 taro conns in each replication of the 'niue' variety, which comprised 85 per cent of the taro planted. Over 4000 conns were harvested from the 0.5 hactare area. Herbage production was measured over the 2nd year period, from the time of planting until the final taro crop harvest. The trees were slashed to an average height of 60 cm on a monthly basis for *Leucaena leucocephala and Calliandra calothyrsus* species and bimonthly for the other species. They suggested that visible evidence of increased growth of taro in plots of certain N.F.T. species (*Gliricidia* and *Leucaena*) versus the control plots (where no trees were planted) shows that this may be a viable cropping system for sustained production of taro at the village level in Western Samoa.

Kass Donald, *et al.* (1987) stated that at Costa Rica an experiment was set-up on alley cropping with *Gliricidia sepium* in

1984. Three of the fourth original treatments, *i.e.* no amendments, *Gliricidia sepium* prunings at 2 kg fresh wt/m$^2$, and *Gliricidia sepium* hedgerows spaced 6 x 1m, remained the same. The fourth treatment was changed from hedgerows spaced 9m x 1m to hedgerows spaced 4.5m x 1m by direct seeding a hedge half-way in between the 9 m rows. Beans (Cv. Talamanca) were planted in December, 1985 and harvested in March, 1986. Maize (Cv. Diamantes-8042) was planted in association with Cassava in June, 1986 and harvested in October, 1986. Beans yields were quite high (1747 kg/ha. no amendment, 1843 kg/ha. mulch at 2 kg/m, alley cropping in 6m. rows 1666 trees/ha, 1333 kg/ha. and alley cropping in 4.5m. rows 1453 kg/ha.) but showed a significant linear response to mineral N in both the control and the treatments receiving 20 t/ha. fresh weight mulch of *Gliricidia* prunings prior to planting. As in 1985, yields were higher with *Gliricidia sepium* mulch, but the difference was not significant at $p<0.05$. Yields in all alley cropping treatments were significantly lower than in corresponding mulch treatments. The significant linear component of the N response observed in the mulch and control treatments was not observed in the alley cropping treatments.

Maize yields were much lower than in 1985 due to the lower population (20,000 plants/ha. instead of 40,000/ha./yr) used and considerable bird damage to the planting. The coefficients of variability were much higher than in the bean yields, and no significant effects were observed. Nevertheless, yields in both the mulch and alley cropping treatments were higher than in the control treatments, but little response to N was observed.

Prunings production was not much higher for the 6 m spacing than for the 9 m rows, but was considerably lower than the 40 t./ha./yr. applied in the mulch treatments. The 1m distance between trees was probably excessive and higher production might have been obtained at a higher tree density. The contibution of the additional row planted from seed was sufficient by October of 1986 to reduce the difference in total pruning production between the 6m and 9m rows by October, 1986.

The Centro Agronomico Tropical de Investigaciones-y-Ensenanza (CATIE) in Costa Rica has been studying *Gliricidia sepium* in alley farming trials since, 1982. *Gliricidia* was chosen for study because of its adaptability in the study areas and farmer's familiarity with the species. A cropping system consisting of

Phaseolus beans followed by maize was used. Gliricidia was planted from cuttings at various spacing : 3.0 x 0.5 m, 6.0 x 1.0 m and 9.0 x 1.0 m. Trees were cut back at crop planting to 1.5 metres and the prunings applied as a surface mulch. Results from 4 years of research carried out at 4 sites, where a total of 12 bean crops and 9 maize crops were harvested showed that a mulch of *Gliricidia* increased bean and maize yields in all cases but one, although the increase did not always attain statistical significance. A positive response to alley farming with Gliricidia was always observed at sites where bean yields in the control plot were less than 800 kg/ha. There was no significant yield increase when bean yields in the control plots were more than 800 kg/ha. There were yield increase with alley farming using *Gliricidia* at sites where maize yields were less than 1200 kg/ha. In only one site where maize yields in the control plots was higher than 1200 kg/ha. was a yield increase associated with alley farming.

There was a tendency of alley farming to provide improved control of grassy weeds, particularly in sites previously in pasture. Differences in yield response are attributed to soil fertility differences between sites and tree spacing. The best results were obtained on sites with low fertility and with trees spaced 3.0 x 0.5 m apart (Nancy Glover, ed. 1987). In 1971, the Mindanao Baptist Rural Life Centre in the Philippines began to develop the Sloping Agricultural Land Technology, known today as SALT. SALT is a form of alley farming system with special emphasis on soil erosion control. *Gliricidia sepium* is one of several species used. *Acacia villosa* and *Flemingia macropylla* are also used. Hedgerows are direct seeded along the contours of sloping land using a double row of trees spaced 50 cm. apart with 4 to 6 metres between hedgerows. Maize, beans, peanuts, potatoes, pineapple and other crops are planted between hedgerows. The hedgerows are pruned about once a month to a height of 0.75 to 1.5 metres producing approximately 10 tons kg (dry weight) of prunings/ha/yr. Prunings are applied to the soil surface at the base of the crops.

Crop residue, branches and rocks are placed against the uphill side of the hedgerows to strenghthen the barrier used for soil erosion control. Maize yields using this system are reported to be double those obtained from traditional production system (Nancy Glove eds., 1987).

Nadagoudar (1988) mentioned that an experiment involving two field crops (ground nut, sorghum, chilli, finger millet, teak, papaya (*Carica papaya*) a fruit crop and two pasture crops (grass and subabool) were started in July, 1984 at the Main Research station, University of Agricultural Sciences, Dharwad, under rainfed conditions.

Anticipating teak would take 3-4 years for inter-row (2m) canopy closure, papaya is planted in between teak plants which has a rotation age of 2-3 years. Along teak and papaya rows, one rows of pasture crop is planted on either side at 60 cm between rows and 60 cm between 2 plants for grass and 20 cm for subabool. Subabool is planted to close, since the object is to harvest fodder leaving stumps in 4-5 cuttings, a thick mat is created. Thus, the two rows of subabool obstruct the rainwater movement serving as "Live-bund". This technique has been worked out at the university and is followed on large areas in wasteland management practice. Even the grass, due to heavy tillering the intra-row space is fully covered in about 6-8 months and serves as live-bund. For field crops, a rotation of 4-years-sorghum, groundnut, chilli and finger millet is followed with sorghum and chilli in the first year. Teak (*Tectona grandis*) at its pole stage is in more demand with the farmers for most of agricultural implements, hand tools, farm house, etc. If left to grow for 30-40 years it will yield very good timber of very high commercial value. In the current experiment it is planted at 2m a part in the rows with 10 and 20 m. in between rows.

Inter-row spacing treatments were allotted to main plots and eight treatments ($T_1$-sorghum, $T_2$-sorghum + teak + papaya, $T_3$-sorghum + teak + papaya + subabool, $T_4$-sorghum + teak + papaya + grass, $T_5$-chilli, $T_6$-chilli + teak + papaya, $T_7$-chilli + teak + papaya + subabool, and $T_8$-chilli + teak + papaya + grass) involving field crops-tree-fruit-pasture to sub-plots.

The combination of all the crops is agrisilvihortipastoral system of hedgerow intercropping system. Field crops are planted during July every year, teak (stump) was planted on 1st August, 1984 ; papaya (variety-Coorg Honey seedlings) on 15th September, 1984, subabool (vatiety k-8 seedlings) on 9th August, 1984 and grass variety-guinea (*Panicum maximum*-root slips) on 5th July, 1984. All the recommended package of practices are followed for field crops, fruit crop and pasture crops. As protective measures, hand watering was done to teak seedlings during first summer (December, 1984 to April, 1985). In the first year papaya did not yield any fruits and moderate yields of pasture did not compensate for the space lost to trees and pasture rows. As a result lower monetary returns were observed in field crops-tree-pasture combinations than entire field crop. During 2nd year higher monetary returns than entire field crops were obtained. Similarly during third year, maximum total income (Rs. 13,153/ha.) was recorded in sorghum + teak + papaya in to 10 m between two rows followed by same combination with 20 m rows. Duguma, *et al.* (1988) stated that a field trials were carried out on an oxic Paleustalf in the humid zone of south western Nigeria with *Leucaena leucocephala*, *Gliricidia sepium* and *Sesbania grandiflora* alley cropped with maize and cowpea. The three leguminous woody species were grown in hedgerows space at 2 m. Trials were carried out one year-after establishment of the hedgerows using a split-plot design with four replications. The *Leucaena leucocephala* trial had twenty pruning combinations consisting of five pruning heights (25, 50, 75, 100 and 150 cm) and four pruning frequencies (monthly, bi-tri and six monthly). The *Gliricidia and Sesbania* hedgerows were subject to nine pruning intensities consisting of three pruning height (25, 50 and 100 cm) and three pruning intensities (monthly, tri- and six-monthly).

For the three woody species, biomass, dry wood and nitrogen yield from the hedgerow prunings increased with decreasing pruning frequencies and increasing pruning height. Biomass, dry wood and nitrogen yields were in the following order *Leucaena*>*Gliricidia*>*Sesbania*.

The various pruning intensities had no effect on survival of Leucaena plants. Pruning frequency had a larger effect than pruning height on survival of *Gliricidia* and *Sesbania* plants with monthly pruning, about 25 per cent of the Gliricidia and all of the *Sesbania* plants died within six months of repeated pruning. Even with lower pruning frequency *Sesbania* plants showed lower survival rates than *Gliricidia* or *Leucaena*.

The various pruning intensities of all the hedgerows species had more pronounced effects on the grain yield of the alley cropped cowpea than an maize grain yield. Higher maize and cowpea yields were obtained with increasing pruning frequency and decreasing height. Alvim (1988) reported that Banana (*Musa sapientum*) and *Erythrina* species are the most common shade trees planted to shade *Theobroma cacao* in Brazil. *Gliricidia sepium* is being planted as an alternative shade species. *Gliricidia sepium* and *Leucaena leucocephala* (k-8) shade trees were established at a 3 x 1.5 m spacing (2,222 plants/ha.) in an oxisol of low native fertility. Alternate rows of the trees were pruned at about 1.2 m. every 8 to 12 weeks, depending upon the amount of shade given to the young cacao planted at a 3 x 3 m spacing. Prunings were left on the soil surface as mulch. This procedure was maintained for four years after cacao seedlings had been planted, and then a thinning scheme was initiated to eliminate the majority of shade trees. Biomass production was measured at the time of pruning in 5 plants per plot, with 3 replicates. *Gliricidia* produced more biomass than *Leucaena*, *i.e.* 18,223 and 15,321 kg/ha/yr. respectively. He has suggested that use of *Gliricidia* hedgerows as a temporary shade for cacao, therefore, seems to be a promising alternatives means of starting new plantations in view of potential for improving soil condition and its ease of establishment and management.

In zonal agroforestry research project at Chalimbana in Zambia a hedgerows-intercropping trial was established in December, 1987 to assess the effectes of *Leucaena leucocephala*, *Fleminingia congesta* and *Sesbania sesban* interplanted with maize. The hedgerows were pruned at heights of 50 to 100 centimetres one year after planting and subsequently on a monthly basis during the growing season. All or no prunings were applied to the maize rows. Inorganic fertilizer was applied at the recommended rate to some plots and not to other. According to preliminary results, the *Sesbania* showed the best height growth, biomass production and wood yield, followed by *Flemingia* and *Leucaena*. Wood yields were highest for all three species when trees were cut at 50 rather at 100 cm. Although some of the results were variable, maize grains yield were highest on plots where tree prunings were applied to the soil together with inorganic fertilizer at the recommended rate. Yields from plots with tree prunings plus fertilizer were higher than yields from plots with fertilizer alone (ICRAF, Annual Report, 1989).

According to Rao and Wesley, (1989) a hedgerow intercropping trial started in 1984 at the Machakos Field station in order to demonstrate alley cropping system and to explore the effects of a number of variables under semi arid conditions. Three tree species *Gliricidia sepium, Cassia siamea* and *Leucaena leucocephala* were planted at 50 cm intervals in rows spaced 3, 4, 5 and 6m apart. Rows of trees were arranged systematically, with four alleys at each interval. In 10 m sections of the different rows, trees were pruned at variable frequencies to height of 30, 60 or 90 cm above ground. The prunings were incorporated into the soil in the alleys and annual crops were grown. After an initial crop of cowpeas in 1985, maize was grown every year during the first rains and cowpeas during the second rainy season. Because the emphasis was to demonstrate the system under favourable condition, the maize was fertilized at a rate of 40 kg nitrogen and 40 kg phosphate ($P_2O_5$) per hactare.

In the later years, all three species produced over 2-tonnes of dry biomass annually per hectare. The *Leucaena* was more competitive with adjacent crops than the other two species particularly during 1987 which was a dry year. Biomass production from the hedgerows increased significantly with higher cutting heights, but this was accompained by significantly lower crop yields, probabaly as a consequence of increased competition for moisture. For this reason, lower cutting heights at around 30 cm above ground are recommended. This strategy allows higher crop yields without drastically affecting biomass yields from the hedgerows.

The effects of alley widths on crop yields in these trials are not conclusive because the systematic arrangement of alley spacing coincided with slope and possible with the fertility gradient. Under these conditions, maize yields increased with alley width, reflecting the fact that wider alleys meant a higher proportion of the total area planted under maize. On plots where hedgerows were pruned to 30 cm above ground, maize yields averaged 2 tonnes per hectare in 1986 and 3 tonnes per hectare in 1988. In 1987, when rainfall was half to normal level for the area, both the maize and cowpea yields were considerably reduced.

They pointed out that this findings indicate that an agroforestry system based on hedgerows intercropping may be productive under semi-arid condition, even using *Gliricidia* and *Leucaena* which have been considered more suitable for humid conditions. Hedgerows management turned out to be an important variable, but results were less clear on the impact of different alley widths. Another question remaining is whether the system is self-sustaining or whether it needs to be supplemented with commercial fertilizers and, if so, how much fertilizer is required.

Shannon, *et al.*, (1990) reported about the alley cropping research in Zaire. The trial was established, January 1986 at

Gandajika station (7° S Latitude, 780 m elevation), on a site which had previously contained cassava. Rainfall average 1400 mm between late August and mid May, permitting 2 maize crops a year. *Leucaena leucocephala* grows very well in the soils (pH 5 to 6) and was choosen as the hedgerow species. The design was a 2 x 2 factorial in a randomized complete block with 4 replicates. Treatment factors were presence or absence of alley cropping ; and presence or absence or fertilizer. Initially, they used the recommended fertilizer at rates of 64 kg N and 46 kg $P_2O_5$/ha to avoid nutrient imbalances ; 30 kg $K_2O$/ha was added as potassium sulphate, because K deficiency symptoms were observed on maize leaves. Hedgerows of *Leucaena leucocephala*, were planted together with maize in rows 4 m apart. The hedgerows were pruned at 50 cm height starting in September 1986 at the beginning of the second maize season. The prunings were applied as mulch. Maize (Var-Salonga 11) was planted in 2 seasons per year at a density of 53,333 plants/ha in rows spaced 80, and later 75 cm apart. In the sixth and seventh season of the trial, 1 Maize row out of 5 was eliminated in plots with hedgerows, in order to reduce competition between the maize and *Leucaena*.

Alley cropping without fertilizer results in stable but low maize yields over the three and a half year of the trial. Yields without alley cropping appeared to decline after the third season, even with fertilizer application. During the first 2 seasons, alley cropping did not benefit the maize crop because insufficient prunings were being produced for mulch. During the third season, the central rows of maize within the alley produced more than the respective rows in plots without hedgerows, but this gain negated by poor yields in maize in the rows adjacent to the hedgerows.

The competition between the hedgerows and the adjacent maize led us to alter the pruning schedule from one every 2 months (except during the dry season) to a 5-6 week pruning cycle, with three and later two prunings effected per season. From the fourth season onwards, maize with alley cropping yielded consistently and significantly higher than without alley cropping. Fertilization increased yields in all seasons, both with and without alley cropping. Nevertheless, fertilizer alone was not able to halt the apparent decline in yield from the third season onwards in plots without alley cropping. The highest yield were obtained from the combination of alley cropping with a low rate of fertilizer. In this treatment, there appears to have been a general increase in yield.

Low yields in second seasons may be attributed to irregular rainfall and possibly depletion of nutrients in the soil. They pointed out that alley cropping alone maintained soil productivity, but integration of alley cropping with moderate fertilizer use may be the best solution to yield stability and increased productivity for farmers in south-central Zaire.

Kokwe (1990), reported that on-farm alley cropping trials were designed for testing on farmers field under farmers condition in 1987/88 cropping season at Luapula Provina, Zambia. Three species were recommended, *i.e., Leucaena leucocephala, Calliandra calothyrsus,* and *Flemingia, congesta.* Three hedgerows of each species and three rows of maize formed a plot (experimental unit) with solo maize as a control.

The trees were planted in 8 m rows at 3 m apart and 1-2 m within rows. The maize was planted in 8 m rows at 20 cm apart and 25 cm within rows. The maize was fertilized at the full recommended rate of 100 kg 'D' compound/ha as a basal dress and 100 kg ammonium nitrate/ha as top dress. Weeding in both the maize and trees was done at the same time (once at 4 weeks after emergence of maize). During the first year, the trees were observed for their establishment capability and yields of maize were recorded. No pruning was done to allow proper establishment. In the 2nd year 1988/89 at the onset of the rains, maize was given only 50 kg/ha 'D' compound as a basal dress across all plots except for the solo maize (control), to which 100 kg/ha. 'D' as basal and 100 AN as top dress was applied. Pruning of trees of all species to 50 cm was done when maize was planted. Each tree hedgerows supplied its biomass to two rows of maize evenly spread on the side by the rows and incorporated in the ridges just before planting. Differences in vigour were observed, those with added prunings looked healthier than the control. The yield results will show the real difference among the treatments. He has concluded that analysis awaits the yield and yield components results. However, initial farmer reaction to the technology has been enthusiatic and co-operative.

Zainabo (1990) stated that a long-term upland rice trial was started in 1987 to study the effect of continuous rice cultivation with or without additions of commercial fertilizers and/or prunings from leguminous trees planted in the rice alleys in the Gambia.

Initially, *Gliricidia sepium*, was planted but later removed due to its very slow growth rate. In 1988 a fast growing leguminous trees, *Cassia siamea*, was planted to replace *Gliricidia sepium*. The study actually started during the 1989 growing season when pruning from the new established *Cassia siamea* could be obtained.

In 1988, *Cassia siamea* was seeded at 10 seeds/m of row in alleys of 4 m apart. The location was the old Yundum tandaka. The design used was a randomized complete block with 4 replications and 4 treatments. The treatments consisted of stripping leaves from the *Cassia* and placing them on the surface of the soil between the rows of rice plants. This was done just after planting (4 July) and eight weeks later (6 September). The treatment used were : (*i*) O-pruning + O-Fertilizer, (*ii*) Prunings + O-Fertilizer, (*iii*) Prunings + half recommended fertilizer rate (75 kg/ha 15:15:15+50 kg/ha urea), (*iv*) Prunings + full recommended fertilizer rate (150 kg/ha 15:15:15+100 kg/ha urea), and (*v*) Virgin land (O-Prunings + O-fertilizer).

The results showed no significant difference between mean yield of treatments. However, the data suggested an increase in yield with increasing nutrient additions (especially nitrogen) from leaf prunings. He has also pointed out that since this is a long-term study, the effects of prunings will be more evident in succeeding years when greater decomposition of the leaves occurs and more nutrient accumulation continues to take place.

**Conclusion**

* More profitable than the traditional bush-fallow rotation.
* Combines cropping and fallow period.
* Increase the productivity of lands and animals.
* No need of extra labour to clear new land in forest area.
* Save capital by eliminating the labour requirements in terrace formation. (Natural terrace are formed).
* Save capital by reducing the need of fertilizer.
* Extra labour required, in trees/shrubs management *i.e.* coppicing, pollarding and pruning.

## REFERENCES

Annual Report (1989). International Council for Research in Agroforestry, (compiled by Sidney, B. Westley), ICRAF., pp. 16-17 and 23-24.

Alvin, R. (1988). *Gliricidia sepium* with cocoa and pepper in Brazil. *Nitrogen fixing tree Research Reports*, Vol.6, pp. 54-58.

Duguma, B., Kang, B.T. and OKali, D.U.U. (1988). Effect of pruning intensities of three woody leguminous species grown in alley cropping with maize and cowpea on an alfisol, *Agroforestry system*, Vol. 6; pp. 19-35.

Kass, D.L. Mario, J.H. and Yael's, C.H. (1987). Second year results of alley cropping with *Gliricidia sepium* (Jacq) Steud. On an oxic dystropept in San carlos, Costa Rica, *Nitrogen Fixing Tree Research Reports*, Vol.V, pp. 42-43.

Kidd, T.J. and Taogaga, T. (1985). Nitrogen fixing trees as green manure for upland Taro in western Samo, Nitrogen Fixing Tree Research Reports, Vol. 3, pp. 67-68.

Kokewe, M. (1990). On-Farm agroforestry trials in Luapula province, Zambia, *AFNETAN, Newsletter*, Vol, 2, No. 2, pp. 6-7.

Nadagoudar, B.S. (1988). Agroforestry systems in the tropical and subtropical parts of India. II structure, function and economics. *Summer Inst. on Social Forestry for rural development*, HA.U., Hissar, pp. 13-26.

Nancy Glover, (Eds.) (1987). Gliricidia production and use NFTA-CATTE workshop, *Gliricidia sepium (Jacq) : Management and improvement*, NFTA publication, pp. 18-21.

Powlick, T.(1989). Coppice with care. *Agroforestry Today*, Vol. I, No.3, pp. 15-18.

Rao, M.R. and Wesley. B. Sidney (1988). Agroforestry for Africa's semi-arid zone, *Agroforestry Today*, Vol.I, No. I, pp. 9-11.

Sacheti, A.K. (1985). Fertilizer and manures, *NCERT Publications*, New Delhi.

Shannon, D.A., Kabaluapa, K.N. and Ngoyi, M.L.(1990). Alley cropping stabilizes maize yields in Gandhajika, Zaire, *AFNETAN*, Vol, 2, No.I, pp. 8-9.

Zainabo, A. Cole (1990). Rice/Cassia alley cropping in the Gambia, *A.F.N.E.T.A.N., Newsletter*, Vol.2, No.2, pp. 6-7.

# 9

# *Agroforestry and Watershed Management as an Interlocked System*

The shortfall of firewood production by 2000 A.D. has been estimated to be 137 million tonnes and we have to produce 240 million tonnes of food grain by 2,000 A.D. Indiscriminate felling of trees and faulty traditional cropping pattern are becoming serious threat to the land. Our present requirements of food, fodder, timber, etc. can be fulfilled by putting all available lands including watershed areas under various cropping system as per their capability class. In watershed areas, productivity on sustained basis depends upon the judicious use of natural resources and applicability of modern agricultural technology.

Tree species are considered important component of agriculture specially in a catchment area. Interaction of trees along with crops plays important role in meeting the requirements of wood and improving the fertility status of soil and also conserve soil and moisture. The vegetative cover to the ground surface reduces the run-off considerably and promotes the maximum rain water infiltration into the ground. The damaging mechanical action of falling rain drops is also greatly reduced. In agroforestry system, roots of trees and crops cover a large area under the soil surface, tree-crop root system loosers the soil, which increases the infiltration rate. In the watershed area, large hilly terrain which are only partially cultivated can be made to yield high returns per unit area by judicious combination of suitable tree species and crops/legumes/fodder along with suitable water management techniques. Selection of tree species having deep roots and several layer of horizontal root systems, is helpful in reducing soil erosion. The different rooting depths in the soil (due to combination of trees/crops/grass) help in increasing moisture retention capicity of soil and minimise soil erosion slips and land slides. Field operations, like ploughing, hoeing, weeding etc., increases the root porocity and permeability, and decreases the velocity of run-off and soil erosion. The hydrologic and soil erosion problem can be better managed under agroforestry system by judcious combination of trees and crops/grasses, etc.

**Fig. 31 :** Many of the slopes are now fallow. Treatment and use of land as per capability

## Watershed and Catchment Concept

Watershed is a geohydrological unit or a place of land that drains at a common point. The natural unit has been evolved through interaction of rain water with land mass and typically comprises of arable lands, non-arable lands and natural drainage lines in rainfed areas. Sustainable production depends on health, vitality and purity of production environment of which land and water are important constituents. Therefore, for scientific utilization of the natural resources, case of land and water, the ideal geographical unit would be the product of interaction of rain with land, *i.e.* the watershed (NWDPRA, 1991).

According to Mohsin (1989) the total area contributing water, which flows through the outlet, is the watershed of that outlet. The outlet may be a small natural drain, a rivulet, a stream or even a river. If the contributing area is very large, of the size of 40,000 sq. km and above, it is called a 'Catchement' wherein multipurpose river Valley Projects, (and medium irrigation projects) like Damodar valley projects, Suwarnarekha Project, etc. are located. These projects are conceived on the basis of the catchments of the river system. For very specific and proper utilisation of land and water resources and for enabling actual micro level planning and implementation, a much smaller area, in comparison to a 'Catchment', is selected which is termed as 'Water-shed'. The specific definition are presented below :

(1) A primary river is a long river having a drainage area of more than one lakh hectares with a number of tributaries of first, second, third, fourth or even fifth order. The primary river generally discharges load to the sea. The drainage area of such a river is called a *Catchment.*

(2) A secondary river is a tributary of a primary river having a drainage area of more than forty thousand hectares. The drainage area of such a river is called a *Sub-Catchment.*

(3) A tertiary stream or river is a tributary of a secondary river having a drainage area of 4 to 40 thousand hectares. The drainage area of such a stream river is called a *Watershed.*

(4) A quarternary stream is a tributary of a tertiary having a drainage area between 2 to 4 thousand hectares. Such a drainage area is called a *Sub-Watershed.*

(5) A pentad streamlet is a tributary of a quarternary having a drainage area between 400-2000 hectares. Such a drainage area is called a *Mini-Watershed.*

(6) A hexad streamlet is a tributary to a pentad having a drainage area of less than 400 hectares. Such a drainage area is called *Micro-Watershed.*

**Objectives**

The conservation strategies in watershed area is based upon ecological balance and production on sustained basis for future welfare of the civilization.

As per NWDPRA guidelines the objectives are as follows :-

(*i*) Conservation, upgradation and utilisation of natural endowments like lands, water, plant, animal and human resources in a harmonious and integrated manner. This will aim at perpetual availability of food, fodder, fuel, fibre, timber and biomass for rural and cottage industries to meet the growing demands of human and livestock population through diversified land use.

(*ii*) Generation of massive employment during the project period and regular employment after the project completion for enhancing the employment opportunities in the backward rainfed areas to ensure livelihood security particularly for under-privileged section of the rural population like small and marginal farmers, landless labourers, tribals, etc.

(*iii*) Improvement of production, environment and restoration of ecological balance through scientific management of land and rain water. In the process, in-situ moisture conservation, introduction of scientific production system, net-work of run-off management structures and devices for recharge of ground water will ensure enhanced availability of water for human and livestock drinking purposes, domestic consumption, life saving irrigation and raising of appropriate cash-crops according to agroclimatic potentials.

(*iv*) Reduction of inequalities between irrigated and rainfed areas. Ultimately, stable production and processing of bio-mass would contribute towards better life in rural areas. This will reduce large scale migration from rural areas to cities.

(*v*) In addition to food, fuel and fodder the project would endeavour to enhance cash flow to the rainfed farmers and landless agricultural labourers through increased casual employment, marketable surplus of agricultural and dairy produce, growing of cash crops like vegetables, etc., in suitable areas.

**Integrated management of watershed areas**

The degradation of catchment watershed area depends upon the several factor and their collective action gradually lead to land degradation and other problems. Management of watershed involved judicious use of land and maintenance of tree, bush or grass cover so as to conserve soil and water for immediate and long term gains along with assured productivity of the land and regular employment to the inhabitants of watershed area. In the past the problems of watershed area were managed through, soil and water conservation means and improved crop husbandry practices were tried on individual basis only. This approach did not solve the problems of watershed area. In the present decade, technologies are introduced in the whole watershed area. Thus, management of watershed has become the important area of planning and development in the agriculture sector.

**Pre-requisites of watershed management**

To introduce any improved technology on the watershed basis, Extension worker should study in depth about the main pre-requisite of the particular watershed area like soil and water conservation. This problem can be managed by mechanical structure and improved farming technology.

**Soil and Water Conservation**

While planning for soil and water conservation measure the climatic (rainfall, etc.), topographic and edaphic factors should be taken into consideration. In each watershed area, lands may be classified into various categories on the basis of land-use capability, etc. To collect the relevant data, soil survey in the watershed area is essential. The capability class does not indicate what is the best use of land. It only indicates range of use to which each group be put. Discretion is left to the scientist or extension worker to select and suggest the best use of land.

**A few words on some specific priority component**

Soil and water conservation and increase in productivity in watershed area is only possible by applying mechanical as well as biological methods together. Biological and mechanical measures are supplementary to each other. To get good result both components should form integral part in the watershed area. This will strengthen the effects of each other as per requirement.

**Biological measure**

The common biological measures to tackle the problems of watershed areas are as follows :

(*a*) Choice of cropping system

(*b*) Suitable crops, varieties sowing time, seed rate and use of fertilizer.

## Classification of Soils in Terms of Potential Use

| Class | Characteristics features and limitation | Use and suggested management practices |
|---|---|---|
| I | Land suitable for tillage, Good land, no need of specific treatment. | Farming, require only ordinary crop management, *i.e.* use of fertilizer, green manure, etc. |
| II | Land suitable for tillage, require moderate conservation measure | Farming/agrisilviculture/ pasture or silvipasture practices. |
| | (*a*) Gentle slope.<br>(*b*) Moderate erosion hazard.<br>(*c*) Some what restricted drainage.<br>(*d*) inadequate soil depth. | Need terracing, contour tillage, strip cropping and crop management like class-I |
| III | Tillage possible after intensive conservation measures.<br>(*a*) Moderate to steep slope.<br>(*b*) High erosion problem<br>(*c*) Low water holding capacity<br>(*d*) Poor fertility status.<br>(*e*) Soil structure less favourable. | Conservation of natural vegetation, crops and pasture production under alley cropping system (Nitrogen fixing tree species, legumes preferred), soil conservation practices required more than class-II. |
| IV | Tillage difficult because of steep slopes, shallow soils, poor drainage, and due to hydrographic peculiarities. | Short term cultivation possible after careful management practices, emphasis should be given to silvipastoral system or agrisilvihorticultural system (alley cropping). |
| V | Land unsuitable for cultivation because of severe erosion problem, frequent flooding, rocky soil, etc.<br>Require mechanical and biological treatments. | Not fit for tillage. Forest and pasture, silvipasture can be introduced. Original vegetation need not be disturbed. |
| VI | Problems more severe than class-V, Require intensive mechanical and biological treatments. | Forest, pasture, silvipasture can be introduced after special measures to tackle the constraints. |
| VII | Problems more severe than class-VI, requires intensive mechanical and biological treatments like class-VI. | Forest, wildlife and conservation of natural ground vegetation. |
| VIII | Land unsuitable due to structure less and hydrographic peculiarities of soil *i.e.* Sandy, beaches, river wash and rock, etc. | Conservation of natural vegetation, recreational site, wildlife, etc. |

Fig. 32 : An eroded area : Need of mechanical structure and perennial grasses and trees as cover crops to stop further deterioration of land.

Fig. 33 : An eroded area kept for tree planting and grasses because the soil is too poor to support any other practice.

## CROPPING SYSTEM

### I. Alley cropping

Objective :
- This system conserve more moisture and soil.
- Improves nutrient status and other properties of soil.
- Increase productivity by optimum utilisation of the land.

Thus, the system involves planting of rows of trees or shrubs along the contour, and food crops and cover crops are grown in between the rows. Trees/shrubs will restrict soil erosion and decrease rate of run-off and tree foliage can be used as mulch or green manure for intercrops. Tree foliage is incorported into the soil before cropping season to improve the fertility similar to the natural process that occurs during the traditional fallow period, thus increasing land use efficiency.

Horizontal distance between the hedgerows decrease with the increase in slope percentage. Under intensive hedgerows system, nitrogen fixing tree species, shrubs, suitable for agroforestry system are planted at a close spacing. To make barrier effective, trees should be spaced densely 5-6 cm within rows. Two rows is considered as the best. In NFT's are planted at larger spacing the gap between trees will facilitate water flow and soil erosion and nutrient loss will be higher. To reduce the run-off, space between trees within rows may be used for growing hedge forming perennial grass or pigeon pea or vetiver grass, etc. If participants are interested to plant fruit trees or tree species having larger rotation period (timber species) the space between the timber tree species within the row can be utilised for growing bush forming grass species or NFT's. In the first year, on the steep slops ; a contour trench may be formed parallel to and behind the hedge rows. These will serve as soil traps, till hedges are established. After 2nd year, soil filtered out of the run-off and run-off oozed through the hedge formed a barrier on the contour. Because of silt load deposited behind the plant, water flows on the slope will be slower. Gradually deposited silt mass forms natural terrace. In the off-season, the twigs and foliage of hedgerows may be incorporated in the soil after chopping and branches may be utilised as fuelwood. The microorganism in the root nodules of nitrogen fixing tree species fixes nitrogen which supports the growth of the intercrops as well.

Leaves used as green manure release humus and minerals to the soil. Further after decomposition, the roots of tree species carry important nutrients from the lower zone of soil due to nutrient pumping action and make it available to crops. In the regular cropping management, roots dye and decomposed roots enriches the soil organic matter. In the hilly terrain hedgerows also holds soil together along the contour. This system facilitates storage of more moisture in the field, thus helping in better growth of the intercrops.

This system is highly flexible. Farmers can modify the model as per their needs, environmental and social condition and can protect the land in the watershed area from degradation and land remains productive on a sustained basis.

**Development of suitable integrated farming technologies for class-II to Class-VI lands in watershed area**

***(A) Agroforestry (alley cropping) on-farm trials to be conducted in Class II to Class IV Land in Watershed area***

*Model-I*

*Intensive alley cropping system*

*Objectives*

- To improve soil fertility,
- To reduce soil erosion
- To increase the productivity of site.

*Perennial hedge forming species*

* *Leucaena leucocephala*
* Any local NFT species
* Perennial pea
* Vetiver grass.

*Pruning management*

At different lopping intensities

*Alley width :*

Alley width will depend upon percentage of slope of the land.

*Plot size* - 15 x 6 m (may be changed)

*Treatments*

T1 = Tree species alone - control

T2 = Agricultural crops/Vegetables-control (without any treatment).

T3 = Agricultural crop two or more/Vegetables two or more + F.Y.M.

T4 = Agricultural crops two or more/Vegetables two or more + Nitrogen application.

T5 = Perennial hedge forming vegetation + Agril crop two or more/Vegetables two or more + Nitrogen application

T6 = Perennial hedge forming vegetation + Agril crop two or more/Vegetables two or more + F.Y.M.

T7 = Perennial hedge forming vegetation + Agril. crop two or more/Vegetables two or more (Tree foliage as green manure).

T8 = Perennial hedge forming vegetation + Agril. crop two or more/Vegetables two or more (Tree foliage as mulch).

T9 = Perennial hedge forming vegetation + Agril. crop two or more/Vegetables two or more

(Branches of trees/Perennial vegetation used as fuelwood).

*Parameters*

* Growth performance of tree species.
* Yield of intercrops.
* Nutrient status of soil
* Nitrogen fixing capacity of the tree involved.
* Percentage decrease in soil erosion, run-off and increase in moisture retention capacity of soil.

*Model-II*

*Alley cropping with perennial grass in watershed area*

*Objectives* :

* To improve soil fertility.
* To reduce soil erosion
* To increase the productivity of site.

**Hedge farming species**

Similar to model-I

**Grass species**

Two or more suitable grass species may be tried

**Pruning management**

At different lopping intensities.

**Alley width**

Varies as per degree of slope of the land.

**Plot size**

As decided by the extension worker.

**Treatments**

T1 = Tree species alone (control)

T2 = Grass alone (control)

T3 = Hedge forming species + Grass + F.Y.M.

T4 = Hedge forming spp. + Grass spp. + Nitrogen application.

T5 = Hedge forming spp. + Grass spp. + Tree foliage as green manure.

T6 = Hedge forming species + Grass spp. + Tree foliage as mulch.

T7 = Hedge forming species + Grass spp.

**Parameter**

* Growth performance of tree species.
* Fuel harvested from the tree spp.
* Yield of grass.
* Nutrient status of soil.
* Soil erosion, run-off and moisture contents of the soil.

*Model-III*

*Wide alley cropping (Agrisilvihorticulture) system in watershed area*

*Objective*

This system will meet requirements of fruits, timber, fuelwood, crop and fodder.

**Design**

Spacing - 5 m x 5m or as suggested extension worker.

Plot size - 30 x 30m or as suggested by extension worker

**Treatments**

T1 = Fruit trees/timber trees + Crops/Vegetables/fruit/grass etc. + Hedge forming tree species (NFT's as fillar within row, space left between the trees.

T2 = Fruit trees/timber trees + crops/vegetable/fruits/grass etc. + Hedge forming species as fillar + F.Y.M.

T3 = Fruit trees/timber trees + crops/vegetables/fruits/grass etc. H.F.S. as fillars + Tree foliage as mulch.

T4 = Fruit trees/timber trees + crops/vegetables/fruits/grass etc. + H.F.S. as fillars + Tree foliage as green manure.

T5 = Fruit trees/timber trees + crops/vegetables/fruits/grass species etc. + H.F.S. as fillars + Nitrogen application.

T6 = Crops/vegetables/fruits/grass/fruit trees.

**Parameters**

* Yield of inter crops/grass.
* Growth performance of tree species.
* Nutrient status of soil.
* Percentage decrease in soil erosion, run-off and increase in moisture retention capacity.

Comparative study of data/economics of different suggested models will be helpful in knowing the best suited model for each soil class of watershed area. With successful introduction of the best model, the tree cover will increase significantly. Crop yield can be obtained on sustained basis in Model-I and II after judicious canopy management.

In Model-III, their will be decrease in the productivity of crop yield due to tree-crops interaction after 4th year. In the long run, farmers will earn more from the fruit trees/timber/fuelwood and also can grow shade tolerant intercrops. The above land use technology will not only help the poor to fulfil their basis needs, it will also help them, to obtain better out put in lieu of their additional labour and will increase the income in the long run.

Moreover, as the poor, farmers, will get regular employment, they will take more interest in watershed management.

**(B) Afforestation of Class V to Class VII land**

In the watershed area to tackle the problem of erosion, reforestation/afforestation of class V to class VII land is essential. These classes requires special and intensive measures against erosion. These classes of land should be managed to conserve the primary vegetation cover and reforested/afforested by fast growing indigenous as well as exotic species with the involving the villagers. Villagers may be permitted to collect fuelwood, grass, minor forest produce. Controlled grass cuttings should be encouraged in closed area. Land may be allotted to the villagers on lease basis. These incentives will prevent pilferage of young pole crop. Need and feasibility of establishing specific firewood plantation will vary from one catchment to another. Where there is serious shortage of fuel in the surrounding villages, and there is no suitable alternative sources in existing forest land, closure and planting with fast growing fuelwood species should certainly be considered. In general, however, multipurpose plantations which would supply fuel, fodder, etc., would be prefarable. There should be provision of firewood depots. Where villagers have been obtaining their fuelwood needs free of charge from surrounding forests, they are not likely to respond favourably into purchasing the fuel from Govt. depots. However, if closure of degraded forest is to be successful, free or heavily subsidised provision of an alternative source of fuelwood might form a component of the watershed project (Mohsin, 1984).

**(C) Suitable crops and cultivers, sowing time, and use of fertilizer**

*Suitable crops and cultivers*

The undulating topography and soil with low moisture retentive capacity restrict the choice of crops or its cultivers in the watershed area. Farmers generally depends upon rains. Thus, it is essential for kharif crops to complete their life cycle within the safe moisture availability limits. In Bihar, Orissa moisture availability is around 80-115 days and we should select crop cultivers that attain maturity at around 85-110 days. In dryland area in Rabi season the crop varieties should be drought resistance and of short duration.

Prasad (1984) have stated that since the crops have to grow entirely or for most of the period on a receding soil moisture regime, the stored soil moisture even in the heavy textured soil is normally not sufficient to grow most of the crops unless it is recharged by the winter rains which arrive mostly in January in Chotanagpur area. Hence, the crop should be either early enough to complete the grain formation before the moisture stress becomes limiting, or should be late enough to be able to capitalise on the winter rains. Example of the former is toria which mature in 70-75 days and of the latter safflower which takes 150-170 days. In the latter case a slow early vegetative phase is desideratum (*e.g.* safflower). Coupled with these characteristics, the crop should have a deep root system (Safflower, Bengal gram, etc.) or should have inherent drought resistance capacity (Lentil, bareley, etc.). Selected crop varieties must also possess proper resistance to diseases and pests.

**Sowing time**

In kharif season monsoon span, is shorter. Land Preparation should be completed in the pre-monsoon showers and sowing should be started soon after receiving the monsoon. In same cases early sowing 6-10 days in advance is advantageous. This helps in proper crop stand establishment, early seedling vigour and moisture and available nutrients are properly utilised. Late sowing of kharif crop may force the grain filling stage to coincide with the moisture stress towards the end of season resulting in low crop yield. Under rainfed condition seed rate must be increased by 15-20 per cent for obtaining an optimum grain yield (Srivastava, 1984).

Sowing of all Rabi crops, should done as early as possible. For this crop planning has to be done in the kharif season, so that right type of land is available in the right time for Rabi seeding. Line sowing is better than broadcasting. Seed should not be below 4 cm from the surface (Prasad, 1984).

In the watershed area, under agroforestry system, foliage of trees or straw should be used as mulch as biological soil and water conservation measures. Pre-monsoon and pre-seedling mulching gave encouraging results in a few experiments.

**Nutrient**

In watershed area, under agroforestry system, best method to improve the nutrient status of soil is use of green manures, harvested from the trees and dung manure etc. It increases the organic contents of soil. Increase in organic matter improves the water holding capacity of soil. It is apparent that use of nitrogenous and phosphatic fertilizer has been found to increase the productivity of cereals and leguminous crops respectively. Response from fertilizer can be greatly enhanced when the fertilizer is placed in the moist zone below the seed. To get better result phosphatic and potassic fertilizers should be applied at the time of sowing in furrows and nitrogenous fertilizer in 2 to 3 splits to coincide with the critical requirement of crops invarious stages (Prasad 1984). However, the application of fertilizers should be done on the basis of soil test and yield target which a farmer envisages in an area.

**II. Essential mechanical measures to tackle the problem of watershed (Sinha, 1993)**

As discussed earlier to strengthen the biological anti-erosion measures application of mechanical means become necessary. The important mechanical means are :

**Terracing**

Principles involved in the design of terraces are : Change the general feature of the land slopes, reduce the total volume of run-off and arrange for the safe disposal of excess run-off.

For this region the following types of terraces have been found suitable.

(*i*) **Bench terrace**

It is perhaps the earliest concept of terracing. Bench terracing is an age old practice in the low lying fields for paddy cultivation in Chotanagpur region. It involves construction of a series of level platforms at suitable intervals by cutting and filling of each, and having a bund at the outerside to arrest run-off.

Bench terraces, levelled or graded longitudinally, are of the following three types :

(*a*) Level of table top.

(*b*) Sloping outward.

(c) Sloping inward.

But for the Chotanagpur region table top or level bench terrace has been found to be most suitable.

For calculating the width of individual fields the general formula is :

$$W = D \frac{(100\text{-}S)}{S}$$

Where W= Width of bench terraced field in ft.

D= vertical interval between terraces in ft.

S= percent slope of land.

*(ii)* **Level or absorption type terraces**

Narrow based bunds of suitable cross section from 6 to 8 sq. ft. is constructed on contours, their function being to intercept the rain and facilitate absorption of water by the soil.

*(iii)* **Graded terrace**

Narrow based bunds and Channels of 8 sq. ft. cross section with uniform grade of 0.1 per cent has been found to be most suitable for plateau rigion. The grade is increased by 0.25 per cent in the last 100 ft length of terrace for better drainage of excess water.

The general formula for calculating the vertical interval of terrace is :

$$VI = \frac{S}{2} + 2$$

Where VI= vertical interval in feet

S= per cent slope of land.

The length of slope in one direction should not normally exceed 600 to 800 ft due to poor maintenance of structures in the field.

*(iv)* **Tati terrace**

This type of terrace was first adopted in the village Tati of Palamau district by D.V.C. authorities and so named after it.

In this method the plots are closed from all sides by bunds of suitable cross section generally varying from 4 to 6 sq. ft., leaving suitable spillway on one. The ultimate objective is to have a levelled field without much cutting and filling of earth. The only technical point kept in view is that the diagonal fall does not exceed two feet. This is a compromise to the cultivator dislike for the graded terrace in which their fields are split into plots of unconventional shape and size.

### (*v*) Conservation Terrace

It will not be out of place to mention here about the "Conservation Terrace" designed by Damodar Valley Corporation (D.V.C.), Instead of having graded bunds as such, what has been, done is that the lower 1/4th to 1/5th part of the terraced plot along the bund in the lower portion has been made level to one foot lower than the general ground surface. All the water from the upper portion of the plot is made to collect in this dugout lower portion. The water remains in it as it is closed at both ends. This water is utilized for raising paddy crop. The data shows that it has given good results specially in medium lands with heavier soils. This needs actual testing in the farmers fields. Conservation terrace combines the principles of watershed and water-harvesting, it may prove a good breakthrough in this direction.

Results of Deochanda (Hajaribagh) and Rahmankhoa (U.P.) have shown that terracing brings about a marked improvement in the productivity of land.

In a systematic study undertaken at the Birsa Agricultural Univ., it was revealed that as a result of the construction of both tati and graded terraces marked improvement in the physio-chemical properties of soil and productivity of land has taken place. The crop yield showed an average increase of 20 to 75 per cent. The analysis of cost benefit ratio, based upon 10 years beneficial life of terraces and calculation of depreciation and interest by the method of "Equated system of loan recovery (Principal + Interest)" followed by Land Mortgage Banks, showed an average increased return of 22 to 30 times of the investment.

### Grassed water ways

In terracing projects sometimes grassed waterways are also constructed for safe disposal of run-off water. Generally it is constructed in trapizoidal shape which takes the shape of parabola in due course of time. The side slope is generally kept much flatter 2 : 1 to 4 : 1. Velocity of flows should not exceed 1 to 5 m/sec. Using this velocity, the carrying capacity of the waterways is calculated as :

$Q = av$, Where $Q$ = Discharge

$a$ = Cross sectional area

$v$ = Velocity.

The design is based on peak discharge which is expected from the area and this is estimated by rational formula.

Q = CIA, where Q= Peak discharge in Cu. ft/Sec.

C= Constant depending upon the watershed characteristics.

A= Area of watershed in acres.

I = Intensity of rainfall in inches per hour for a duration equal to the time of concentration.

The grassed waterways is generally constructed a year earlier than the construction of terraces, so that grass is established well before the water is turned into it. Afterwards it should be properly maintained. Any negligence may transform the waterways into a gully.

**Gully control and reclamation**

Gullies are a symptom of a functional disorder of the land as a result of improper landuse. But gullies are not the first symptom of land misuse, they are generally a secondary symptom, indicating misuse of land over a long period of time.

The most important work in gully reclamation is to reduce the run-off through adopting water conservation measures in its watershed. In case of small gullies planting some vegetation is essential to stabilise it.

In case of active gullies diversion of the run-off through diversion channels avoids further damage. Gullies develop in three directions—elongate along head, expand in width and increase in depth.

Thus control works are also of three types

(*i*) Gully side control

(*ii*) Gully head control

(*iii*) Gully bed control.

**(*i*) Gully side control**

Periferal bunds should be constructed to check the entry of water from the sides. The side slopes should be eased out and vegetation should be established.

(*ii*) **Gully head control**

If the drop is small sod flume will serve the purpose. In case the drop is more or volume of incoming water is more, sod flumes will not work and masonry structures are required. In such cases Chute spillway of suitable size is required to be constructed. This will safely lead the water from the higher level to the bottom of the gully.

(*iii*) **Gully bed control**

For stabilization of the gully bed, construction of check dams may be needed at suitable places to reduce the velocity and force of moving water and harvest the soil coming with the water. These structures may be temporary in case of small gullies or permanent in case of bigger ones. The temporary check dams can be constructed with any locally available materials like rock, timber brushwood, wiremesh etc. In addition, some sort of vegetation may be established.

In case of bigger, active gullies and where the intake of run-off is more, permanent structures like earthen dams with suitable masonry spillways are to be constructed. If the drop is less than 10 ft. and there is less chance of deposition of silt pipe spill way or drop spillway may be the choice. But if the drop exceeds 10 ft and there is chances of accumulation of silt then drop inlet spillways should be preferred. Apart from this gully bed should be levelled and vegetation should be established.

In case of considerably wider gullies the beds may be levelled and put under cultivation if possible.

**Silt detension cum water harvesting dams**

At suitable sites where the grass waterways end or there is an accumulation of water from the project area silt detension cum water harvesting dams are also constructed to arrest whatever silt is being washed out and at the same time harvest the excess run-off for use in raising crop after rainy season.

In the construction of such dams, proper design of spillway and corewall is of prime importance. The spillway should be designed at proper place. On the basis of peak discharge rate keeping in view storage capacity or the reservoir. Proper construction of corewall is necessary to check seepage losses which is quite much is earthen dams. Corewall should invariably be made of clayey material. Sides of the earthen dams should be given proper slope depending upon the soil type for its stability. For average quality of soil material, it should not be steeper than 3 : 1 on the upstream side and 2 : 1 on the downstream. Coarse soils may need side slopes of 3 : 1 to 4 : 1. The upstream side should be protected with riprap against wave action, whereas down stream should have good turf for protection. The height of the dam should be determined after fixing the height of spillway. The dam height should be at least 2 ft above the spillway height. The top width may be kept anything between 8 to 12 as per requirements.

**Spurs and a Levees**

Spurs are constructed as projections from a bank of a stream so to deflect the flow of water and check stream bank erosion. The head of spur must be well built into the natural bank of the stream and placed at some angle down stream and at interval all along the vulnerable spot. It may be temporary of boulders, brushwood etc. or permanent of masonry work.

Levees are earthen dams constructed along the flow on the banks of the stream to protect the adjoining areas from flood and check stream from changing its course. The in-side wall of levees are kept quite flat 6 : 1 to 8 : 1 and protected by rip rap. The outside wall is kept a bit steeper 2 : 1 to 4 : 1 to give stability to the structure and protected by sodding.

**Mass mobilisation and peoples participation**

For successful implementation of watershed development programme, first step is group mobilisation. Trainers and extension workers should create environment by proper training communication and learning process so that farmers can realise that successful execution of the programme is in their interest and so it is their responsibilities. The extension workers should always create environment to sustain the continuous interest in the programme. When desire to achieve some standard of excellence will appear from farmers side, they are ready to accept latest technology and their participation will be permanent features.

**Constraints in watershed development project**

* Lack of integrated watershed development technologies. Most of the technologies are tested on individual farmers land. Technologies should be developed for integrated development of entire watershed areas.

* Lack of co-ordination between forestry, agriculture, animal husbandry, State co-operative departments and other funding agencies. A joint co-operative body should be formed with representatives of all sections of rural society.
* Lack of state level multidisciplinary team for watershed Development.
* Lack of management plan of watershed area.
* Lack of active participation of farmers.
* Weak linkages between research, training and extension programmes.
* Lack of infrastructure, like road, communication, market education, etc.
* Delay in administrative approval, staff sanction.
* Adequate object is not provided timely.
* Lack of input supply in the watershed area.
* Lack of short term and long term credit facilities for the participants.

**Suggestions**

* In the hilly area still there is no shortage of land. Govt. should allot land permanently to the weaker section of the watershed area, individually or on community basis, so that they can realise that protection and reclamation of the land by improved technology is in the interest of their own development.
* Watershed management is multidisciplinary integrated programme. The personnel of agricultural, forestry, animal husbandry and fisheries should work together.
* Separate training for development and management of watershed is essential.
* Vigilence cell consisting of farmers representative, should be formed at remote village of the watershed area, so that participants may avail timely benefits granted by the Government.

**Data on the following attributes are needed for proper planning of watershed**

| Sl. No. | Attributes | Need |
|---|---|---|
| 1 | 2 | 3 |
| (A) | *Socio-economic Profile* | |
| | (*a*) Land use pattern | * To finalise appropriate land use planning. |
| | (*b*) Occupational pattern & economic condition. | * To study economic activities of farm activities.<br>* Consumption habits.<br>* How they like to improve their economic condition ? |
| | (*c*) Educational | * Level of understanding of the participants. |
| (B) | *Soil and Land use status* : | |
| | (*a*) Land Qualities | * To know texture, slope, drainage, depth, moisture level and extent of erosion. |
| | (*b*) Land use pattern | * To know the present land use system, constraints for future planning. |
| | (*i*) **Forest status** : | |
| | * Phytogeographical importance of Catchments. | |
| | * Forest and other vegetation cover | |
| | - The past history | |
| | - Main mode of livelihood (hill farming, shifting cultivation etc. | |
| | * Present status of forest | |
| | (*a*) Types of forest area in per cent | |
| | (*b*) Legal status and ownership pattern of forest. | |
| | (*c*) Plantation activities under different schemes. | |
| | (*ii*) **Agricultural status** : | |
| | * Soil texture, slope, drainage, depth, extent of erosion measure level. | |
| | * Problems in agricultural practice (Topography, Rainfall, Temperature, Technology, Farm implements, Input supply, Communication, etc. | |

*(Contd.)*

| 1 | 2 | 3 |
|---|---|---|
| | * Present status of agriculture in the watershed area. (Crop sequence, Crop yield, Potentialities for development of agriculture | |
| | (*iii*) **Horticulture status** : | |
| | * Name of the village. | |
| | * Area under orchard | |
| | * Facilities available for horticultural development (marketing, processing and storage). | |
| | (*iv*) **Live stock status** : | |
| | * Existing status | |
| | * Livestock resource | |
| | * Infrastructure support | |
| | * Existing animal husbandry and veterinary Institute | |
| | (*c*) Proposed land use as per quality of land. | * For allocation of land under various cropping system. (Crops, pasture, Horticulture, silvipastoral, agrisilvipastoral, agrilsilvihorticulture, forestry, etc.) |
| (C) | *Geological and Geophysical setting* - | |
| | (*a*) Lithostratigraphy | * To Know geology of catchment area. (Age, group formation, thickness and lithology) To know seismicity (Location of eplicenter Intensity) To know joint pattern, nature of exposed rock surface, soil mineralogy economic potential of ground water, road metel, etc. |

*(Contd.)*

| 1 | 2 | 3 |
|---|---|---|
| (D) | *Hydrometrology* :<br>Rainfall analysis :<br>- 6-weekly and monthly<br>- Number of rainy days in each month<br>- Annual rainfall values<br>- precipation. | To know moisture availability period for irrigation planning and proper cropping system. |
| (E) | *Geomorphic parameter* :<br>- Watershed shape, area, length, perimeter, length of main stream, stream order contour interval. | To identify causes of erosion and run-off |
| (F) | *Geomorphic factors* :<br>(*a*) Watershed shape, form factor, compactness co-efficient, elongation ratio, circulatory ratio and rotundity factor.<br>(*b*) Drainage pattern<br>Bifurcation ratio, drainage density, stream frequency, stream grade, length of overhead flow and time of concentration.<br>(*c*) Topographic factors<br>Watershed relief, relief ratio, relevant relief, ruggedness number, channel maintenance factor. | To know the hydrological behaviour of watershed. |

## INDICATIVE MANAGEMENT PLAN

### (A) Strategies proposed

* Development programme.
  - Training programme.
  - Provision for long term and short term credit to the participants.
  - On-farm trial in colloboration with farmers :

(*a*) Agriculture development

(*b*) Livestock development

(*c*) Orchard development

(*d*) Agroforestry development.

- agrisilviculture system
- agrisilivihorticulture system
- Silvipastoral system
- Multistoried cropping system
- Silvisilkculture
- Silvilacculture

(*e*) Forest development etc.

**(B) Benefits**

* Employment potential
* Additional income to the participants

  (Crop, vegetables, fruits, firewood, timber, fodder, silk, Honey, etc.)
* Ameliorative gains

  (Improvements of nutrient status, soil organic matter content, soil structure reduction in soil temperature, etc.)
* Protective gains

  (Minimization of soil erosion, landslides, surface run-off, nutrient loss, etc.).

**(C) Componentwise and yearwise financial outlay**

(*i*) *Outlay for developmental programmes*

| Sl. No. | Item | 1st | 2nd | 3rd | 4th | 5th | 6th | 7th | 8th | 9th | 10th | Tot-al |
|---|---|---|---|---|---|---|---|---|---|---|---|---|
| 1. | Agriculture | | | | | | | | | | | |
| 2. | Horticulture | | | | | | | | | | | |
| 3. | Live stock | | | | | | | | | | | |
| 4. | Agroforestry | | | | | | | | | | | |
| 5. | Forest development | | | | | | | | | | | |
| 6. | Biological and Mechanical means of soil conservation. | | | | | | | | | | | |
| Total | | | | | | | | | | | | |

*(ii) Outlay for salary, consumable and miscellaneous expenditure*

| Sl. No. | Item | 1st | 2nd | 3rd | 4th | 5th | 6th | 7th | 8th | 9th | 10th | Total |
|---|---|---|---|---|---|---|---|---|---|---|---|---|
| 1. | Staff | - | - | - | - | - | - | - | - | - | - | - |
| 2. | Training Programme | - | - | - | - | | - | - | - | - | - | - |
| 3. | Misc. | - | - | - | - | - | - | - | - | - | - | - |
| | Total | | | | | | | | | | | |
| | | | Grand Total :- i + ii | | | | | | | | | |

## REFERENCES

Mohsin, M.A. (1984). Forestry and watershed management as interlocked system, *Proceeding Social Forestry & Watershed Management, Muhammad, S.* Eds., B.A.U., Ranchi, pp. 110-115.

Mohsin, M.A. (1989). Watershed and its management, K.V.K. Souvenir, Ramkrishna Mission, Ranchi, pp. 33-39

National wasteland Development Project for rainfed areas—Guidelines (1991). Ministry of Agril. Dept. of Agril. and Cooperation, Govt. of India, New Delhi, pp. 1-103.

Prasad, R.B. (1984). Technology for Dryland agriculture on watershed basis, Proceeding, Dryland Farming and watershed Management, Mohsin, M.A. and Prasad, R.B., Eds. B.A.U., Ranchi, pp. 72-83.

Sinha, A.K. (1984) Approved practices of soil conservation for watershed management. In : *Agroforestry - Indian perspective*, Jha, L.K. and Sen Sarma, P.K. (eds.); Ashish Publishing House, New Delhi, pp. 159-172.

Srivastva, V.C. (1984). Water Managements : its importance in dryland agriculture, *Proceeding Dryland Farming and watershed Management,* Mohsin, M.A. and Prasad, R.B. Eds., B.A.U., Ranchi, pp. 162-176.

# 10

# *Agroforestry an Alternative to Shifting Cultivation*

Shifting cultivation, the main mode of livelihood of hill people, was simple most acceptable, self sustained and environmentally sound practice in the past centuries.

The limited population, vast land areas with little needs, never forced them to search alternative practices. All these promoted the practice of shifting cultivation with long fallow period, allowing the abandoned shifting cultivation lands longer period till the natural fertility of the soil is restored. In this way, this system was providing social security to the hill people and was considered highly evolved natural technology for the forest dwellors.

In the present situation poverty in Jhum cultivation area can be attributed to the population explosion and insufficient production of food grain. Their material needs are also high. They have to meet their own requirements and farmers has to produce more in order to get other produce what they are not able to grow themselves. Due to increase in needs, larger forest areas are put under shifting cultivation and jhum period is shortened. The reduced fallow periods is decreasing production potential and creating other environmental hazards.

**Extent of Area**

This practice is common in different socio-economic and environmental conditions in Africa, Central and South America, Southern Asia, Oceania, etc.

In India this is common practice among the hill dwellers of Assam, Arunachal Pradesh, Andhra Pradesh, Bihar, Karnataka, Manipur, Meghalaya, Mizoram, Madhya Pradesh, Nagaland, Orrissa and Tripura.

According to Roy Burman (1989) the task force on shifting cultivation in India during the seventh plan has drawn the attention to the absence of reliable statistical data on the extent of area under

**Fig. 34 :** Shifting cultivation has led several ecological and social problem due to our negligence and carelessness.

shifting cultivation and the number of person and families involved. According to N.C.A. only 0.54 million hectare of land in India was under shifting cultivation. According to the task force set up by the Ministry of Agriculture, the corresponding figure was 9.95 million hectare.

By the visual interpretation of landset imaginary pertaining to the years 1975 and 1984 the Forest Survey of India has come to the conclusion that during the period of 9 years, the area affected by shifting cultivation in the North Eastern hill states has reduced from 73,410 sq.km. to 62,854 sq.km. (Table 10.1).

**Table 10.1 :** Area affected by shifting cultivation in the North Eastern Hill States

| State | Geographical area (sq km) | Extent of forest affected by shifting cultivation (sq km) | | % Change in 1984 as compared to 1975 | |
|---|---|---|---|---|---|
| | 1975 | | 1984 | | |
| Arunachal Pradesh | 83590 | 7940 | 8521 | (+) | 7.3 |
| Assam | 78520 | 4160 | 7276 | (+) | 74.9 |
| Manipur | 22360 | 17770 | 13846 | (−) | 22.1 |
| Meghalaya | 22480 | 10240 | 6812 | (−) | 33.5 |
| Mizoram | 21090 | 16110 | 12442 | (−) | 22.8 |
| Nagaland | 16520 | 10970 | 10641 | (−) | 3.0 |
| Tripura | 10490 | 6220 | 3315 | (−) | 46.7 |
| Total : | 255050 | 73410 | 62854 | (−) | 14.4 |
| % | 100.0 | 28.8 | 24.6 | | |

**Source :** The State Forest Report, 1989.

The report is encouraging but decrease in practice may be due to some difficult situation. In fact, the main reason may be as follows :

- Today one can see long stretches of barren hills.
- The families once dependent on Jhum are now facing problem in maintaining their family due to various ecological and socio-psychological problems. At present, they prefer to work in the urban areas. This problem can be only solved by the integration of traditional land use system with modern practices. If in the existing jhum land, trees could be grown

Fig. 35 : Site clearance (Shifting cultivation).

along with food crops without affecting the yield and fertility status of soil, it would be acceptable to rural populace.

**Steps involved in shifting cultivation**

The steps/activities involved in shifting cultivation practice are as follows. The period of operation varies according to area, availability of man power and nature of crops. (See Chart No. 5)

**Comparative analysis of shifting cultivation and improved Technology**

The shortening of the Jhum cycle does not allow the soil fertility to build up. Thus, farmers needed larger areas for cultivation for the same quantity of produce. Many Research Centres have been

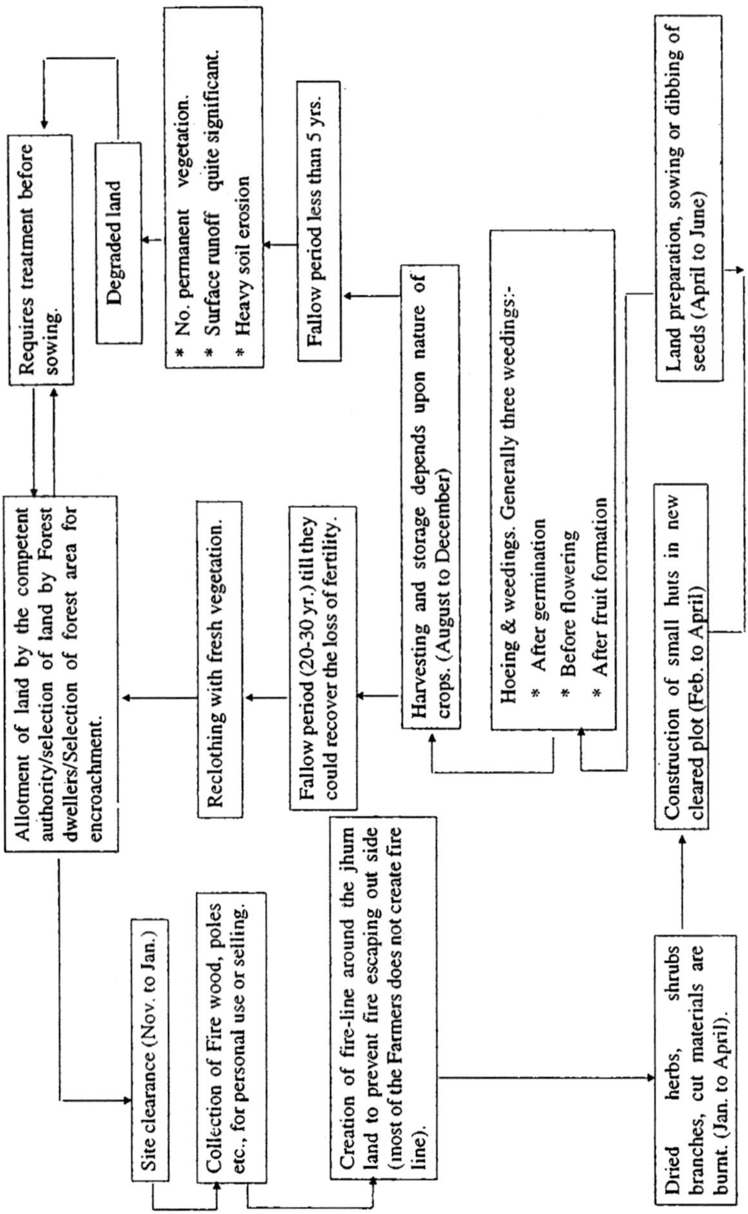


Chart No. 5 A Schematic Diagram showing the Shifting cultivation cycle

carrying on experiments to tackle the problems of shifting cultivation. In recent decades, continuous cropping have accelerated the soil erosion and condition is progressively deteriorating. In most of the shifting cultivation areas, wrong soil tillage practices also played important role in increasing rate of run-off and soil erosion. In these areas, luxurious vegetation (grass species) misleads people to believe, that the soils are extremely rich. Infact, soils are poor and heavy rain destroy them gradually.

**Fig. 36 :** Tharlam (Small House) construction in Jhum land by using bamboos.

## Soil loss, surface run-off and water conservation

ICAR Publication (1983) specially mentions that average soil loss due to shifting cultivation has been recorded to be 40.9 tonnes/ha in a study in Meghalaya. The highest sediment load however, was 76.9 tonnes/ha/yr. The annual loss of soil was computed by taking into account an average soil loss of 40.0 tonnes/ha/yr. Report also indicated that annual soil loss due to shifting cultivation in the region was about 15.5 million tonnes.

Figs. 37&38 : Paddy cultivation in a Jhum Land, at Aizawl, Mizoram.

Jha & Rathore (1981) mentioned that erosion ratios of Orissa soils under shifting cultivation in India were 18.03 and 18.78 in the surface and 6.5 and 11.4 in the sub-surface as compared with erosion ratio of soils not under shifting cultivation which were 10.7 and 10.8 in the surface and 3.7 and 11.7 in subsurface soils. Erosion ratios were higher in the upper layer than in the lower layers. Soils under shifting cultivation had high erosion ratios. While, soils not under shifting cultivation are within the safety limits of erodibility. Borthakur *et al.*, (1978), studied that surface run-off under shifting cultivation was about 11.40 mm. Partial terracing reduced surface run-off further to 3.28 mm. This study indicates that physical barriers considerably reduces surface run-off.

Mishra *et al.*, (1975) are of the opinion that with appropriate soil and water conservation measures the rate of sediments can be reduced from 80 tonnes/ha/annum to 6-7 tonnes/ha/annum within 5-6 years of treatment. In general, soil erosion and waterloss are the most critical phenomena throughout nation. Trees control erosion. This is achieved in two ways - with trees acting as barriers or as cover. The barrier function in the conventional approach to erosion control by checking run-off of water and the tree cover reduces the raindrop impact (Young, 1989).

Reynolds *et al.*, (1988) pointed out that on slopping land, when leguminous tree species planted along the contour, the trees limit soil erosion. Tree foliage can be used as much and fertilizer for food crops.

**Nutrient status**

Mishra and Ramkrishnan (1983) studied the effect of Jhum on soil fertility at high elevations of Meghalaya, North Eastern India using 15-, 10- and 5-year Jhum cycle and terrace system. Soil nitrogen concentration under a 5-year Jhum cycle was significantly lower than under the 10- and 15-year cycles. The concentration of this nutrient at the surface layers declined sharply after the burn and was attributed to volatilization. The degree of volatilization is dependent on the intensity of the burn and, therefore, the nitrogen decline is lower in a 5-year cycle than in longer cycles with a greater fuel load. Nitrogen concentration in the soil is expected to improve at the end of one year under all cycles, but such an improvement was actually observed only under 10- and 15-year cycles.

Under terrace cultivation, nitrogen losses continued up to the end of two years. Available-P was generally lower under a 5-yr. cycle than under 10- and 15-year cycle. Available-P under the three jhum cycle declined significantly after the burn. Recovery occurred after 30 days under 10- and 15-year cycle and after 100 days for the 5-year cycle with subsequent decline during cropping. In the terrace system, the available-P declined during cropping after an initial increase. The concentrations and total quantities of all cations on the surface soil layers under the three jhum cycles improved markedly after the burn owing to releases through ash, but subsequently declined sharply. The decline is partly due to absorption by the developing crop; but due more to through sediment and water loss. Large quantities of cations were released under 10- and 15-year jhum cycles owing to greater quantities of slash burned than that under a 5-year cycle. Starting from a lower level in a 0-year fallow, the recovery of C, N and available-P occurred in older fallow of 5- and 10-year, with a slight decline in a 15-year fallow. Longer fallows gave greater improvement in humus and nutrients in order to sustain slash and burn agriculture. A jhum cycle of 5-year is definitely short as far as nutrient recovery is concerned. Longer cycle of 10 years are preferred. A terrace system does not offer a better alternative to a short cycle because it cannot sustain continuously cropping without heavy fertilizer inputs.

Chauhan (1990) in his paper also referred to results of ICAR Research complex (1983). When the forest canopy of virgin land is slashed and burnt for shifting cultivation, there is very likelihood that organic matter of the surface soil is also oxidised. This may cause a decrease in the level of soil organic matter. The decrease in organic matter will be more pronounced particularly when the temperature during burning exceeds 150°c. Both erosion and cultivation cause a decrease in the organic matter content of the soil. According to Chauhan (1990) the annual loss of soil and crop yield in the North Eastern region, due to shifting cultivation was estimated to be 15.5 million tonnes and 52.23 thousand tonnes respectively. The on-site cost of soil degradation was computed to be Rs 980.6/ha/yr (Rs 37.9 crores/yr) which was 36.6 per cent of the value of the crop produce harvested. The crop yield loss, plant nutrient loss and land depreciation contributed 48.2, 39.6 and 12.2 per cent, respectively towards the on-site cost of soil degradation. Among the plant nutrient, nitrogent contributed maximum (29.2) followed by phosphorus (7.8%), potassium (1.6%) and sulpher (1.0%).

Under agroforestry system, hedgerows may be utilised as green manure or mulch. Nutrients are released when it is incorporated into the soil. Under alley cropping system moderate fertilizer or F.Y.M. may be best solution to yield stability and increased production under modified shifting cultivation models.

In the jhum cultivation, villagers cut and clear a strip of forest, three to four metres wide, all round the jhum plot as fire line before they set fire to the area and scrupulously guard against fire escaping outside the area. With the lowering of vigilence, now more damage to vegetation is caused by the escaping jhum fire than the actual practice of jhuming cultivation itself. The sprouting young vegetation covering the left over jhum area is destroyed year after year till the land becomes completely devoid of vegetative cover worth the name (Status Report, AFCL, 1989).

In this practice, increase in temperature is also likely to destroy beneficial microorganisms. Besides this there are some positive effect of control burning. The smoke liberated in the slash and burn cultivation kills pathogens and pests.

**A Look into Future**

Jha and Singh (1993) opined that the causes of the jhum cultivation were socio-economic, though, there is dearth of factual material with regard to the psycho-socio-cultural perspectives with special reference to social organisation, social behaviour, socialization and social cognitions relating to forestry, and their relationships (as the social causes of jhuming), on the one hand, and the underlying motivational and cognitive characteristics associated with jhuming and permanent agricultural settlement and forestry on the other. They also suggested that jhuming is a socially learned behaviour which affects social organisations, and on the higher level upon the organisation of the entire community and subsequently the populations. This process of learned socialization and associated perceptual, motivational, other cognitive characteristics (including learning and thought processes) ultimately determines the adaptability and change proneness of individual(s) and groups of people for achievement, motivation and orientation as sources for the origin of social motivation, and the need hierarchy of people. This kind of social motivation is related to the general phenomena of social facilitation, and the factors which may produce social facilitation may be the arousal by other numbers of the groups, reduction of fear and conflicts between the traditional and modern technology, and the coping of the individuals between the systems. An empirical study on these counts would only provide empirical foundations for a major breakthrough for cognitive and behaviour modifications of people for successful implementation and follow up programmes of development strategies relating to productivity and sustainability of the jhum land, and restoration of the forest.

Thus, before implementation of modern technology, we should study in depth the psycho-socio-cultural profiles, particularly the social organisation, social behaviour and social cognitions associated with shifting cultivation and permanent settlement patterns by employing standard psychometric and sociometric parameters. To increase productivity on sustained basis in the jhum land, below mentioned study should be carried out in each agroclimatic zone :

* An inventory of general physiographic, edaphic, climate and socio-economic conditions,
* Study on existing land use pattern in the shifting cultivation zone,
* Evaluation of productivity of the system and the components.

The finding of the above would provide empirical foundation to formulate strategies for cognitive behaviour modification of individual(s) and community orientation programmes for planned elimination (minimization) of jhuming and to fecilitate a permanent settlement pattern relating to agriculture and forestry, hence meeting the objectives means to replace shifting cultivation suitably for the rehabilitation of the people.

It is apparent from above discussion that the approach of shifting cultivation should be need based and location oriented. This practice can be only modified by studying existing agricultural practices in each shifting cultivation zone and utilising its natural resources, keeping in mind the psycho-socio-economic problems of the area.

Several designs/models have been suggested to solve the problem of shifting cultivation. The three tier system suggested by the ICAR Shillong *i.e.* cultivation of trees at the hill top, cultivation of horticulture crops in the middle portions of the hill slopes and opening of terraces at the lower portion for cultivation of field crops. Singh and Pandey (1989) have suggested the tree shifting cultivation rotation. Each of these methods has limitations.

In my opinion, any alternative model can only be successful, if adopted by local people, suitable to local environment and meet the needs of farmers. Besides this, the model should be similar to traditional practices with minor alternations. The above problem can be tackled by developing an appropriate agroforestry models for each agro-climatic zone. Introduction of all the three components i.e., agronomy, forestry and animal husbandry will help in minimising the soil erosion, conservation of moisture, increase in inflitration rate, decrease in evaporation from soil and balance nutrient status and finally help in land remaining productive on a sustained basis. Keeping in mind problems of jhumias, we can suggest the following proposals under agroforestry programme. The key features of the proposal are :

1. Maximum fruitful utilisation of unculturable jhum land.
2. Production of resources of local demand as well as national needs.
3. Economic rehabilitation of jhum farmers by creating perpetual income opportunities through quick and sustained return from timber/agril. crop/fruits etc.

Thus, Agrisilviculture, Agrisilvihorticulture or Agrihorticulture practices may be best solution for rehabilitation of jhum land.

**Agrisilviculture**

As is known, Agrisilviculture implies cultivation of agricultural crops in conjunction with forestry crops over the same unit of land to maximise productivity and sustainability of the land.

Here, I would like to point out that most of the trees/shrubs are not suitable for agroforestry system. As a general rule, a good agroforestry species should have following characteristics (Getahum and Reshid, 1988)

(*i*) Multipurpose *e.g.* fuel wood, fodder, poles, timber, green leaf manure, etc.

(*ii*) Fast growing.

(*iii*) Deep rooted, with narrow root zone.

(*iv*) Noncompetitive.

(*v*) The ability to grow back after cutting.

(*vi*) Nutritious and palatable leaves for livestock feed.

(*vii*) Nitrogen fixing.

(*viii*) Produce economic products and by products which can be used or sold.

(*ix*) Light canopy to allow sunlight penetration.

In North Eastern part the trees are generally planted on the margin of terraces, which are still planted and proctected and this could be pronounced as traditional Agrisilvicultural system. Contour system is generally followed where there is greater danger of erosion and irrigation of the area is difficult. The main purpose of this system is to minimise soil erosion and to conserve soil moisture so as to make the slope fit for growing fruits. The width of contour terrace varies according to the nature of the slope. If the slope is stiff, the less width of terrace, the less the chance of erosion (Das, 1985). In high rain fall area, soil is generally found deficient in nitrogen which tends to move deeper in soil having more sand and gravel content.

Timber/tree species including fruit trees take many years to generate income. Farmers do not get immediate return from the monoculture or mixed plantation area. If farmers adopt agroforestry practices *i.e.* intercrops are gown in the space between the rows of young trees, participants will get some income even in the initial years.

In selecting the crops to be grown in the interspace, attention must be given to the following points.

(1) The crop should not exhaust the soil.

(2) Spacing of the main tree crop and intercrop should be planned according to their requirements.

(3) Combinations should be able to maintain sufficient soil nutrients and check soil erosion.

(4) Intercrop may be utilised as green manures for maintaining organic matter content of the soil. This will also check weed, control soil erosion and help in enriching nitrogen content of the soil.

(5) Intercrop may be utilised as mulch also. This is very common in tropical and sub-tropical conditions. The mulch acts as a surface barrier preventing evaporation, suppresses weed growth and decreases soil temperature.

The best mulch will absorb little moisture and permit the rainfall to move downwards rapidly into the soil. This help in improving the soil structure.

During summer months, soil moisture is considerably reduced which causes severe fruit drop in citrus, if trees are not irrigated frequently. Mulch may be possible solution to this problem. Mulching with leaves should be adopted as standard practice in citrus orchards to tide over high soil temperature and rapid evaporation of soil moisture during high atmosphere temperature and low humidity. This is practiced in many states.

(6) Farmers should keep in mind that grain forming timing of the crop should not synchronise with flower bud differentiation in fruit trees. In non - bearing orchards farmer can grow any crop.

## SILVIPASTURE SYSTEMS

Silvipastoral practices can ensure high yield of good quality fodder, besides maintaining the productivity of such sites. The trees tap the lower layers of the soil, which will otherwise not be tapped by grasses. Silvipastoral practices will thus ensures better utilization of the site.

Pasture land may be developed in the degraded jhum land by closing the area for 5 years and introducing good palatable grass. The area should be fenced and be divided into 4 parts. From 2nd years, villagers may cut grass free of cost for their bonafied use only. Controlled grazing can be managed by rotation of each part.

If all the pasture area are managed properly the yield will be multiplied every year and full requirements of grass fodder will met. Grasses conserve soil and moisture and improve fertility status of soil.

### Models for the rehabilitation of jhum land

I have not discussed about the slope and nature of crops to be tried in particular zone, because farmers generally practice jhum cultivation within permissible level of slope. Percentage of slope will assist in computing the alley width under agroforestry system. Extension workers and, scientists should be careful in computation of the alley width. Agricultural crops, tree species varies in different agroclimatic zone. In each agroclimatic zones models must be tried with local species as well exotics. The proposed models of agroforestry system may be tried to rehabilitate the jhum land.

**Model-I : Intensive hedgerow (Alley) cropping on the Jhum Land**

**Objective**

* This model improves nutrient status and other properties of soil.
* Conserve more moisture and decrease soil erosion.
* Maximum utility of site by increasing productivity.

Alley cropping involves planting of rows of trees or shrubs, along contours, and food crops and cover crops are grown in between the rows. The trees will restrict soil erosion, decrease rate of run-off and tree foliage can be used as mulch or fertilizer for agricultural crops/or as animal feeds. Trees foliage is used to maintain and improve soil fertility, while cropping is in progress, similar to natural process that occurs during the traditional fallow period, thus increasing land use efficiency.

**Perennial vegetation (Hedge forming species)**

* *Leucaena leucocephala* (K-8, K-28, etc).
* *Gliricidia* species
* Perennial pea
* Vetiver
* Any other local Nitrogen fixing tree species suitable for agroforestry system.

**Pruning management**

* At different lopping intensities.

Alley width = 6m or may be changed according to percentage of slope of the jhum land.

Plot size = 15 x 6 m (may be changed)

**Treatments**

$T_1$ = Perennial vegetation (1-5 No.) + Agricultural crop two or more/vegetables two or more.

$T_2$ = Perennial vegetation (1-5 No.) + Agril. crop two or more/vegetables two or more (Tree foliage as mulch).

$T_3$ = Perennial vegetation (1-5 No.)/+Agril. Crop two or more/vegetables two or more (Tree foliage as green manure).

$T_4$ = Perennial vegetation (1-5 No.)/Agril. crop two or more/vegetables two or more.
(Farm yard manure)

$T_5$ = Perennial vegetation (1-5 No.)/Agril. crop two or more/vegetables two or more.
(Nitrogen fertilizer application)

$T_6$ = Agril. crop two or more/vegetables two or more (Nitrogen fertilizer application).

$T_7$ = Agril. crop two or more/vegetables two or more (F.Y.M.).

$T_8$ = Agril. crops/vegetables - control
(without any treatment).

$T_9$ = Tree species alone - control.

**Parameters**

* Yield of annual crop.
* Nutrient status of soil.
* Nitrogen fixing capacity of the trees involved.
* Growth performance of tree species.
* Percentage decrease in soil erosion, run-off and increase in moisture retention capacity of soil.

**Model - II Alley cropping with perennial grass**

**Objective :**

Planting of nitrogen fixing tree species with grasses to improve the soil fertility, reduce soil erosion and improve the production of fodder in previous jhum land.

**Hedge forming species (H.F.S.)**

* *Leucaena leucocephala*
* *Gliricidia* species
* Perennial pea
* Vetiver grass
* Any other local nitrogen fixing tree species.

**Grass species**

Tow or more grass species suitable for the particular Agro-climatic zone.

**Pruning management** : At different lopping intensities and intervals.

Alley width = 5 m or may be changed according to percentage of slope.

Plot size = 15 x 5 m (may be changed).

**Treatments** :

$T_1$ – H.F.S. (1 to 5 No.)+ Grass two or more species.

$T_2$ – H.F.S. (1 to 5 No.)+ Grass two or more species (Tree foliage as mulch).

$T_3$ – H.F.S. (1 to 5 No.)+ Grass two or more species (Tree foliage as green manure).

$T_4$ – H.F.S. (1 to 5 No.)+ Grass species (Nitrogen fertilizer application).

$T_5$ – H.F.S. (1 to 5 No.)+ Grass species (F.Y.M.)

$T_6$ – Grass alone

$T_7$ – Tree species alone.

**Parameter**

* Yield of grass
* Growth performance of tree species
* Nutrient status of soil
* Soil erosion, run-off and moisture contents of the soil.

**Model - III : Introduction of agril. crops/vegetable under different spacing of tree species including fruit trees on the jhuming land or degraded jhum land.**

**Objective**

* To find out best combination of tree species and intercrops.
* To study impact of trees/intercrops on the fertility status of soil.
* To reduce soil erosion and increase water conservation.

**Tree species to be tried**

* Important socially accepted tree species
* Important fruit trees.

**Inter crop**

* April. crop– Two or more
* Vegetables– Two or more (Rabi and Kharif Season).

* Spices – Two or more
* Fruits – Pineapple, sweet potato, etc. (Selection of species depends upon farmers choices and agro-climatic zone).

*Spacing* – 3 x 3m, 4 x 4m, 5 x 5m or may be changed according to percentage of slope.

Plot size – 24 x 24m or 12 x 12m (may be changed).

**Treatment**

$T_1$ – Tree species + Agril. crop two or more/vegetables two or more/spices two or more/fruit two or more species. (Control).

$T_2$ – Tree species + Agril. crop two or more/vegetables two or more/fruits two or more species (nitrogen application).

$T_3$ – Tree species + Agril. crop two or more/vegetables two or more/fruits two or more species (F.Y.M. application).

$T_4$ – Tree species + Agril. crop two or more species/fruits two or more species/vegetables two or more species. (Mulching by natural grass).

**Parameter**

* Growth performance of tree species
* Yield of intercrops
* Nutrient status of soil
* Increase or decrease in soil erosion, run-off and moisture content of soil.

**Model - IV : Horti-Silvi-Agriculture system (wide alley cropping) in the jhum land**

**Objective**

* To meet the requirement of fruits, wood crop and the fodder.
* To conserve more moisture and minimise soil erosion.
* Maximum site utility by increasing productivity.

(Fruit Trees, nitrogen fixing hedge forming tree species or vetiver grass may be integrated to meet the diverse requirements of the farmers. The space between fruit trees will be utilised by *leucaena leucocephala* or any other NFTs as fillers to form hedges.)

Spacing – 5 x 5 m (may be changed according to percentage of slope).

Plot size – 30 x 30 m (may be changed).

**Treatment**

$T_1$ = Fruit trees (Citrus/Aonla/Guava etc.) + Hedge forming tree species as fillers between trees + Agril. crops two or more/vegetables two or more/fruits two or more (Ananas/banana etc.).

$T_2$ = Fruit trees + Hedge forming tree species between trees + Agriculture crop two or more/vegetables two or more/fruits two or more

(Hedge foliage as mulch).

$T_3$ = Fruit trees + Hedge forming tree species + Agril. crops two or more/vegetables two or more/fruits two or more.

(Hedge foliage as green manure).

$T_4$ = Fruit trees + Hedge forming tree species + Agril. crop two or more/vegetables two or more/Fruits two or more (Nitrogen/F.Y.M. application and hedge foliage used as fodder).

$T_5$ = Agril. crop/vegetables/fruits (control).

$T_6$ = Fruit trees (control).

**Parameter**

* Growth performance of fruit tree species.
* Yield of intercrops.
* Nutrient status of soil.
* Soil erosion, run-off and moisture contents of soil.

**Block Plantation**

Block plantation of fuel/timber species in fallow jhum land may improve rural economy and meet the requirements of fuelwood timber. This will indirectly help in decreasing pressure over traditional forests.

**Parameter**

* Growth performance of tree species at different spacing.
* Nutrient status of soil.
* Soil erosion, run-off and moisture contents of soil.

Comparative study of data/economics of different model will be helpful in knowing best model for jhum cultivators and after successful introduction of the best model, the tree cover will increase significantly, although the grain yield/vegetables etc., may go down to some extent at later stages because of tree shade which can be solved through proper lopping managements. In the long run farmer will earn more from the fruit trees/timber/fuel wood and shade tolerant intercrops.

In the above model, farmers are required to put in additional labour to maintain trees and to harvest the produce. The inter-cultural operations will further result in soil and water conservation and will improve fertility status of the land.

This land use technology will not only help the poor to fulfil their basic needs, but it will also help them to get better output for their labour which will increase their income in the long run. Further, poor jhumia's will get regular employment in existing jhuming field. Naturally they will prefer to adopt settled-land use pattern.

## CONSERVATION OF SOIL AND MOISTURE AND MAINTENANCE OF FERTILITY STATUS UNDER THE PROPOSED AGROFORESTRY MODELS

At present forests are removed at a faster rate due to jhum cultivation. Jhum farmers prefer monocropping. Generally they do not care about growth performance or regeneration ability of tree species within the jhum land. Jhumia's life style is thus going to disturb normal environmental equilibrium and so much, they may not obtain needed tree products easily.

The rate of run-off and soil erosion between two Jhum cycle have recorded higher. Sheet erosion is the crucial stage. The accelerated soil erosion influence water cycle within its range.

Under this condition lands are considered as most degraded, unsuitable for future agricultural practices, as soil is without vegetative cover which decreases water absorption capacity and increases rate of run-off.

In some parts, conscious farmer form soil bund in his plot and a channel for the run-off to prevent erosion. As shown in Figs. 41 and 42, bund is made up of soil of the same plot. After heavy rains, heaped soils gets eroded within a year due to increase in run-off.

**Fig. 39 :** Diversion of forest land towards non-forest use.

**Fig. 40 :** Misuse of land and water resources (faulty cultivation).

This measure brings short term relief. The practice of mono-cropping of unscientific mix-cropping accelerates erosion problems and the soil can no longer absorb much rain water. In addition, this practice requires much labour.

As apparent in Fig 43, some farmer use gunny bag filled with soil to form terrace or stones piled up or branches of trees are used to form natural wall across the bed. The height is raised, above the level of field. This structure reduces soil erosion, silt gets deposited near the stone wall and water oozes out, thus facilitating increased infiltration of the rain water that falls in the crop filed. This may be treated as good and effective method to control soil erosion for a short period. However, after few years, this structure also looses its viability, fails to check soil erosion and runoff etc. (See Figs. 43 and 44).

We have discussed all the traditional approach which help the farmers in solving their problems for a short span.

As shown in Figs. 45 and 46 problems of jhum land *i.e.* soil erosion, increased run-off, loss in nutrient status can be solved by adopting suitable agroforestry models. My suggestion is to introduce intensive hedgerow or alley cropping or wide alley cropping system (as discussed earlier). Horizontal distance between the hedgerow depends upon the slope percentage. The distance between the hedge decreases with the increase in slope percentage. The most important feature of this system is that the farmers can do the whole work themselves. Labour requirements is less. The hedgerows must be established across the slope (Figs. 45 to 48) on the contour line. Contour line passes round the slope and it is line of equal elivation. If hedgerow are established in straight line, the hazards of shifting cultivation increases. Under intensive hedgerows system, nitrogen fixing tree species *e.g.*, *Gliricidia* species, *Leucaena leucocephala* or any N.F.Ts. suitable for agroforestry system are planted at close spacing. To make the barrier effective, trees should be spaced densely, 5-6 cm with in rows. Two rows may be considered as the best. If nitrogen fixing multipurpose tree species are planted at larger gap, the gap between trees will facilitate waterflow and soil erosion and nutrient loss will be more. To overcome the above difficulties, the space between the trees can be used for growing perennial grass or pigeonpea, vetiver grass, etc. In case farmers intend to plant fruit trees or tree species having longer rotation period (timber species), spacing between the trees will be

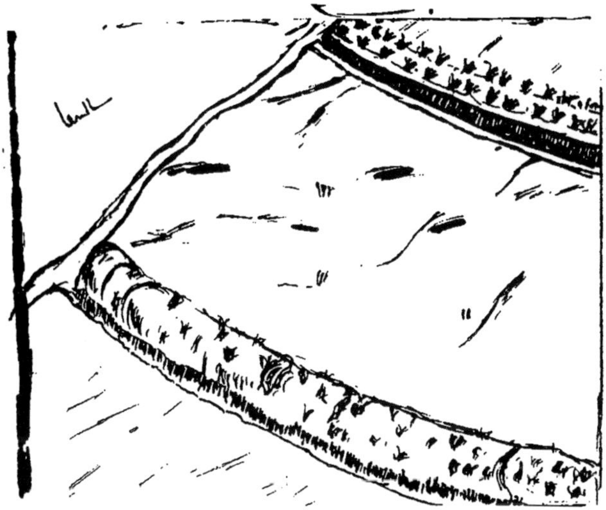

**Figs. 41&42 :** Traditional soil and water conservation practice (Soil bund on contour line).

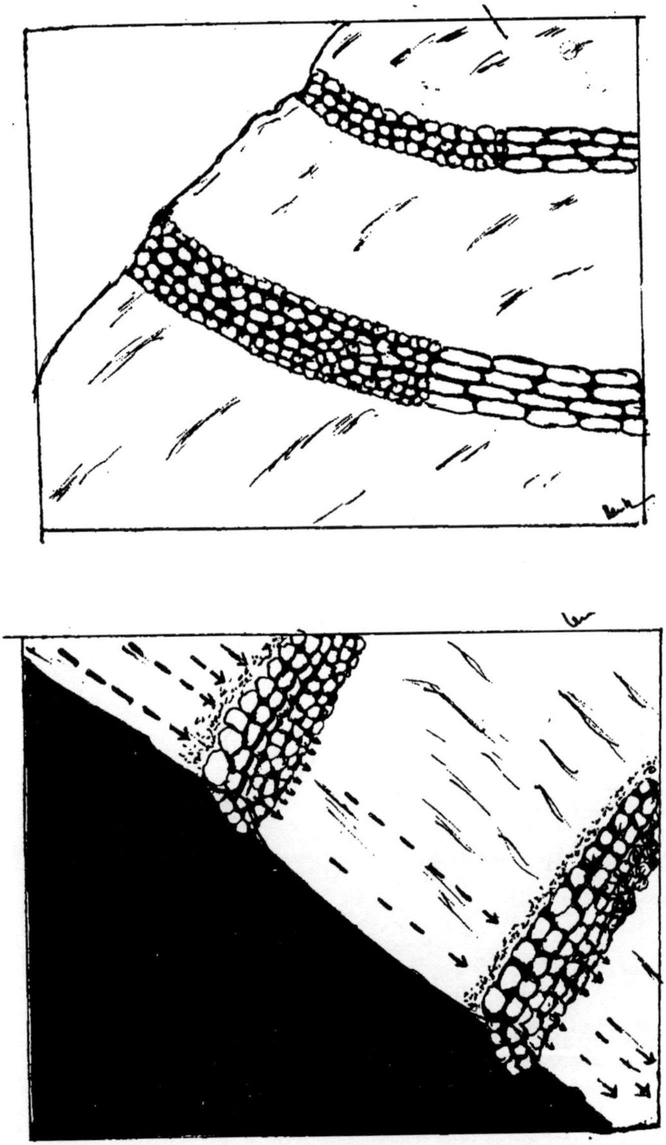

Figs. 43&44 : Traditional soil and water conservation practice (Stones piled up on contour lines).

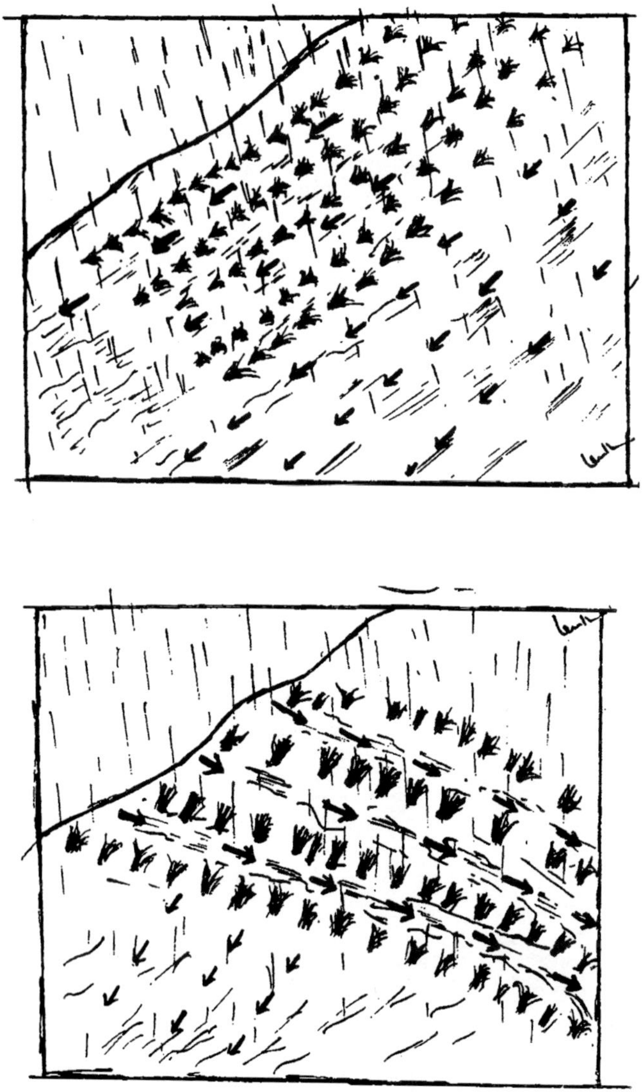

**Figs. 45&46 :** Faulty hill farming: deterioration of land.

**Figs. 47&48 :** Intensive alley cropping to tackle the problems of Jhum land.

more. *Leucaena leucocephala*/vetiver grass/any suitable perennial grass/pigeon pea etc. may be used as fillers within the rows. The hedge forming tree species and grasses should be harvested after proper establishment. Within two years, hedge forming species will have well established root system. After that hedge species may be managed under coppice system. Tree species (hedge forming) are generally cut 10-15 cm above from the ground level. Vetiver or perennial grass can be cut down to ground level.

In the initial year a contour trench may be formed parallel to the hedge rows (behind each hedgerows). This will serve as soil trap. After establishment of the hedgerows soil filters out of the run-off and run-off oozes through the hedges formed on the contour. Silt is deposited behind the plant and water flows on the slope with slow speed with very low quantity of silt load. Thus, a natural terrace is formed. Farmers have to take special care of hedge in the initial year, till the establishment of natural hedge rows at the edge of terrace (with deep root system).

The small twigs and foliage of nitrogen fixing tree species of hedgerows may be incorporated in the soil and poles and branches may be used as fuelwood. The micro-organism in the root nodules of nitrogen fixing tree species fixes the atmospheric nitrogen, which supports the growth of the intercrops. Leaves when used as green manure, release humus and minerals to the soil after decomposition. The roots of nitrogen fixing tree species due to nutrient pumping action carries important nutrient from the lower zone of soil and makes it available to agricultural crop. In the regular coppicing management, roots die, decompose and add organic matter to the soil. The hedgerows hold soil together along the contour. The hedgerows also help in reducing the wind speed. The foliage of N.F.Ts. may also be utilised as fodder or mulch. Soil nutrient status improves in a manner similar to fallow period. Higher moisture stored in the filed, also helps in better growth of the inter-crops. to protect the inter-crops from shade effect, pollarding or pruning is done.

We can conclude that this system is highly flexible, farmers can modify the model as per their needs, environmental, social and economic condition and can protect their jhum land from degradation and help jhum land remaining productive on a sustained basis. These models have been designed to simulate traditional practice.

**Fig. 49 :** Intensive alley cropping system (Top storey- silvipastoral ; Middle storey- silvihortipastoral and Lower storey- agrisilvipastoral).

**Suggestions**

(1) Government is trying to change the jhum cultivation to settled scientific cultivation. In the hilly areas their is no shortage of land. Therefore, Government should allot land permanently to the jhumia's, individually or community basis as has been provided in the New Land Use Policy of the government of Mizoram, so that they consider that permanent reclamation of jhum land is in their own interest or their development is directly related with reclamation of jhum land and they will not get new areas for above practice.

(2) Establishment of a network of alternative models in each zone in conformity to traditional practices which should be demonstrated to the jhumia's.

(3) As the system is multidisciplinary, separate eduction and training for development, management and extension of agroforestry system at different level are required.

(4) Most of the jhumia's are below the poverty line. Moreover, in spite of the efforts made during our five years plans, infrastructure in our jhum belts is not up to the mark. The facilities for education, drinking water, transport and communication, institutional credit etc. are lagging far behind. Under these circumstances we cannot expect voluntary involvement of jhumia's to settled agriculture.

(5) Vigilence cell consisting of jhumia's, representative, should be formed at remote village, so that jhumia's may avail timely benefits granted by the Government. They must be protected from the practice of greating palm of corrupt officials.

(6) Frequent transfer of senior officers, who are directly involved in development programmes including rehabilitation of jhumia's should discouraged. Such practice adversely affects the executive of rehabilitation programme.

(7) *Jhumia's co-operative society:* Needless to say Co-operatives can play a pivotal role in accelerating the

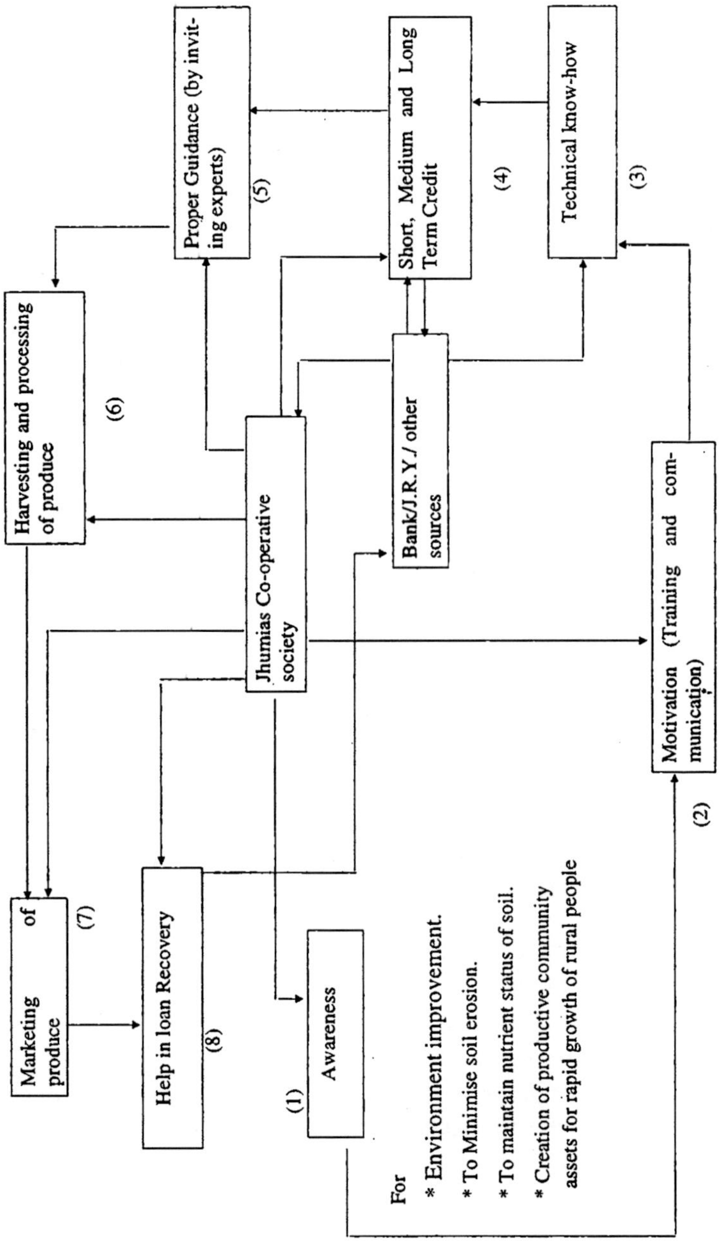


Chart No. 6. A Schematic Diagram Showing Role of Jhumia's Co-operative Society to Tackle the Problem of Jhum Farmers.

tempo of rehabilitation of jhum land. The jhumia's belong to the weakest section of our Indian society. It is apparent that proper functioning of jhumia's Co-operatives Society would reduce their economic exploitation and will raise their socio-economic status. It is satisfying that at present utmost emphasis is laid for the rehabilitation of jhum land to increase productivity and sustainability.

The Co-operative Society will create awareness among the farmers and motivate them to implement rehabilitation programmes of jhum land (Chart No. 6.). The society can invite technical experts/extension workers to guide and provide latest technical know-how for successful rehabilitation of jhum land. The jhumia's Co-operative Society should arrange timely disbursement of short, medium and long term of credit. Credit will be provided for consumption purpose during the lean period of the year to eliminate exploitation of professional money-lenders or in the initial years of settled cultivation. The jhumia's will get technical guidance from the experts through the Society. Marketing of produce of members will be exclusively done by primaries. Consequently, exploitation by the intermediaries will be eliminated. Processing and marketing of the produce of the members by the Society will also facilitate the timely recovery of loans and hence the problem of overdues will be automatically solved. Thus, it is apparent that Co-operative Society can solve the problems of jhumia's and help in improving the overall quality of life of jhum cultivation, leading to the formation of an egalitarion social order.

(7) Thus, in my opinion to reclaim jhum land, 1st phase is permanent allotment of land to jhum cultivators and 2nd phase is creation of awareness, motivation, guidance about latest technology, financial assistance etc., to jhum farmers. Farmers may be provided these incentives through Co-operative Societies or any suitable agencies. The farmers can utilise available space and resources by introducing latest technology. Suppose farmers have been alloted 10 hac. land, planning should be such that after providing land for construction of house, kitchen, pig and cow shed, compost pit, Gobar gas plant, water harvesting tank etc., remaining area be put under different cropping systems as per need of farmers. Boarder area of the allotted land should be planted with bamboos. One portion of plot may be put under high density of plantation *i.e.* timber species at 10 x 10 metre spacing and fuel wood/nitrogen fixing tree species at 2 x 2 metre spacing or as suggested by the

experts. The hedge rows with perennial grass species may be established to protect the field from soil erosion. In the 2nd plot farmers can plant fruit trees at 8 x 8 m or 10 x 10 m spacing or as suggested by extension worker and in the interspace agricultural crop, vegetables may be grown. Hedgerows should be formed at required intervals. In the third plot, farmers can plant only tree species to fulfil the requirements of fuel wood or silk host plant may be planted. This will also assist in raising the income of farmers after 4 years and lops and tops of the host plant can be used as fuel wood. Farmers can put their third plot under multicropping system. Fruit trees at larger spacing, 12m x 12m space or as suggested by experts, in between them, papaya, banana or any other fruit species may be planted. In the interspace agricultural crop, spices, vegetables etc., can be grown as per their requirements and demand of local market. Farmers can grow perennial flower or seasonal flower in a small part of his field and can practice agriculture to supplement income. The hedgerows foliage can be utilised as green manure or fodder. As I discussed earlier, under Government financial help, farmer should be provided cow/pig/goat etc. The grass species and foliage of nitrogen fixing tree species can be used as fodder. The farmers can use farm yard manure of his compost pit in his crop field and rearing of cows/pigs/goats etc., will improve their supplementary income. It is apparent that under agroforestry programme, collaboration and cordination of all the three departments *i.e.*, Agriculture, Forestry and Animal Husbandry are essential. Finally this system will improve the economic status of jhum cultivators and loans provided by the Government can be recovered in instalments. In the initial year jhum cultivators will maintain their family from the loan provided by the Government. Concerted efforts are to be made by all concerned to enthuse jhumia's for voluntary and mass involvement, so that, economic stability may be a reality.

## REFERENCES

A.F.C.L. (1989). *Status Report on Dhaleswari Catchment,* Mizoram, Department of Agriculture, Government of Mizoram.

Borthakur, P. *et al*, (1978). Cited by R.K. Gupta (1981), Ecological consequences of irrational land uses in loss of productivity soil, Tingsanhall, T. and Eggers, H. Eds. *Southest Asian Regional symposium on problems of soil erosion and sedimentation.* Bangkok, Thailand, pp. 27-29.

Chauhan, B.S. (1990). On site degradation due to shifting cultivation in North Eastern Region, NEHU, *Journal of Social Sciences and Humanities,* Vol. VIII No. 3, pp. 5-8.

Getahum, A. and Rashid, K. (1988). *Agroforestry in Kenya, A Field Guide,* Rural afforestation Extension, Forest Department, Ministry of Environment and Natural Resources, in collaboration with the Swiss Development Co-operation.

I.C.A.R. (1983). *Shifting cultivation in North-Eastern India,* ICAR, Research Complex for North-Eastern Region, Shillong, Meghalaya, India, pp. 35-40.

Singh, J. & Pandey, M.C. (1989). Some measures to improving shifting cultivation, NEHU, *J. of Social Sciences,* Vol. V, No. 4; pp. 9-12.

Jha, L.K. and Singh, L.N. (1993). An empirical study on the psycho-socio-cultural perspectives of jhuming; *Tribal Research Bulletin,* (communicated).

Jha, M. and Rathore, R. (1981). Erodability of soil in shifting cultivation areas of Tripura and Orissa, *Indian Forester,* Vol. 107(5), pp. 310-313.

Jha, M. and Singh, M. (1980). Effect of various stages of shifting cultivation on soil erosion from steep hill slopes, *Indian Forestor,* Vol. 107(5), pp. 319-313.

Mishra, B.K. and Ramkrishnan, P.S. (1983). Slash and burn agriculture at higher elevation in North Eastern India, II, Soil fertility changes, *Agric, Ecosystem and Environ.,* 9. 83-96. Cited by R.E. de La Cruz and N.T. Vergara, Protective and ameliorative roles of agroforestry, An overview, *Agro Forestry in the Humid Tropics,* pp. 18-19.

Roy Burban, B.K. (1989). Critical appraisal of shifting cultivation farming technology and of various approaches, *J. of the N.E.C.,* Vol. X, No. 2-3, pp. 1-4.

Reynolds, L., Atta-Krah, A.N. and Francis, P.A. (1988). *Alley-farming with live-stock Guidelines,* Humid Zone Research Site, I.L.S.G. for Africa, Nigeria, pp. 1-30.

Singh, A., and Singh, M. (1980). Effect of various stages of shifting cultivation on soil erosion from steep hill slopes. *Indian Forester,* 106(2); pp. 116-121.

The State Forest Report (1989). Government of India, F.S.I., 25, Subhash Road, Dehra Dun (U.P.).

Young, A. (1989). 10-hypothesis for soil agroforestry research, *Agroforestry Today,* Vol. 1(1), pp. 13-16.

11

# On-Farm Research : A Case Study

Indian farmers with the help of modern technology under supervision of a scientist/an extension worker have done commendable work in pushing our food grain production from 52 million tonnes to 152 million tonnes in about 30 years. If we consider about increase in production and productivity in other developing countries like, China, Philippines, North Korea, South Korea etc., we can realise that our achievement is not an extra ordinary feat.

In the present context, we have not only to increase the productivity of the land but we have also to think about the rehabilitation of adivasis, weaker sections, people living in hill sides and those villagers having poor communication with insignificant education or even literacy.

Extension Department of Ministry of Agriculture/State Department of Agriculture and Extension education system of ICAR are playing crucial role in this tasks. When we visualise the objective of ICAR Extension Education system, it clearly highlight, role played by ICAR and State Agricultural Universities in conducting research and extension works :

* Conducting research in different agroclimatic zone and testing the technologies in the different psycho-socio-economic condition of the farmers.
* Demonstration of most successful model in the farmers field under operation research/On-farm research programme.
* Conducting training for different level of functionaries, with especial emphasis to train the farmers and farm women.

In India Operational Research Projects of Indian Council of Agricultural Research/State Agricultural Universities have now been adopted by the many state Agricultural Departments to increase the productivity. Prasad (1989) has stated that the future model of Extension Project of the ICAR is going to be the Krishi Vigyan Kendras. All other projects namely: National Demonstration Projects, Operational Research Projects, Lab-to-land Programme etc., will be merged with the Krishi Vigyan Kendra Scheme and the functional structure of the K.V.K. will be enlarged in terms of :—

(*i*) Farm advisory service — demonstration of latest technologies,

(*ii*) Vocational training of the farmers, farm women, young farmers and field level extension functionaries,

(*iii*) On-farm research/ operational research.

To fulfil the above objectives K.V.K. model has been established by ICAR in 100 districts and efforts are being made that this model will be introduced in each agro-climatic zone for strengthening vocational training for the farming communities in the country.

Therefore, K.V.Ks. will have to be strengthened so that these become true harbinger in transferring the technology to the farmers who will, in turn increase food productivity by adopting higher production technology.

Shkula (1989) has mentioned that if sincere efforts are made in transforming the existing Krishi Vigyan Kendras into integrated Centres for "Knowledge dissemination and input distribution" the country can surmount the present challenge of increasing agricultural production and generating surplus grains after meeting the food requirements of growing population.

**What is on-farm research ?**

The environment at on-stations differ greatly from the on-farm experiments. We conduct research under controlled conditions under on-station trial. Most of the research station are equipped with all facilities. This help in conducting various types of experiments. In on-farm trial we have to face many problems. On-farm trials are conducted on farmer's land having difficult terrain, small plots, degraded land, soil erosion problems etc.

On-farm trials help in eliminating the farmers problems and increase the productivity. The activities of on-farm trial may be divided into two steps :—

(*a*) Diagnosis and design study conducted by extension workers, plays an important role in recognizing various bio-physical and psycho-socio-economic constraints of the farmers. Extension worker examines closely the precise requirements and identify improvement required for future on-farm trials.

(*b*) In the final step, best on-station models are introduced in the farmers field. The model may be modified as per psycho-socio-economic requirements and biophysical constraints of a village.

(*c*) These village laboratories will be a continuing feature on long term basis.

**On-farm research in Agroforestry**

Farmers of different parts of the world are practising agroforestry since time immemorial. The Hindu, a national newspaper of India in its issue of 6 Nov. 1937 published that a scheme known as agroforestry, which was a combination of agriculture and forestry, was adopted at the meeting of Central Fodder and Grazing Committee of the Imperial Council of Agriculture research (renamed Indian Council of Agricultural Research), which met at New Delhi. It is apparent that Indian Scientists was associated with agroforestry research since long ago. In the recent decade, ICAR has established All India Co-ordinated Project on Agroforestry to cover all agroclimatic zones. Similarly under Mission oriented work of National Wastelands Development Board, Government of India has created technology extension based on agroforestry in different part of country.

Agroforestry research is conducted in the agricultural experimentation centres, but generally the results of the experimental stations remain unused or one should say that newly proposed models tested in on-station trial is not accepted by the farmers. Farmers never accepts an imposed technology. Under above condition impact remains for short period. Our latest technology in the field of agroforestry can only be accepted by the farmers, provided our extension workers work along with farmers to examine requirements of the farmers and/accordingly tested models/designs are modified. If scientist/extension worker follows all steps during transfer of on-station model in farmers field (on-farm trial), the farmers acceptance of the model as a continuing feature on long term basis can be ensured. During the last ten or

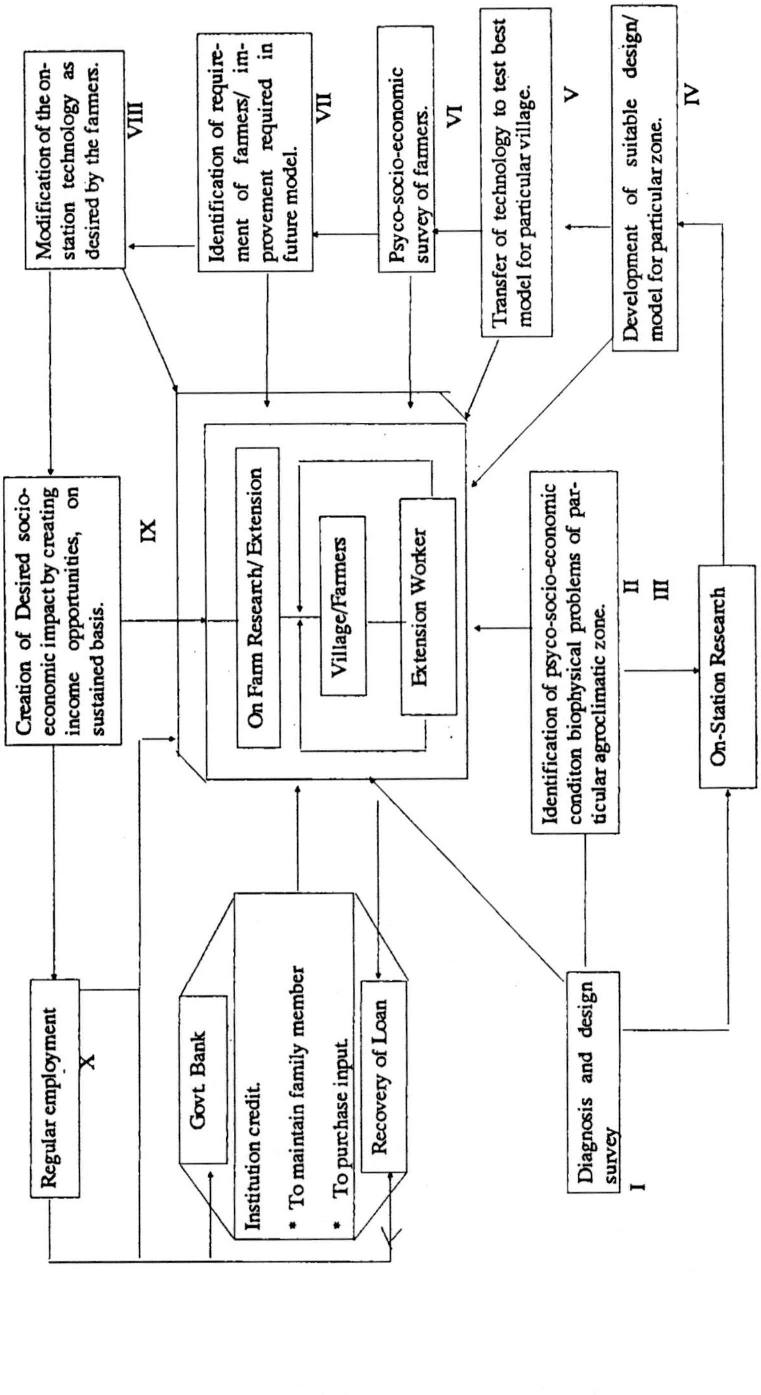


**Chart No. 7.** A schematic diagram showing steps involved in transfer of on-station research model in farmers land

more years most of the agroforestry research had been conducted under controlled condition. Now, it has become imperative to transfer the proto-type system trial to compare the productivity of the agroforestry technology to traditional systems.

**Experiments (on-farm)**

On-farm studies focussing on discrete environmental factors, practices or interventions may be designed to yield precise results on the effects and interactions of selected key variables. Comparative data might be obtained on an existing agroforestry system, with and without a specific intervention, for instance, lopping or not lopping established hedgerows between farmer's fields. Data might also be obtained on the relative impact of a limited number of ecological variables on (approximately) identical production system. For instance the productivity of several identically planted sites could be compared in terms of site-specific ecological factors such as soil, shelter, aspect (direction) or influence of pre-existing vegetation.

An experiment could be designed to determine the impact of a single agroforestry practice, such as the use of tree-litter mulch, or a small number of discrete or interrelated practices, such as the use of a particular method of soil tillage with or without the addition of mulch. Different levels of a single practice might also be tested, However, distinctions between levels should be obvious. In a study on the use of mulch, for instance results could be obtained using a lot of mulch, a little or none.

The amount of useful information obtainable from any experiment or trial depends largely on the initial planning and design stage. The spatial arrangement of plant components of various kinds, the range of management practices and the variability of environmental factors within the site must be carefully considered so that a well defined set of experimental situations can be sampled and reliable data gathered. Clearly the researcher working together with the farmer, must choose or design each study unit carefully with a particular set of well specified objectives in mind.

The proposed approach begins with the identification of a larger number of observation units of approximately identical size, probably spread over several farms. Each unit is then classified according to all relevant environmental variables.

**Fig. 50 :** Cleaning of wasteland, to put land under agroforestry practice.

**Fig. 51 :** Land preparation for sowing of seeds in rainfed condition.

Within groups (blocks) of similar units, an appropriate number of units is selected at random. These may be compared with a second set of units, from a different block which shares different environmental variables, or some units may be subjected to one or more interventions and compared with other untreated units from the same block. The crucial considerations are that all units in the same block are as similar as possible and that an adequate number of study units is choosen as a random sample of all units sharing the same characteristics.

Given the large number of ecological variables and production factors characteristic of any agroforestry practice or system, the accurate identification of similar production units is a challenging, but important, aspect of experimental design. Whether or not researchers are able to select truely homogeneous sites for experimentation is an important subject of testing in itself.

To expedite this task, the researcher should work closely with the farmers who would provide sites for experimentation. The farmer's classification of factors such as soil fertility at different sites, based on experience will require less time and resources than scientific measurements and will also be more closely linked to practical management considerations and expectations of productivity.

If research blocks are selected with closely similar biophysical attributes, if within these blocks test units are selected at random and if enough units are selected for adequate replication, then information will be obtained that is valid both biologically and statistically. Furthermore, the small, scattered test units are the ready sites on farmer's fields, and research sites on which productivity is demonstrably enhanced may easily be observed, expanded and replicated by farmers (I.C.R.A.F., News letter, 1988).

## ON-FARM AGROFORESTRY RESEARCH IN CHOTA-NAGPUR PLATEAU OF BIHAR : A CASE STUDY

The present study was conducted in Kumharia village of Chotanagpur plateau of Bihar. The people of this district are economically backward. In Kumharia village, total geographical area comes around 509 acres. Out of this 40 per cent land is considered unproductive and barren wastelands. The percentage of scheduled tribe, scheduled castes and others in the village is 70 per

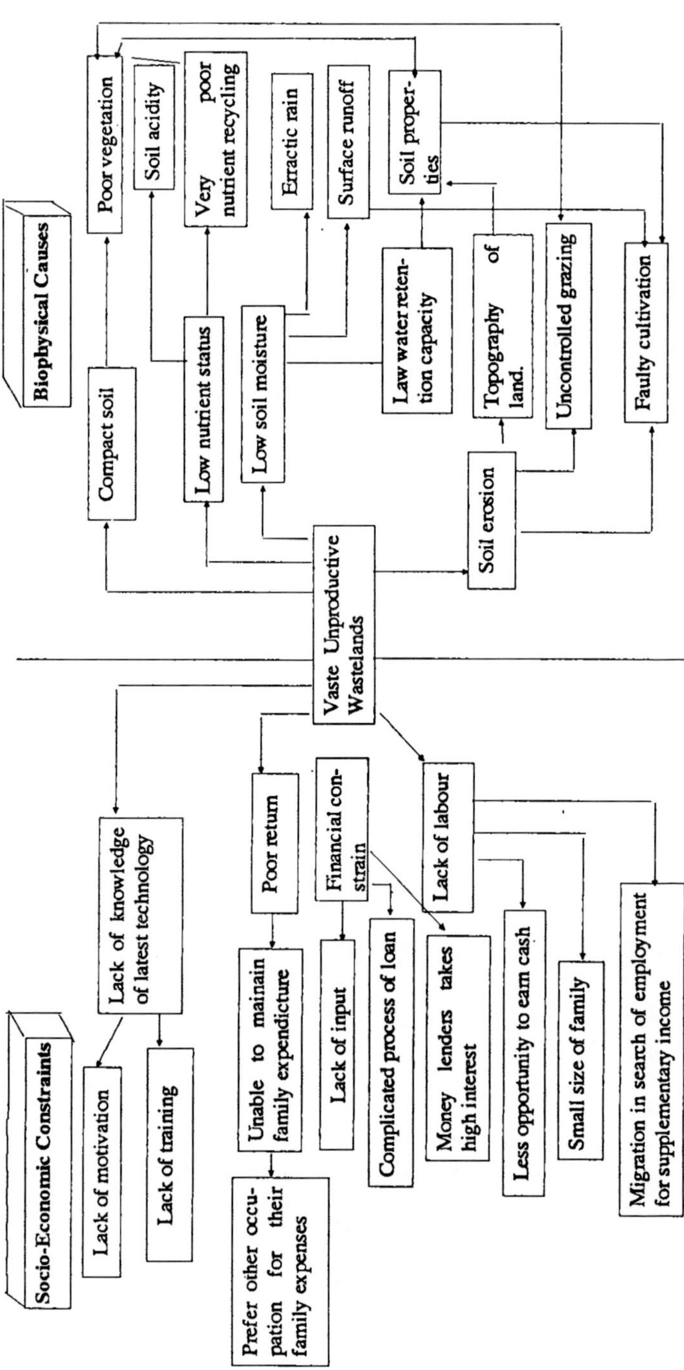


Chart No. 8 : A schematic diagram showing the farmer's problem tree of Kumharia village.

cent, 5.0 per cent and 25 per cent respectively. The distribution of household according to land holdings are landless 11.6 per cent, less than one acre 5.16 per cent, 1 to 5 acres 77.41 per cent and more than 5 acres 5.8 per cent. The average land holding per house hold is 3.2 acres and literacy rate is 25 per cent. The socio-economic survey was conducted before implementation of agroforestry. On-farm research in Kumharia village and farmers problems tree was drawn and central point of study was farmer's vaste area of unproductive barren wastelands. (See Chart No. 8)

The primary biophysical causes were compactness of soil due to uncontrolled grazing and poor vegetative cover. The age old faulty cultivation practice, poor nutrient recycling and acidity in soil, low moisture level, soil erosion problem progressively converted agricultural land into unproductive wasteland. Financial constraints was the main problem and they can't purchase input. They are also not willing to cutivation on the land because of low productivity of the site. Moreover, the farmers were not aware of latest farming technology to improve the productivity of these wastelands.

Previously, most of the farmers specially women used to collect fuel, fodder and timber from the adjacent forest area and many farmer specially women folk also used to sell it in the local market to supplement family income.

In this village, there was no organised system of fodder collection activity earlier. The entire system was primitive. Only straw was the valuable fodder. Farmers were also not selective regarding the types of fodder. Total annual requirements of fodder and fuelwood were 799.36 tonnes/yr, 1532.40 quintals respectively. From the response of participants it was apparent that the local community use wood as the major source of fuel both for cooking and other purposes. Fuelwood consumption was the highest during winter and the lowest in Monsoon when the consumption rate declined. In this village also problem of collection of fuelwood revolved round the distance, hazards in the process of cutting the wood, stacking it, preparing bundles and then carrying it to the village. Distance was another problem in fetching the fuelwood. Most of the roofs of the houses were constructed with the help of small size poles. The roofs required frequent repairing and replacement of wooden poles. Up till now, villagers used to collect the poles from the forest area. Since this is a sal (*Shorea robusta*) growing belt, most of the farmers could collect this valuable trees pole crop for their poles requirement. The problem of collection of timber also revolves round the distance hazard as well as the process of cutting the wood and stacking. It was clear that perambulating in the forests and carrying headload to the house might took considerable time and energy.

Generally, the process of transfer of on-station technology includes the following : —

(*a*) release of innovations from research stations,

(*b*) efforts to popularise them by extension experts, and

(*c*) response of farmers.

Each segments has its own groups of factors interacting with each other and simultaneously with the factors of other group, which ultimately affect the process of transfer of technology. As process starts from research station, an innovation is released to the farmers for adoption from these (Sinha, 1984).

In the Kumharia village, approach was different. After analysing the problems of the villagers, Directorate of Extension Education directed Department of Extension and Social Forestry, Birsa, Agril. University, Ranchi to start on-farm, agroforestry experiments in phases and experiments were conducted in the year 1984, in collaboration with farmers and researchers. However, farmer/participants were left free to make own decisions while adopting integrated programme.

The key objectives of on-farm agroforestry experiment were :

1. Utilisation of wasteland.
2. To generate resources for use by farmers as well as local demands.
3. Economic rehabilitation of poor farmers by creating regular income opportunities through quick and sustained return from agril. crop/fruit/vegetables etc., in short term and from trees in long term.
4. Recycling of available resources by judicious management practice.
5. Creation of regular employment opportunities.

Input, partial support in labour wages and technical know-how were provided to the participants by Department of Extension and Social Forestry, Faculty of Forestry, Birsa Agril. University, Ranchi.

The motivation and training programme of the farmers were conducted in a congenial atmosphere.

Keeping in mind the requirements of the participants a small village nursery was developed in the year 1984, by the farmers in their own land, under technical and partial labour assistance provided by the Department of Extension and Social Forestry, Faculty of Forestry, B.A.U. Ranchi.

After great pursuation, 24-marginal farmers offered 17 acres wastelands where silvipastoral system was introduced by the participants. The participants formed a society to protect the silvipastoral plot from the stray cattle and due to their pressure the villagers gradually gave up habbit of uncontrolled grazing and started stall feeding, (*Social Fencing*). Farmers started harvesting grass from 1986. The tree species planted were teak (*Tectona grandis*), Gamhar (*Gamelina arborea*), Sissoo (*Dalbergia sissoo*), *Acacia auriculiformis*. The spacing was 4 x 3, 4 x 4 m and grass species grown was *Cenchrus*. On an average farmers harvested 4.0 t/ha fodder in the year 1987 and 4.1 t/ha in the year 1989.

Maximum average height was recorded in *Acacia auriculiformis* and minimum in teak. In the year 1989, the plots were handed over to the farmers. In 1989, the farmers uprooted, burned root slips of grasses and started growing cash crop. Introduction of silvipastoral trial improved the soil fertility so much-crop cultivation was possible. Change in the nutrient status encouraged farmers to grow cash crop along with tree species, the farmers also harvested 150 trees for timber and fuelwood and replanted the site by new seedlings in 1989.

In the year 1987, 40 farmers of the same village voluntarily offered 50 acres of land for agroforestry, under the overall supervision of scientist of Birsa Agril. University. The participants whose lands were adjacent agreed to form common boundary with mutual understanding in order to protect the entire area from the biotic pressure collectively and to share produce among themselves as per their land area.

**On-Farm Experiment**

The entire area was divided into three parts and three demonstration cum experiment plots were laid in collaboration with farmers and scientist. Farmers collected seedling from the village nursery. The On-Farm experiment on the following aspect were laid down.

***Silvi-agriculture plantation***

Tree species planted were *Dalbergia sissoo, Eucalyptus* hybrid and *Gmelina arborea*, and in the interspace Gorra paddy (specially suitable for Chotanagpur plateau of Bihar) and arhar were grown by the participants in the year 1987-88 and 1988-89. The height and diameter measurements of tree species were generally taken in January every year, after harvesting of agricultural crops. Paddy seeds were sown at 20 cm row to row and muriate of potash was added at 75 kg per hectare, in addition T.S.P 28.13 kg and urea 95.5 kg per hectare. Arhar seeds were sown at 25 cm apart in a row. Muriate of potash 26.3 kg/ha ; T.S.P. 35.3 kg/ha, Urea 167.4kg per/ha were used. In control plot in the 2nd year average height and girth of *Dalbergia sissoo* were 105.5 cm and 7.12 cm respectively. Along with gora paddy average height was 362.4 cm and girth 12.6 cm and along with arhar average height was 417.6 cm and dia. 17.2 cm recorded.

Growth of *Eucalyptus* hybrid was very poor due to attack by termites. Performance of *Gmelina arborea* was also good. In control plot average height was 140.9 cm and average girth 12.6 cm. In the agrisilviculture plot average height was 377.4 cm and average girth was 16.1 cm and in conjunction with arhar average height was 387.4 cm and average girth 16.9 cm.

It is apparent that average height and diameter growth of tree species was better in plot with crops than in the control. Growth of tree species along with arhar was more in comparison to Gorra paddy. It is apparent that the raising of crops accelerated the rate of growth of tree species as well. Total area under the above experiment-cum-demonstration plot was 20 acres. The plot was left under direct control of farmers in the year 1990.

In the another 20 acres of land, natural grass was allowed to grow along with *Cassia siamea, Acacia auriculiformis, Pongamia pinnata* and *Gmelina arborea.*

In the 2nd year participants harvested natural grass at the rate of 4.5t/ha/yr and maximum height growth 172.10 cm was recorded in case of *Acacia auriculiformis*, followed by 167.82 cm, 163.76 cm and 143.30 cm respectively in *Cassia siamea, Gmelina arborea and Pongamia pinnata* and maximum diameter growth was 4.52 cm for *Gmelina arborea* followed by 3.66, 3.64, 3.52 cm for *Cassia siamea,*

*Acacia auriculiformis* and *Pongamia pinnata*. The growth performance of tree species were not satisfactory due to lack of cultural operation (hoeing, weeding and manuring). Farmers were only interested to grow natural grass. As the area was protected from biotic interference, the growth of natural grass was satisfactory compared to adjacent barren plot.

In order to motivate the farmers for silk worm rearing, *Terminalia arjuna* was planted at 2.5m x 2m spacing in 10 acres land. It is known that one hectare of *Terminalia arjuna* plantation can support a family of landless, if only tasar worm rearing is done. After establishment of trees (approximately 5 years) tasar worm can be reared.

In the year 1988 again 36 farmers put 40 acres of land adjoining to previous areas under agroforestry system. A co-operative society (unregistered) headed by head of the village panchayat of Kumharia village was formed under guidance of I/C Extension, Faculty of Forestry, Birsa Agril. University.

In 1988, 8 acres of land were also put under agrisilviculture experiment with twelve forest tree species. The total number of replications was 5 under randomized block design. The tree species were *Eucalyptus tereticornis, Melia azedarach, Acacia auriculiformis, Leucaena leucocephala, Terminalia arjuna, Shorea robusta, Gmelina arborea, Tectona grandis, Azadirachta indica* and *Pongamia pinnata*. (See Figs. 52, 53, 54 and 55).

Agricultural crops (Niger and Horse gram) were grown along with tree species by the farmers. Partial financial assistance was provided by the Faculty of Forestry, Birsa Agricultural University, Ranchi. The crops grown by the farmers were distributed among the land owners after harvesting as per the land holding. The data recorded so far indicated that *Eucalyptus tereticornis, Acacia auriculiformis, Tectona grandis* and *Gmelina arborea* showed the best growth both in average height and average diameter. Height and diameter of *Shorea robusta* and *Pongamia pinnata* were recorded the lowest.

The observation indicates that during the first year of seedling out planting, no significant difference in grain yield of Niger or Horse gram was recorded. However, the yield was the highest in combination with *Shorea robusta* (8.8 q/ha. Horse gram and 5.8 q/ha Niger). However, grain yield seemed to have been affected in the second year of tree growth (See Figs. 56 & 57).

Fig. 52 : Kulthi grown along with *Azadirachta indica* at Kumharia Village, Ranchi.

Fig. 53 : *Sesamum indicum* grown under *Gmelina arborea* at Kumharia Village, Ranchi.

Figs. 54&55 : Kulthi grown along with *Acacia auriculiformis* at Kumharia Village, Ranchi.

**Fig. 56 :** Kusum grown along with *Albizia lebbeck* at Kumharia Village, Ranchi.

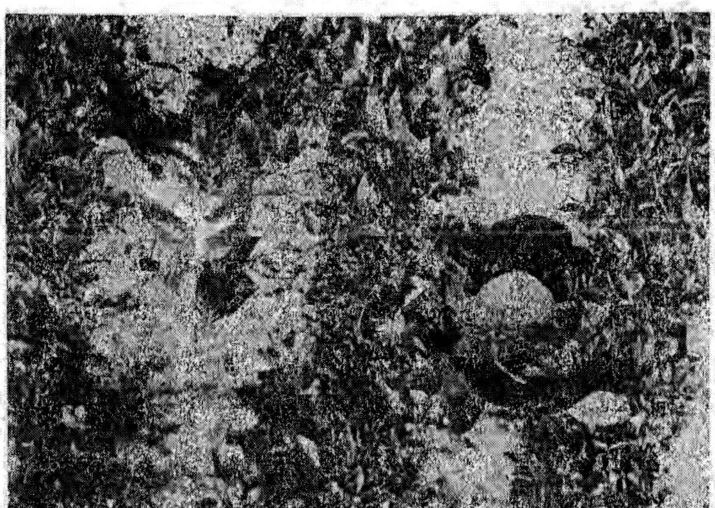

**Fig. 57 :** Leguminous crops grown with *Shorea robusta* at Kumharia Village, Ranchi.

In 20 acres of barren wastelands, nearly 10 tree species were planted at different spacing (2.5 x 2.5m, 3 x 3m and 4 x 4m) and natural grasses which showed rapid growth due to fencing, were harvested by the farmers for stall feeding of the cattles.

In the year 1989 N.W.D.B., Govt. of India, sanctioned a project in which author was the Associate Investigator. This project provided partial assistance to the participants in term of inputs and labour wages. In 1990, nearly 100 acres of wasteland of marginal farmers were put under agroforestry system. In 40 acres farmers started intensive alley cropping where the alley width was 3m and 4m excluding boarder area.

The tree species planted was *Leucaena leucocephala* (k-8 and k-29). The treatments were use of *Leucaena leucocephala* leaves as mulch, as green manures and as fodder. When leaves were harvested for fodder, FYM and NPK was used as alternative source of manure and change of nutrient status of soil was studied. Two vegetables/grass/crops were intercropped.

In the experiment-cum-demonstration plot No.II, 30 acres of land were put under wide alley cropping system. The treatments and intercrops were the same as adopted in Experiment No.I. In the experiment I and II pruning height and pruning intervals were studied to ascertain the maximum productivity of hedgerows and the optimum intervals for pruning.

In Expt. III, nearly 30 acres of land were put under wide alley hedgerows cropping along with 4 fruit trees, aonala, sarifa, guava and jack fruit. *Leucaena leucocephala* used as fillers. The treatments were similar to in all the tree experiments.

In this way, author, as a head, of Extension and Social Forestry Dept., Birsa Agril. University, Ranchi and Associate Investigator of O.R.P. wasteland project, Ranchi could cover 210 acres wasteland in collaboration of farmers under different systems of agroforestry in a single village. This village is henceforth referred to as forest village-Kumharia. Jha and Singh (1991) again surveyed village Kumharia to ascertain the degree awareness of farmers regarding agroforestry. The data indicated that 71.7 and 51.7 per cent of farmers of Kumharia village were aware of single row bund plantation and double row bund plantation. When respondents were asked regarding growing of grasses along with timber/fruit trees, 63.4% responded positively. 95% of respondents showed awareness about traditional practice of agroforestry system. In general farmers showed awareness regarding farmforestry programmes. But among the all villages surveyed in Chotanagpur plateau maximum awareness was recorded in Kumharia village where the farmers were acquainted with most of the components of the agroforestry systems.

They have also suggested that under present circumstances it is the duty of the experts/extension workers to transfer latest technology related to agroforestry practices. Besides this, scientists/foresters should conduct experiments in the farmers field to ascertain the suitable tree species ; suitable combination of trees and grass for each agroclimatic zone. In Kumharia village, percentage of farmers adopting various alternatives were 8.30 per cent bund plantation, 10.70 per cent for fruit tree, 20 per cent Block plantation of misc. species, 23.0 per cent silvipastoral system and 38 per cent agrisilviculture. In Kumharia village extent of adoption is maximum in comparison to other surveyed village (Jha and Singh, 1991).

It is apparent that the extent of adoption of farmforestry largely depends upon appropriate technological inputs. This is particularly very significant in rained farming.

Infact, this programme cannot be implemented rapidly with the help of private source of credit with very exorbitant rate of interest prevailing in rural India. The participants, particularly poor, requires credit for purhasing physical inputs, farm implements, fencing of the plot, etc. Physical labour may be done by several participants, but for the maintenance of their families, provision of consumption credit is absolutely necessary. Credit in cash may be misspent on household requirements instead of inputs. Therefore, credit should be provided, as far as it can be practicable, in the form of physical inputs.

## REFERENCES

Jha, L.K. and Singh, S.K.(1991). Social Forestry/Agroforestry in important component in the development of tribal villages : A case study, *Mendal*, 6(3-4), pp. 487-490.

I.C.R.A.F. (1988). On farm research in agroforestry old problems and new approaches, *News letter and Agroforestry Review*, pp. 1-5.

Prasad, C. (1989). Extension Education of the ICAR. A first line mechanism, *Souvenir, VI national seminar-cum-workshop*, K.V.K., Ramkrishna Mission, Ranchi, pp. 18-24.

Sinha, H.S.P. (1984). Transfer of technology on rainfed and dry farming, *Proceeding Dryland watershed management*, B.A.U., Ranchi, pp. 287-289.

Shkula, A.N. (1989). Krishi Vigyan Kendra. A futuristic view, *Souvenir, VI National Seminar-Cum-Workshop*, K.V.K., Ramkrishna Mission, Ranchi, pp. 46-47.

# 12

# *Potential of Medicine, Essential Oil and Insecticide Producing Crops in Agro-forestry System*

The familiarity with plant species producing medicine, essential oil and insecticides dates back to the beginning of civilisation. It has been found that most of the plant species having above qualities grows in the forest in the wild. A scientific survey of these plants and their ethnobotanical studies has been done by Botanical Survey of India. A number of detailed ethnobotanical explorations were conducted in different tribal areas of the country. More than 800 plant species of ethnobotanical interest were collected at different centres (Annon, 1990). These plant species are over exploited since time immemorial and many of them are now referred to as endangered species. There is need of domestication of economically important plant species to decrease pressure over natural source and to fulfil national need.

Introduction of these species in farming system would increase production, villagers may become self sufficient and earn money. We have basic information about the most of the economically important species. In the present situation to conserve them, propagation and cultivation of these species in the farm land on commercial basis is essentially required. From last two decades we have conducted on-station and on-farm trials to know the production potential of various crops and grasses in agroforestry system. This is getting popularised progressively.

Study related to growing tree species along with medicinal, aromatic and insecticidal crops in different agroclimate zone is in infancy at present. Keeping in mind, national and international demand and supply, State Agricultural Universities, State Government Research Institutes and Central Government Research Institutes should conduct study on cultivation of plants producing medicine, essential oil, insecticide under agroforestry system.

**Figs. 59&60.** Medicinal plants suitable for agrisilviculture system in Chotanagpur plateau of Bihar.

**Scope Of Cultivation**

Due to over exploitation several medicinal and aromatic plants are facing extinction. There is tremendous scope of growing important medicinal plants in agroforestry cropping system in the hills as well as in the plains. The R & D activities at Central Institute of Medicinal and Aromatic Plants and other research institute have studied major medicinal plants in depth and foreign exchange from export of medicinal and aromatic plants is also earned. However, few important medicinal and aromatic plants which losses its properties soon after harvesting cannot be exported. But export of green folliage/fresh roots of these can be done under conditions of refrigeration, proper packing and direct flight to location without any transhipment. Such facilities are not available at present and air freight charges are very high. Air freight should be reduced so that exports of medicinal and aromatic plant complete successfully in the international market.

Maithani and Nautiyal (1987) have suggested the important species which can be initiated for commercial cultivation are *Aconitum chasmanthus* (Balnag), *A. ferox* (Bish), *A heterophyllum* (Atis), *Adhatoda vasica* (Vasingar), *Atropa acuminata, A. balladona, Berberis* species, *Digitalis purpurea*, *Dioscorea deltoides* (Tairu), *Ocimum basilicum* (Bentulsi), *O. sanctum* (Tulsi), *Carum carvi* (Kala Zira), *Mentha arvensis* (Jangli Pudina), *M. piperita* (Vilaiti pudina), *Picrorhiza kurroo* (Kukki), *Podophyllum hexandrum* (Ban Pakri), which occur in mountain ecosystem.

Husain (1988) has listed the medicinal plants in great demand in country and abroad. These are cultivated mints, aromatic grasses, rose, lavender, clary sage, opium poppy, tropane alkaloid-bearing plants, sapogenin bearing yams, senna, pysllium husk and seeds, cinchona and ipecac.

At present, India has monopoly over exports of few medicinal plants, like opium, pysillium husk, etc. To increase the productivity and upgrade the quality of medicinal and aromatic plants, their domestication and incorporation in the agroforestry production system will help in making country self sufficient and to supplementing foreign currency earning. This will also reduce the pressure on the natural resource (forest). Introduction of medicinal and aromatic plants in agroforestry system may be possible after identification of varieties and ecotypes for different edaphic and climatic condition and developing improved agronomic practices to boost production.

Figs. 61&62 : Medicinal plants suitable for agrisilviculture system in chotanagpur plateau of Bihar.

Figs. 63&64 : Medicinal plants suitable for agrisilviculture system Chotanagpur plateau of Bihar.

Jha and Gupta (1991) have intercropped sixtyfour, mainly herbaceous, medicinal plants, collected from local as well as distant forest from hills along, with two year old *populus deltoides*. The species which did the best under poplar are *Acorus calamus, Adhatoda vasica, Aloe barbadense, Artimicia vulgaris, Cymbopogon winterianus, Cyperus scariosus, Mimosa pudica, Piper longum, Solanum indica, Vetiveria zizanoides, Canabis sativa, Centella asiatica* and *Curcuma amada.* Other species like *Amomum subulatum, Calotropis gigantia, Cymbopogon jwarancusa, Ocimum sanctum, Plantago avata, Valaerina wallichii* and *Zingiber officinale* closely followed by species listed above. 'Very good' performers as 'companions' are *Allium sativum, Anethum graveolens, Catharanthus roseus, Dioscorea bulbifera, Dioscorea deltoidea, Euphorbia royleana, Rauvolfia serpentina, Viola odora,* and *Vitex negundo, Achyranthus aspera, Berginia ligulata, Boerhavia diffusa, Ocimum americanum* and *Ocimum gratisimum* turned up as good performers.

Many other workers (Bhattee, 1983 ; Lahiri, 1972 ; Pareek & Gupta, 1984 ; Lahiri, 1983a & b ; Sharma, 1985 ; Singh *et al.*, 1985 ; Gupta, 1987 ; Maheshwari *et al.*, 1988 ; and Singh *et al.*, 1990) have reported successful intercropping of medicinal and aromatic plants in the agroforestry system.

The high yielding strains of medicinal and aromatic plants mentioned below may be cultivated in agroforestry system in different agro-climate zone on commercial basis.

## (A) MEDICINAL PLANTS SUITABLE FOR AGROFORESTRY SYSTEM

### (*a*) Sub–Temperate and Temperate Zone

(1) *Atropa belladonna*

(Luckmunee)

*Atropa acuminate*

Family—Solanaceae

*Atropa belladonna* is an erect, glabrous herb with long pointed ovate leaves. *A. belladonna* has purple flower, tinged with yellow colour whereas *A. acuminate* has yellow coloured flowers. Among the all available species *A. belladonna* has highest alkaloid content. The plant contains drugs atropine and hyoscyamine. Various preparations are used in the pharmaceutical industry. Technology for cultivation of belladonna has been developed. The State Forest

Department in Himachal Pradesh has made an attempt to introduce belladonna in forest blanks of deodar and fir region. It is noted that belladonna does not stand weed competition during monsoon rains and need frequent weedings (Gupta, 1987). When *Atropa belladonna* is cultivated along with tree species, weed problem is greatly reduced. Farmers will earn both from medicinal plants as well as from tree species. Gulati *et al.,* (1982) have given the cultivation technology of belladonna. Berries of belladonna are crushed gently to separate seeds. The healthier seeds after drying are treated with a fungicide like Dithane M-45 to minimize mortalities at nursery stage due to damping-off disease. The seed is sown (broadcast) in nursery from 15 May to end of July at the rate of 1kg/100 sq m. The nurseries are irrigated frequently to keep the soil moist throughout the period of germination. Before transplanting the seedlings are again treated with a fungicide like agallol. Autumn transplantation is now preferred. Spring transplanting is also advisable and is practised at such place where irrigation is available. The leaves are cut with the help of a cutter, about 30 cm above ground level. Immediate irrigation and fertilization help in quick and fast emergence of several secondary shoots which in turn are harvested at about the same height from the ground level. Harvested leaves are sundried. If belladonna is cultivated in large scale under agroforestry system, our entire requirements of dry leaf of belladonna can be met domestically and the country will be self-sufficient.

(2) *Swertia chirata*

(Chirayata)

Family—Gentianaceae

Chirayata is a perennial, small, erect herb with robust stems and green yellow flower, tinged with purple colour. Swertiamarin found in *Swertia* species is 5-hydroxy-gentopicroside (Shah, 1982). Gupta (1987) advocated domestication of *Swertia chirata* and *Valaeriana jatamansi* under agroforestry programme in Himachal Pradesh. It is also cultivated in Darjeeling and other sub-temperate zone. The medicinal herb found in the market, consists of bundles of dried twigs of brownish colour, which is very bitter, but have a pleasant taste. The whole plant is used medicinally, but root is reported to contain maximum swertiamarin. It is a tonic, stomachic and febrifuge, and prescribed in decoction or infusion in the quantity of a small tea cup ful twice daily (Kirtikar and Basu, 1980).

(3) *Dioscorea floribunda*
(Sapogenin – bearing Yams)
Family—Dioscoriaceae

This is a herb with slender twining stems and round or oblong tubers. The tubers contain diosgenin, basic raw material for synthesis of steroidal drugs. Although a large number of industries in the world have switched over to other raw materials such as stigmasterol from soyabean oil, diosgenin obtained from various species of *Dioscorea* is still one of the major raw materials for the manufacture of these drugs and the diosgenin requirement will go on increasing, especially for the use as oral contraceptives (Husain, 1988).

*Dioscorea* species are successfully grown along with poplar in U.P. Due to slow growing nature, cultivation of *Dioscorea deltoides,* on commercial basis is not profitable. However, *Dioscorea floribunda* is suitable for cultivation on commercial basis in Assam, Goa, Karnataka etc. (Bammi and Rao, 1980). The crown portion of tuber is the best planting material. It is planted in the field in the sand bed when sprouting is about 70 per cent complete. The *Dioscorea* vines need support for their optimum growth. A two year crop is more economical. On an average, a two year old plant will yield 2.5 to 3 kg of tubers, having 3.0 to 3.5 per cent diosgenin content.

(4) *Cephaelis ipecacuanha*
(Ipecac)
Family—Rubiaceae

Ipecac is the native of Brazil. This plant was introduced in India in 1866 and its cultivation is confined to lower foot hills of Eastern Himalayas. The dried rhizome and roots of *Cephaelis ipecacuanha,* commercially recognised as Rio or Brazilian ipecac, are the best source of emetine used in the treatment of amoebic dysentery. There is acute shortage of ipecac and India can seize the opportunity to increase its production for export (Husain, 1988). According to Chatterjee *et al.,* (1980) nursery preparation involves considerable expenditure and results of preliminary studies to grow this crop 'open' and under *Crotalaria* shade are encouraging.

(5) *Papaver somniferum*
(Opium poppy)
Family—Papaveraceae

This is an erect, branchless (rarely branched), shrubs with ovate-oblong, dentate or serrate leaves. Flowers are bluish white. It attains height upto 150 cm. The poppy growing tract in India covers nearly 23,000 ha mainly in the three adjoining states of U.P., M.P. and Rajasthan. India, the largest and the only legal producer of opium, harvested about 865 tonnes in 1986-87. Central Institute of Medicinal and Aromatic plants has developed and released two improved varieties of opium poppy namely Shweta and Shyama. These varieties give 40-45 per cent higher latex yield than the existing local cultivar. The varieties have been released to growers in collaboration with the Narcotics Department. Another new variety, 'Sanchita', rich in morphine (1.2 per cent) and suitable for processing alkaloids from poppy straw, was developed in 1986-87. The variety yield 9-10 qtl. seeds per ha. This was done to avoid lancing of capsule so that social abuses are checked in the near future (Husain, 1988). This plant can be introduced in agroforestry system ; Maize-opium, chillies-opium, groundnut-opium are commonly practised in one year rotation. Dabral and Patel (1976) suggested that best time of sowing appear to be the second fortnight of October to the first fortnight of November. However, cultivation of this plant requires permission and supervision of the narcotics Dept.

(6) *Valeriana wallichii*

(Dipana)

Family—Valerianaeceae

This is a small perennial herb. Stems are thinner than the root stock, flowers white. It contains monoterpene derivatives called valepotriats. In India, it is a good substitute for *Valeriana officinalis.* The root is bitter with a flavour. Its used in epileptic fits, headaches pains in the joints and disease of the liver and kidney. Jha and Gupta (1991) claimed successful intercropping of *valeriana wallichii* along two year old *Populus deltoides.*

(7) *Saussuria lappa*

(Pachak, Kuth)

Family—Compositae

Kuth is now cultivated in Northern Himalayas. This is tall, erect herb. Leaves are membranous and toothed. The root contains essential oil, an alkaloid saussurine and resin. The roots are aromatic and bitter in taste. This is used in the treatment of blood disease, bronchitis, headache, chest pain, etc. In India it is used on a limited scale, but it has demand in international markets.

(8) *Podophyllum emodi*
(Pilijati)
Family—Berberidaceae

This is small, succulent, erect herb with creeping rootstock. The rhizome and the root contains podophyllotoxin and uncrystallizable podophylloresin. It is used in controlling various skin diseases.

(9) *Nardostachys jatamansi*
(Jatalasi)
Family—Valerianaceae

This is a small, erect, perennial herb with woody root stock. The roots are fragrant and bitter in taste.

Jatalasi has all properties of valerian. This plant is used as a substitute for valerian. The roots are employed in the treatment of convulsive afflictions, hysteria, epilepsy and palpitation of heart.

(10) *Berberis aristata*
(Daruhaldi)
*Berberis lycium*
(Chitra)
Family—Berberidaceae

A tall evergreen shrub reaching upto 2 m height. Twigs are yellowish brown and bark pale brown rough and deeply furrowed. The roots and bark contains alkaloids and have long been used in the treatment of various diseases. Alkaloids of *B. aristata* heals ulcer, disease of the mouth like toothache, urethral discharges etc., whereas roots and barks of *Berberis lycium* are used as remedial therapy to cough, throat troubles, piles etc.

(11) *Glycyrrhiza glabra*
(Jethimadh)
Family—Papilionaceae

Jethimadh is perennial plant, reaching upto 1 m height. The dried roots and underground stems have medicinal properties. The root has sweet flavour. Plant roots are used after the 4th year of planting.

This plant is introduced in farming system of Himachal Pradesh and Jammu & Kashmir. For proper growth irrigation is essential. This plant is the source of commercial liquorice. It contains glycyrrhizin upto 14 per cent. It is used in the treatment of coughs, consumption and chest complaints.

(12) *Asparagus adscendens*

(Hazarmuli, Safed musli)

Family—Liliaceae

This is a suberect prickly much branched shrub, occurring over the Western Himalayas. The white tuberous roots cure dysentery and diarrhoea.

(b) **Tropical and Sub Tropical**

(1) *Cinchona ledgeriana*

(Cinchona)

Family—Rubiaceae

This is a small to large evergreen shrubs or trees. It has good coppicing ability. It gives the "yellow bark" which is the richest in quinine content (4 to 5 per cent, occasionally as high as 14 per cent). One tree in the West Bengal plantation showed as high as 19 per cent (Chatterjee, 1980). This species is highly suitable in alley cropping system. Inter-space may be utilised for growing crops, vegetables, etc.

(2) *Costus speciosus*

(Custus)

Family—Zingiberaceae

The plant is growing in a wide range of agroclimatic condition. It contains 2.12 per cent of diosgenin out of a total of 3.86 per cent of total sapogenins present in the glycosidic form in the alcoholic extract. Diosgenin is the main component of *Costus speciosus*. *Costus* cultivation is cheaper than cultivation of *Dioscorea* Spp., as *Costus* does not need supports. The crop can be harvested in 16-18 months compared to 24-36 months in the case of exotic *Diascorea* species introduced in India (Sarin *et al.*, 1980).

(3) *Rauvolfia serpentina*

(Sarpagandha)

Family—Apocynaceae

This is a small, erect herbaceous shrub. It is characterised by leaves having bright green colour above and pale colour beneath, occurring in whorls of three two five. Flowers are white often tinged with violet colour. Outer portion of tap root is corky with longitudinal fissures. Root contains alkaloids like ajmaline, ajmalicine, ajmalinine, rescinamine, reserpinine, reserpine, serpentine, serpentinene and yohimbine. Roots are harvested during winter months as alkaloidal content is higher during this period (Sarin, 1980). The total alkaloidal content of the root usually varies from 0.8-1.3 per cent.

The alkaloids of *Rauvolfia serpentina* cures hypertension or also to its sedative properties. However, its demand has decreased on account of availability of synthetic substitute.

(4) *Datura innoxia*

*D. metel*

Family—Solanaceae

This is an annual or perennial shrub. *Datura innoxia, D. stramonium* and *D. metel* are important medicinal plants. *D. innoxia* is a bushy annual shrubs. It attains 1 m height. It is characterised by dense pubescence, ovate leaves. Fruit bears long weak spines. It occurs in the hilly regions in Western Himalayas.

*D. metel* plant is densely clothed with greyish tomentum. It occurs throughout India. The dried leaves and flowers have medicinal values. *D. innoxia* and *D. metel* are rich in scopolamine. The alkaloid scopolamine is used as pre-anaesthetic in surgery and child-birth, in opthalmology and in prevention of motion sickness (Sobti and Kaul, 1980). RRL Jammu has developed and released two improved varieties of *Datura metel,* namely RRL-purple and RRL-Green. RRL-purple contains 0.24 to 0.36 per cent and 0.12 to 0.19 per cent of alkaloids in leaves and seeds respectively, whereas RRL-Green contains 0.24 to 0.28 per cent and 0.098 to 0.128 per cent alkaloids in leaves and seeds respectively.

(5) *Catharanthus roseus* (syn. *Vinca rosea*)

(Medagascar Periwinkle)

Family—Apocynaceae

The species is a perennial erect herb. It attains height of 0.6 m. Leaves are green, opposite, oval, obovate or oblong and flowers are pink or white. It grows all over the country. It contains more than 100 alkaloids. The most important are raubasine from roots and vinblastine and vincristine from leaves. Integration of this plant in agroforestry on commercial scale would increase foreign exchange. The plant is propagated by direct sowing of seeds. The plant is drought tolerant. The roots are harvested after one year. Leaves can be harvested thrice annually. Raubasine is used to check the fragility of capallaries and vinblastine and vincristine are used to cure hodgkin's disease and acute leukemia in children respectively (Datta, 1980).

(c) **Arid and Semi-Arid Regions**

(1) *Cassia angustifolia*

(Senna)

Family—Caesalpiniaceae

Senna is an erect shrub with pale ascending branches, reaching a height of about 70 cm. The leaves are pale green colour, flowers yellow. The pods are thin, flat about 2 cm., in width and 7 cm in length. Senna is an important export item. It is used as a laxative, the laxative principles being two glycosides-sennoside A and B. In 1984-85, about 4,000 tonnes of leaves and pods worth more than Rs. 37 millions were exported. There is considerable scope to increase the production and export (Husain, 1988). It has been grown as 130 days crop under rainfed condition in Jodhpur and Delhi under irrigation whenever rainfall is low or erratic (Gupta, 1984).

The dried leaves of the plant known as seena of commerce. It is chiefly employed in constipation, liver complaints, abdominal troubles, typhoid, jaundice, etc.

(2) *Catharanthus roseus*

Vide 'under medicinal plants for sub-tropical region.'

(3) *Solanum viarum*

(Khasi ketaris)

Family—Solanaceae

This species can be grown in marginal land in rainfed condition. During kharif season it produces higher berry yields. The dry berries contains solasodine and 16 DPA. In India, there is considerable scope to increase the production and export. A number of varieties are evolved for commercial cultivation but there is need to evolve high yielding cultivars and to develop improved agronomic practices.

(4) *Plantago ovata*

(Isapgol)

Family—Plantaginaceae

This species is a small annual herb, nearly stemless, leaves appearing in whorl. It generally reaches a height of about 30-45 cm. Isabgol husk is a membranous covering of the seeds, white to light pink in colour, it is 2 to 3 mm x 0.5 to 1.0 mm in dimension, translucent and odourless. It absorbs moisture and forms a tasteless mucilaginous substance which constitute the drug. The plant comes in flowering stage at about 60 days of sowing and the flowering is completed in next 8 to 17 days in a spike (Gupta, 1980). India enjoys, a monopoly in production and export of Isabgol. There is scope to increse the productivity by cultivating *Plantago ovata* on a large scale under agroforestry system.

(5) *Withania somnifera*

(Ashwagandha)

Family—Solanaceae

This is an erect branched shrub. Branches are covered with mealy stellate heary tomentum. It reaches a height of about 0.3-1.5 m. Flowers are greenish or lurid yellow, usually about 5 together in a sessile or nearly sessile umbellate cyme. Berry red, smooth and seeds are 2.5 mm diam ; yellow (Kirtikar and Basu, 1980). *Withania somnifera* contains alkaloids as well as steroid lactones. This is used as tonic and cure bronchitis, asthma, ulcers etc. Seeds are used to coagulate milk.

(6) *Mucuna pruriens*

Family—Leguminosae

Mucuna is twindling annual leguminous plant. This is a hardy species which withstands drought. It is common in almost all parts of India. The July-August is the best planting season and crops are harvested in January-February. To get the best yield 1-2 weeding in initial stages and artificial support are essential. The pods are harvested thrice or four times in a cropping season. After drying the pods, seeds are collected and stored for use. The seeds contain L-Dopa (L-3, 4 dihydroxy phenylalanine). Dopa is a precursor in biosynthesis of catecholamines like nor-adrenaline, dopamine and 3-methoxy tryptamine. Recently L-Dopa has been used successfully in the treatment of Parkinson's disease (Shah, 1980).

## (B) AROMATIC PLANTS SUITABLE FOR AGROFORESTRY SYSTEM

### (1) *Mentha arvensis*

(Japanese mint)

Family—Labiatae

This is a perennial herbaceous plant with bad taste and smell. The stems are short, branched covered with short hairs. Leaves are ovate, oblong, lanceolate and toothed. The Japanese mint was introduced in 1952 in Jammu & Kashmir. The improved agrotechnology for growing the plant was developed by Regional Centre of CIMAP, Pantnagar and Jammu. Two superior varieties *i.e.* CIMAP/Hybrid-77 and CIMAP/MAS-1 are available. India has not only become self sufficient, but has been saving foreign exchange of about Rs. 600-1200 lakhs a year depending on the market price, for the last 15 years. The crop covers an area of approximately 10,000 ha; provides employment to more than 50,000 people in the rural area. Several hundred small and mini-rural industries for distillation of oil and production of menthol have been set up (Husain, 1988). Crops are harvested before flowering stage and left for drying for 2 hours in sunny weather. Semi-dried crops are again dried in the shade area. By steam distillation, a yellow volatile oil is obtained. From last few years, India is in a position to export menthol to the developed countries.

### (2) *Mentha spicata*

(Spearmint)

This is mostly used for flavouring food products, tooth paste, chewing gums etc. To increase its productivity new clone was developed from U.S. spearmint (CIMAP/MSS-1) known as CIMAP/MSS-5. With the release of this new variety, production of 20 tonnes per year is feasible. In India, there is considerable scope to increase the production and export.

### (3) *Mentha cardiaca*

(Scotch spearmint)

Scotch spearmint (U.S.) introduced by CIMAP in India and an important clone was developed which is better and superior to its parental clone. This clone is released for cultivation. Its cultivation on commercial basis would lead India to be a major producer of the oil in the world.

(4) *Mentha viridis*

(Pudina)

This is commonly known as garden mint. This is a herbaceous perennial having pungent smell. It attains height up to 0.9 m CIMAP has developed a superior strain, which has very good flavour, more oil yield and can be grown throughout the year. Its essential oil contains carvone. Leaves are used in 'chutney', in fever and bronchitis. The leaves are also used in preparation of cosmetics.

(5) *Cymbopogon flexuosus*

(Lemon grass)

Family—Graminae

Most of the cymbopogon grasses are perennial, densely tuft and aromatic. *Cymbopogon flexuosus* are commonly grown in south India. The lemon grass oil contains Citral 45-65 per cent. The farmers left cultivating lemon grass due to low content of Citral. The CIMAP variety of lemon grass possesses 80-90 per cent, Citral content. The new CIMAP variety also produces more oil. The main use of lemon grass is in perfuming soaps and cosmetics. Lemon grass oil is also used in production of Beta-ionone which used in the synthesis of Vitamin-A.

(6) *Cymbopogon martinii*

(Palmarosa)

The essential oil is obtained on distillation from inflorescence and leaves of Palmarosa. Two high yielding improved varieties of Palmore grass *i.e.* CIMAP/CMP-1 and CIMAP/PRC-1 are introduced in U.P., Assam, Karnataka, etc. In the year 1984-85 India exported 16 tonnes of oil worth about Rs. 6.5 millions. Palmorosa oil is used in high grade toilet soaps and in other cosmetics.

(7) *Cymbopogon winterianus*

(Citronella java)

Citronella Java is mainly cultivated in Assam, Andhra Pradesh, Karnataka, Tamil Nadu and West India. India is self sufficient and an exporter too. The productivity of this species has been increased through the development of new high yielding variety – CIMAP/Bio-13 and CIMAP/C-22 (Mandakini) and CIMAP/C-73

(Manjusha). CIMAP/Bio-13 produce 50 per cent more oil content and 100 per cent higher herb yield. C-73 is suitable for North Indian plains and C-22 is suitable for terai belt of U.P. Essential oil is mainly used in the production of chemical citronellol, citronellal, Geraniol and hydroxy citronellol.

(8) *Vetiveria zizanioides*

(Khus)

Family—Gramineae

This is a densely tufted, awnless, wiry, glabrous, perennial grass, is a "shy breeder" and is considered sterile rhizomes, no stolons, propagated by root divisions or slips. The plant grows in large clumps from a much branched "Spongy" root stock with erect culms 0.5-1.5 m high. The leaf blades are relatively stiff, long and narrow up to 75 cm long and 8 mm or less in width, glabrous, but "downward rough" along the edges (Bredero, 1987). This grass is now considered as one of the best species, suitable in alley cropping system of agroforestry. It is found throughout the plains and lower hills up to 1,500 m. It can be grown in a variety of problem soils. Khus oil is used in perfumery, cosmetics and chewing tobacco industries. CIMAP has developed two superior clones of *V. zizanioides,* known as 'KS-1' and 'KS-2'. It is an excellent soil conservator and increases the water retention capacity of soil.

(9) *Lavandula angustifolia*

(Lavender oil)

Presently to fulfil requirements, India is importing lavender oil from Europe. This is one of the valuable oils used in cosmetic and perfumery industry. CIMAP has introduced high grade lavender in Kashmir Valley. CIMAP has also developed a new variety designated 'Sher-e-Kashmir', yielding 80-100 per cent more oil than the previous Bulgarian variety (Husain, 1988). To be self-sufficient in future and to export, its cultivation to newer areas is needed.

## (C) IMPORTANT INSECTICIDE PRODUCING PLANTS SUITABLE FOR AGROFORESTRY SYSTEM

There is no gainsaying the fact that plant protection works as an important tool for growing more and more food. However, hard one may work in the field to grow crop, fertilise it and irrigate it, one cannot hope for desired output of crop, if the pest are not kept under control. The growth of these inimical insects can be kept under control in a number of ways : chemical control, biological, insect pest suppression and integrated pest management. Many chemicals (Insecticides) are extracted from plant material. The demand for plant insecticides would increase indefinitely in the future. Insecticides producing plants may be grown successfully along with tree species to increase the farmers' income and to help the nation in earning substantial income from export.

(1) *Chrysanthemum cinerariaefolium*

(Pyrethrum)

Family-Compositae

Chrysanthemum is perrenial herb, with daisy-like flower head. About 95 per cent commercial pyrethrum is obtained from the flower heads. Pyrethrum acts as insecticide in six different forms, esters. They are Pyrethrum-I, Pyrethrum-II, Cinerin-I, Cinerin-II, Jasmolin-I and Jasmolin-II. These esters are derived from two cyclopropane carboxylic acids. In relation to other compounds, pyrethrum is non-poisonous to warm-blooded animals (Jha, 1987). Kenya in Africa is the principal exporting country of this insecticides, which produces not only large quantity but also of distinctly improved quality. In 1942, it was tried in different parts of India whereas in Kashmir it was introduced earlier in 1931. Its productivity can be increased by identifying suitable agroclimatic zone and scientist should extend latest agrotechnology and high yielding clones etc., to the farmers to whom Government should provide short term credits to cultivate this plant.

(2) *Nicotiana tabacum*

*N. rustica*

(Tamaku)

Family—Solanaceae

Nicotine is obtained from the tobacco wastes. It is growing in all most all states of India. Andhra Pradesh, Bihar, Gujarat, Karnataka and Tamil Nadu produces in large quantity. The main tobacco crop cannot be used as raw material for nicotine extraction due to its high costs. Main tobacco crop is used for normal consumption. From the commercial point of view, Nicotine is prepared by extracting coarse parts of the tobacco plant, *Nicotine tabaccum* (2-5%) and *Nicotiana rustica* (5-14%) through distillation. When freshly distilled, they turn into colourless and almost odourless liquid. Nicotine alkaloids easily mixes with water at atmospheric temperature and is also miscible in the organic solvents. It is marketed as nicotine sulphate with 40 per cent nicotine in it (Jha, 1987). Nicotine sulphate is solid and its application is done either in liquid or dust carrier. Nicotine in this form assumes stability and is less volatile. The effect of nicotine after its application is of short duration. It is generally used as a killing agent for aphids and other soft bodied insects. The waste product of bidi industry, etc. can be potential source of raw material for production of good quality of nicotine sulphate.

**Suggestions**

* Seed multiplication, storage and marketing will produce quick result. Formulation of a National medicinal, aromatic plant seed policy with the objective of a successful supply of high yielding seeds of medicinal and aromatic plants.
* A series of trials with local and exotic medicinal, essential oil, insecticide producing plants should be conducted in different agroclimatic conditions in the agroforestry system, to identify the best methods of planting, months of planting, months of harvest to get maximum produce, etc.
* Development of improved variety and improved agronomic practices to boost production of internationally acceptable quality in different agroclimatic condition.
* To study integrated nutrient management for medicinal and essential oil, and insecticide producing plant in agroforestry system.
* Strengthening research, training and extension for successful transfer of technology in the farmers land.
* Provision of credit facilities at concessional rate to extend the area under cultivation of medicinal and aromatic plant etc.
* Development of sound marketing facilities.

## REFERENCES

Anonymous (1990). *India 1990. A reference mannual,* Publication Division, Ministry of Information and Broadcasting, Govt. of India, pp. 1-929.

Bammi, R.K. and Rao Gangadhara, G. (1980). Commercial cultivation of *Dioscorea floribunda,* In : *Cultivation and Utilization of Medicinal Plants,* Atal, C.K. and Kapur, B.M., Eds., CSIR Publication, pp. 77-86.

Bhattee, S.S. (1983). Agroforestry in Arunachal Pradesh, In : *Agro Forestry in India,* Mathur, R.S. and Gogate, M.G., Eds. pp. 22-26.

Chatterjee, S.K., Nandi, R.P. and Ghosh, N.C. (1980). Cultivation and utilization of Ipecac in West Bengal, In : *Cultivation and Utilization of Medicinal Plants,* Atal, C.K. and Kapur, B.M. Eds,. C.S.I.R Publication, pp. 295-301.

Chatterjee, S.K. (1980). Cultivation of cinchona in Darjeeling hills, West Bengal. In *Cultivation and Utilization of Medicinal Plants,* Atal, C.K. and Kapur, B.M. Eds., C.S.I.R Publication, pp. 22-224.

Dabral, K.G. and Patel, O.P. (1976). *Indian Farming.* 27 (15), pp. 17-18 (Cross reference).

Datta, S.C. (1980). Cultivation of *Catharanthus roseus* in West Bengal, In : *Cultivation and Utilization of Medicinal Plants.,* Atal, C.K. and Kapur, B.M. Eds. C.S.I.R. Publication, pp. 279-283.

Gupta, R. (1980). Recent advances in cultivation of Isabgol (*Plantago ovata Forsk*) in India, In : *Cultivation and Utilization of Medicinal Plants.* Atal, C.K. and Kapur, B.M. Eds., C.S.I.R. Publication, pp. 406-417.

Gupta, R. (1984). Cultivation of medicinal and aromatic plants in new afforestation areas with particular reference to Rajasthan. In *Proc. Application of Science and Technology for Afforestation,* Department of Science and Technology, Jaipur, pp. 155-164.

Gupta, R. (1987). The role of medicinal plants in agroforestry in India, In : *Agroforestry for rural needs,* Khosla, P.K. and Khurama, D.K. Eds. *I.S.T.S,* Solan, pp. 297-301.

Gulati, B.C. and Qureshi, N.A. and Taj-ud-Din (1982). Cultivation of *Belladonna* in Kashmir (An appraisal), In : *Cultivation and Utilization of medicinal plants.* Atal, C.K. and Kapur, B.M., Eds., C.S.I.R., Regional Research Lab. Jammu-Tawi, pp. 262-269.

Husain, A. (1988). Medicinal and aromatic plants solid base for export thrust, *The Hindu Survey of Indian Agriculture,* pp. 139-145.

Jha, K.K. and Gupta, C. (1991). Intercropping of medicinal plants with poplar and their phenology, *The Indian Forestor,* Vol. 117 (7), pp. 535-544.

Jha, L.K. (1987). *Applied agricultural entomology.* New Central Book Agency, Calcutta, pp. 1-258.

Kirtikar, K.R. and Basu, B.D. (1980). *Indian Medicinal Plants* Vol-III, In : Blatter, E., Caius, J.F. and Mhaskar, K.S.; Eds., (Revised and rewritten), Bishan Singh Mahenra Pal Singh, Dehra Dun, pp. 1593-2393.

Kirtikar, K.R. and Basu, B.D. (1980). *Indian Medicinal Plants.* Vol-IV, In : Blatter, E; Caius, J.F. and Mhasker, K.S., Eds., (Revised and rewritten), Bishan Singh Mahendra Pal Singh, Dehra Dun, pp. 2395-2761.

Lahiri, A.K. (1972). Intercropping trials with *Turmeric* in North Bengal, *Indian Forestor,* Vol. 98 (2), pp. 109-115.

Lahiri, A.K. (1983a). Agroforestry in West Bengal, In : *Agroforestry in India,* Mathur, R.S. and Gogate, M.G., Eds., pp. 218-231.

Lahiri, A.K. (1983b). Citronella grass intercropping in West Bengal, In : *Agroforestry in India.* Mathur, R.S. and Gogate, M.C., Eds., pp. 138-146.

Maheshwari, S.K.; Dhahatonde, B.N.; Yadav. S. and Gangarade, S.K. (1988). Intercropping of Rauvolfia, *Indian J. Agril. Sci.,* Vol. 55, pp. 332-334.

Maithani, G.P. and Nautiyal, S. (1987). Mountain ecosystem of Garhwal Himalaya and its development strategy, *Himalayan Research and Development,* Vol. 6, pp. 4-9.

Pareek, S.K. and Gupta, R. (1984). Exploratory study on yield and comparative economic of medicinal plants based cropping system in India, *Ann. Agric. Research.*, Vol. 5, pp. 169-177.

Sarin, Y.K. (1980). Cultivation and utilization of *Rauvolfia serpentina in India, In : Cultivation and Utilization of Medicinal Plants,* Atal, C.K. and Kapur, B.M., Eds., C.S.I.R., R.R.L. Jammu-Tawi, pp. 288-294.

Sarin, Y.K., Singh, A.; Bedi, K.L ; Kapur, S.K., Kapathi, B.K. and Atal, C.K. (1980). *Costus speciosus* Rhizome on commercial source of Diosgenin, In : *Cultivation and Utilization of Medicinal Plants,* Atal, C.K. and Kapur, B.M., Eds., C.S.I.R., R.R.L. Jammu-Tawi, pp. 107-119.

Shah, C.S. (1980). Recent development of some natural products, In : *Cultivation and Utilization of Medicinal Plants,* Atal, C.K. and Kapur, B.M., Eds., C.S.I.R., R.R.L., Jammu-Tawi, pp. 24-48.

Sharma, B.S.K. (1985). Agroforestry in Manipur, In : *Agroforestry in India,* Mathur, R.S., and Gogate, M.C., Eds., pp. 138-146.

Singh, K.; Singh, V.K.; Hussian, A. and Kothari, S.K. (1990). Aromatic plants as efficient intercrops under poplar, *Indian Forestor,* Vol. 116(3), pp. 189-193.

Singh, K; Singh, V; Singh, J.P. and Kothari, S.K. (1985). Cultivation of medicinal and aromatic plants along agroforestry, *Indian Farmer Digest,* Vol. 18 (9), pp. 34-36.

Sobti, S.N. and Kaul, B.L. (1980). Cultivation of *Datura innoxia* and *D. metel* in India, In : *Cultivation and Utilization of Medicinal Plants,* Atal, C.K. and Kapur, B.M., Eds., C.S.I.R ; R.R.L.; Jammu-Tawi, pp. 24-48.

Vetiver grass, A method of vegetative soil and moisture conservation, (Ed. by Fran Bredero (1987), *World Bank,* New Delhi.

# 13

# *Developing Economic Produces in the Agroforestry System*

A sizeable section of our society living in and around forest requires the raw material for the cottage industries.

## (A) AGROFORESTRY AND APICULTURE

Apiculture is nothing but an organised system of bee-keeping or the system of keeping bees for the production of honey is known as apiculture.

In Indian condition the potential rewards of bee-keeping are great but apiculture is not achieving their full potential mainly due to shortage of flora and qualified personal to train farmers properly. In monocropping system, honey bee gets nectar and pollen for a short period within a year. According to Powlick (1989) agroforestry offers some solutions, based on symbiotic relationship between bees, field crops and trees. The increased presence of bees could lead to better pollination of both tree and field crops and thus to greater yields of timber, fruit and crop. The latter, along with greater honey yields, would bring farmers increased cash income. According to Dr. Baumer, woody perennials are important sources of nectar and/or pollen for honey bees, and bees are important as pollinators. A system that combines apiculture and agroforestry apiforestry, in other words - is therefore likely to improve the quality of life and the revenue of many small farmers who would harvest hive products and would also probably have better yields from their trees as a result of better pollination.

If four well managed bee hivess (per hectare) are placed in a field of sunflower, the productivity of sunflower seeds increases 15 to 20 per cent and each seed will give up to 2 to 3 per cent more oil. The presence of bees and their pollinating activity are solely responsible for it. Powlick has also pointed out that by growing appropriate trees, the period of nectar and pollen availability can be extended. Woody plants are generally preferred to herbaceous species for this purpose because they are less affected by climatic veriations at moderate scale. Best results are achieved by planting trees which are actually somewhat ill-suited to the environment. Tree species those are not at their ecological optimum, which are slightly marginal to local condition, will often flower a time when others are in flowering stage. Some trees under these conditions even react by producing more flowers than normal.

Thus, farmers should grow combination of tree species under agroforestry system, so that honey bees will get nectar and pollen round the year in the same unit of land.

**Criteria for selection of tree species**

Rajan (1986) has suggested the following criteria for selection of tree species in agro/farmforestry system to boost-up apiculture.

1. The tree species should be fast growing like *Eucalyptus* hybrid (mysore gum,) *Casuarina equisetifolia, Acacia auriculiformis, Cassia siamea, etc.*
2. It should be good coppicer like mysore gum, acacia, syzygium, etc.
3. It should respond well if pollarded *e.g.* Mysore gum, *Morninga oleifera,* etc.
4. It is advantageous if it grows from cuttings *e.g., Moringa oleifera.* Such a plant develops a crown quickly and produces flowers much earlier than those raised from seedlings.
5. As far as possible all the selected tree species should not put forth blossoms at one time. They should flower in different months of a year, preferably, they should keep on flowering over a longer period in a year, like mysore gum.
6. Since land available to grow tree species is generally limited in a farm, preference may be given to those which produce both nectar and pollen, preferred by honey bees instead of any of them.

**Fig. 65 :** Silvi apiculture

**Tree Species Suitable for Apiculture in Agroforestry System**

| Sl. No. | Botanical and Local name | Family | Description | Flowering months |
|---|---|---|---|---|
| 1 | 2 | 3 | 4 | 5 |
| 1. | *Michelia champaca* (champa) | Magnoliaceae | A tall, evergreen tree with dark greybark. Flowers pale, yellow, very fragrant. | April to June (Occasionally throughout the year.) |
| 2. | *Bombax ceiba* (Semal, silk cotton tree) | Bombacaceae | A Large deciduous tree. Flowers—fleshy, crimson or orange. | January to March |
| 3. | *Magnifera Indica* (Mango) | Anacardiaceae | A large evergreen tree with rough thick dark-grey bark. Flower—yellowish green. | February to April. |
| 4. | *Zizyphus mauritiana* (Indian plum) | Rhamnaceae | A moderate-sized tree, almost ever-green, bark dark, grey or nearly black, with long vertical cracks Flower—greenish-yellow. | September to November. |
| 5. | *Litchi chinensis* (Litchi) | Sapindaceae | A medium sized evergreen tree, bark is greyish brown. (Flower—Yellow) | February to April. |
| 6. | *Schleichera oleosa* (Kusum or Ceylon Oak) | Sapindaceae | A large tree; bark grey Flower—yellowish green. | March to April. |
| 7. | *Butea monosperma* (Palas) | Papilionaceae | A small tree, bark brown. | March to May. |
| 8. | *Dalbergia sissoo* (Shishum) | -do- | A fairly large tree, bark grey Flowers—pale white. | March to May. |
| 9. | *Ougeinia dalbergioides* (Ougeinia) | -do- | A middle sized tree with a thin ashy-grey of light-brown bark. Flowers—light pink or white. | March to April. |

*(Contd.)*

| 1 | 2 | 3 | 4 | 5 |
|---|---|---|---|---|
| 10. | *Pongamia pinnata* (Karanj) | Papilionaceae | A large sized tree, with grey bark; Flowers—White. | January to April. |
| 11. | *Sesbania* species | -do- | A soft wooded shrubs. Flowers—pale yellow. | August to September. |
| 12. | *Cassia siamea* | Caesalpinaceae | A moderate-sized evergreen tree, with greyish brown bark. Flowers—bright yellow. | February to May. |
| 13. | *Tamarindus indica* (Imli) | -do- | A large tree, bark dark grey, with longitudinal fissures. | May to June. |
| 14. | *Hardwickia binata* (Anjan) | -do- | A moderate sized tree. | February to May. |
| 15. | *Albizia lebbeck* (Siris) | Mimosaceae | A large deciduous tree with a spreading crown, bark dark grey, irregularly cracked. | April to May. |
| 16. | *Leucaena leucocephala* (Subabool) | -do- | A medium sized fodder cum fuel tree, contain enormous root nodules for nitrogen fixation. | Throughout the year. |
| 17. | *Prosopis cineraria* (Kheajri) | -do- | A medium sized evergreen tree | February to July. |
| 18. | *Pyrus malus* (Apple) | Rosaceae | A medium sized fruit trees, leaves glabrous above, tomentose beneath, flower pink-tinged. | March to May. |
| 19. | *Psidium guajava* (Guava) | Myrtaceae | A medium size fruit-cum-fuel wood tree, soft brown bark. | April to May and August to September. |
| 20. | *Syzygium cumini* (Jamun/black plum) | Myrtaceae | A tall, evergreen tree, generally grown as shade tree or wind break on farm bunds. Flower—light yellow in colour. | March to April. |
| 21. | *Terminalia arjuna* (Arjun) | Combretaceae | A large deciduous tree, smoothen grey bark; Flower—dull yellow. | April to May. |

*(Contd.)*

| 1 | 2 | 3 | 4 | 5 |
|---|---|---|---|---|
| 22. | *Anthocephalus cadamba* (Kadam) | Rubiaceae | A large handsome tree, Flower—yellow. | April to May. |
| 23. | *Celtis australis* (Kharak) | Urticacea | A medium-sized deciduous tree, bark bluish grey with narrow horizontal wrinkles. | March to May. |
| 24. | *Broussonetia papyrifera* (Paper mulberry tree). | -do- | A medium sized deciduous tree with smooth grey bark. Flower—yellowish. | March. |
| 25. | *Emblica officinalis* (Amla) | Euphorbiaceae | A small tree with smooth grey bark. | March to May. |
| 26. | *Jatropha curcus* (Safed anand) | -do- | A shrub or small tree with short irregular trunk, bark grey, smooth, Flower—yellow. | Throughout the year. |
| 27. | *Sapium insigna* (Khirni) | -do- | A medium-sized deciduous tree with grey cork bark. Flower—with green bracts. | November to February. |
| 28. | *Sapium sebiferum* | -do- | A small gnarled tree. Flower—greenish yellow. | June to August. |
| 29. | *Salix tetrasperma* (Laila) | Salicaceae | A moderate sized tree; bark greyish brown blackish | February to March & October to November. |
| 30. | *Salix acmophylla* | -do- | A small handsome tree, bark rough with vertical cracks. | February to March. |
| 31. | *Salix daphnoides* (Bashroi) | -do- | A large shrubs or small tree, branches dark green or brown shining. | March to April. |
| 32. | *Populus* species (Poplar) | -do- | A lofty tree, bark grey smooth in young plants, with vertical wrinkles or fissures on old stems. | March to April. |
| 33. | *Alnus nepalensis* | Cupuliferae | A moderate size tree, bark compact silvery grey. | October to December. |

**Bee-keeping (Jha, 1987)**

*Systematic position*

Order - Hymenoptera

Family - Apidae

Genus - Apis

One need not bother much for raising capital to start bee-keeping. Even a courtyard or the parlour can be a fit place for keeping bee-hive. The popular kinds of bees are :

*Apis cerana indica, Apis cerana dorsata, Apis cerana florea* and *Apis cerana mellifera.*

*Apis cerana indica*

*Apis cerana indica*, constructs a number of combs side by side in perfect parallel to each other. It may also reside in hides or hollows of trees. An average of yield offered by their colony weigh 3 to 5 kg of honey at higher altitudes and 1 to 3 kg on plains.

*Apis cerana dorsata*

Contrary to *Apis cerana indica, Apis cerana dorsata* is too uncultivated and tremendous a variety to thrive in the hive. To rear it, is to cry in wilderness. It constructs a comb on tall trees and ceilings. It frequently shifts its colony at its sweet will. Its migration is correlated to the variation of climate. It constructs cells in the comb of similar shape and size. The queen is almost equal in size to other doers but she can be easily singled out of the swarm for the dark colour of the abdomen. They accumulate surplus honey in the upper most layer of the comb. The capacity of a single colony for yield is estimated upto 25 to 38 kg. of honey per year.

*Apis cerana florea*

In the third category of bees, comes *Apis cerana florea,* contented within a single small comb similar in length and width to the palm of the hand. It does little in the name of producing honey. So it is not a beneficent bee from the commercial point of view.

*Apis cerana mellifera*

Now to come to the fourth and last one, *Apis cerana mellifera* bees are more commonly found in the land of Europe. True to the nature of Indian bees, they are also adept in building several combs side by side. The queen is more prolific. It hoards honey in rich quantity.

**Individuals of a colony**

The life of honey-bees is an interesting and amazing tale of a well-formed society consisting of well-behaved individuals. Like people of one nationality of occupation living together in noisy place in different quarters, they too have a colony of their own in which all of them live up to their rights and duties. Categorically, they are of three types—queen, worker and drone.

**Queen**

Each and every colony has a queen of its own. Only the queen is sexually potent female. She has to think of nothing but to lay eggs. At most, she lives for three years. When she is understood to have become incapacitated to lay eggs, she is put to death by the workers. In season, the queen of an Indian bee lays 200 or more eggs a day.

**Drone**

No colony can do without drones, the important members of the hive, not less than four hundred of them live together in a hive. They are all males always busy in mating with the queen to add to the fertility. The drone lie lazily inside the hive.

**Worker**

Every colony is destined to have a group of worker-bees which are always up and doing. They are underdeveloped females. They make a majority answerable for the up-keeping and advancement of the colony.

In course of their ramblings, they never miss to collect nectar in their honey stomach and pollen in their pollen basket. It has been calculated that these bees do visit not less than 22 haunts a day. It is their simple nature to secrete Royal jelley from the glands in their head and apportion bee-bread of their own doing to the grubs. They always keep the hive neat and clean and look after the honey already stored up.

**Life cycle**

The Queen lays up to 800 eggs per day. It takes three days to hatch the larvae out of the eggs. Among the hatched eggs some are

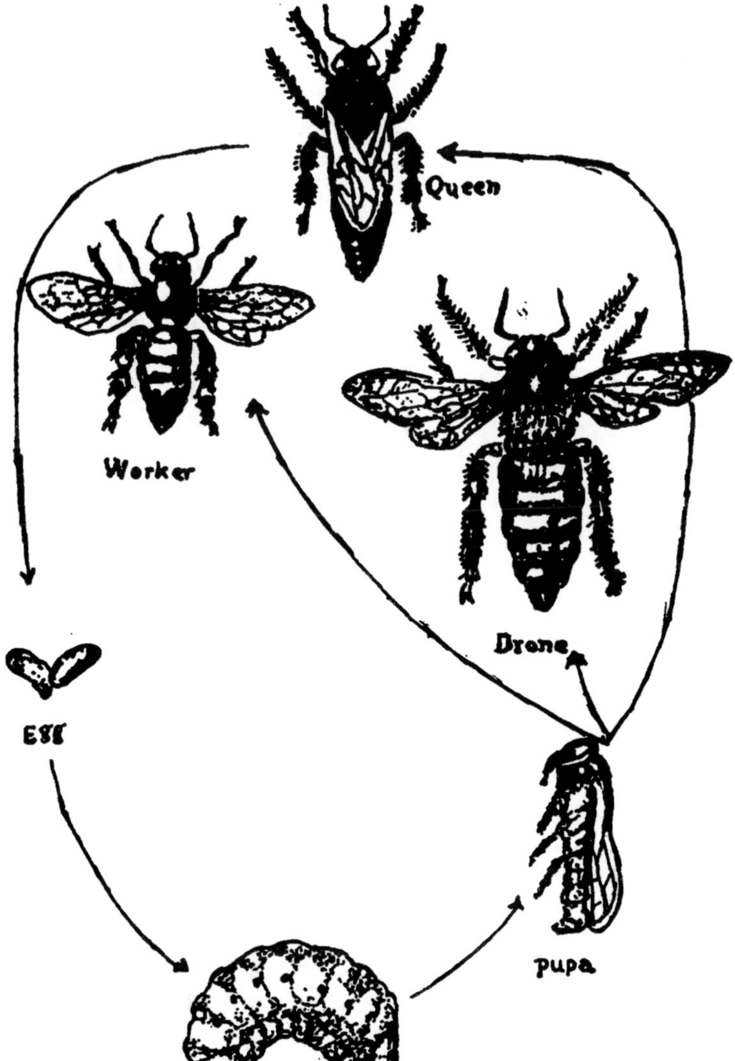


Fig. 66 : Life-cycle of Honey Bee

unfertilised and some are fertilised. Those unfertilised become drones and fertilised become either females or workers. Females are always diploid and males hapoid. Males develop parthenogenetically.

Female (2ZX) — Meiosis — ZX (Ova) — Parthenogensis —> Male (ZX)
ZX (Ova)
Male (ZX) — Mitosis — ZX (Sperm) ——> Female (2ZX).

The honey bee develops in an indirect way and it has three stages of its development—egg, larva and pupa.

The queen take 16 days to have full growth, worker bees 20 days and drones about 24 days. During first 50 to 74 hours, all the larvae are fed with special food called Royal jelly secreted by the salivary glands of worker bees. This very special food constituted the diet of the queen. While she is in her larval life, the others are fed with coarser food called bee-bread which are made up of mincing pollen and honey.

**Tools**

The bee-hives are of various type but lengstroth frame-hive is common. The lengstroth bee-hive consists of double storeys having following parts in them.

**Stand** *:* It has four-legged stand supporting the bottom board.

**Bottom-board** : It is provided with an entrance rod ahead of it. On this bottom board, the brood chamber rests.

**Brood chamber**

It is rectangular box having neither top nor bottom. It is set with movable frames. This chambers is only for queen and her workers.

**Super**

At the head of the brood chamber, there is an analogous chamber known as super, which is inaccessible to the queen. Egg-laying is prohibited in this chamber. It is laid on the brood chamber and left there until the season of honey lasts and bees collect sufficient honey. There is an inner cover attached to the top layer of the super chamber. The hive has to be kept well-ventilated and painted exteriorly, so that they may remain weather proof. With such a painting over it, the bees have no difficulty in locating their hive. The colour mostly used in painting is white yellow or green. The U.P. Government's Bee-keeping station has evolved a tip-top type of hive called jeolikote Standard hive, which can be used with equal case on the hills and plains.

**Queen excluder**

Consisting of a frame furnished with wire or zinc plates with holes made into them, it is so built as to facilitate the worker-bees to move in and out, but, at the same time debar the queen and drone from entering.

**Honey extractor**

It is a machine, whose work is based on the theory of centrifugal force. It sends out honey from the cells, while keeping the comb intact.

**Wire entrance guard**

It resembles the queen excluder. It is laid on the opening of the hive.

In addition to the foregoing equipments, a few others that need mentioning are smoker, hive tool, bee evil, bee gloves, bee-brushes, uncapping knife and the rest.

**Catching**

With the setting in of swarming season, the swarm emanate from the hives. A swarm of them may be seen and heard buzzing in the cluster of trees. From such locations, bees can be stored into a bag. The swarm, so trickishly accumulated, should be placed in a hive protected from light. In the mean time, some sugar syrup should be poured into them. Tempted by the sweet scent of the liquid, the scattered bees come darting into the hive. When all the bees are well within the hive, the door of exit should be closed.

**Control of swarming**

Swarming is a reproductive division of the honey bee colony. A lust for swarming entices the old queen and others to desert the hive and find other haunts entrusting the hive to the authority of a young queen. Swarming does not absolutely dilapidate the hive, as does absconding. Generally, two factors bring about swarming: (1) When the hive becomes too full of bees and (2) Prior to the oozing out of honey.

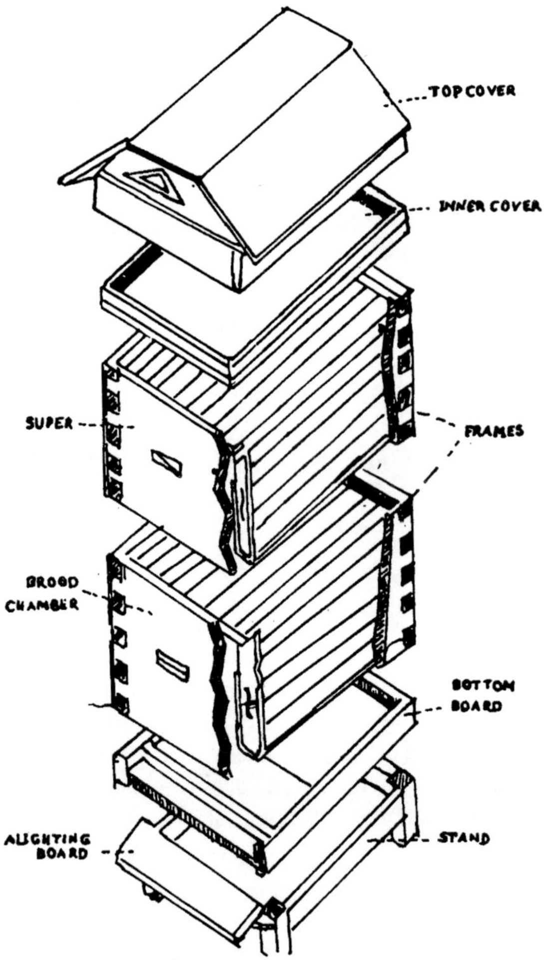


**Fig. 67 :** Multiple-storey Hive.

**Contents of honey**

Biochemists have made a study of the percentage composition of different components of honey which are the Levulose - 38.19%, Dextrose - 31.28%, Maltose and other Sugar 8.81%, Enzymes pigments 2.21%, Sucrose 1.35%, Acid proteins and Ash 1% and Water 17.20%.

**Economic importance**

Honey produced by honey-bees is said to be the elixir of life. Its nourishing and medicinal value can not be under-rated. It has been practically examined that 200 gm of honey is no less wholesome than 1.15 litres or 1.60 kg of cheese or 330 gm of meat or 8 orange. A table spoon of honey measuring 21 gm. affords as much as 67 calories of energy.

It helps to build up haemoglobin. It is taken as a preventive against cough and cold as an expectorant. Honey can be taken tastily with tea, coffee, milk or lemon etc. Honey is made use of in making alcoholic drinks. Bee-wax is useful for candle making.

**Pest of bees**

Like other insects, bees are also assailed by enemies. Two types of moth (*i*) *Galleria mellonella* (*ii*) *Achroia grisella* are their sworn enemies. The female moth lay eggs inside the hive or open portion of the comb or in the cracks or crevices of the wall. The egg hatches in the comb. After hatching, larva makes tunnel into the comb and eat their wax. If its growth be not controlled, all the bees make away from the hive for life. Watchfulness and killing of parasitic eggs and larvae keep the bees from the onslaught of pest. Apart from the pests spoken of above, a few other may be cited : ants, wasps, lizard, snake, birds etc.

The nymph and adult of *Acanthospis siva* pierce the body of the workers and suck the body fluid.

Frequent cleaning of surrounding of hives help in keeping away this bug from attacking honey bees.

Protection from ants can be granted to them by placing them on raised stands and spraying dieldrine and chloridane on its legs.

The weeds and grass must not be permitted to rise over or near the hive. Ant nests, if any, adjacent to the hive, should be unbuilt. The hive must be kept out of the reach of wasp.

**Diseases**

Bees are also found to have been suffering from one or the other disease which are mainly of two types.

I. American foul brood.

II. European foul brood.

American foul brood is caused by *bacillus* and European foul brood is caused by *Streptococus pluten* and *Bacterium eurydice.*

They also suffer from dysentery and nosma. But Indian bees enjoy the privilege of the providence and are free from such diseases.

## (B) AGROFORESTRY AND TASAR CULTURE

Tasar culture is an ancient cottage industry. India has the unique distinction of being the only country in the world producing all the four commercially known varieties of natural silk, *viz.* Mulberry, tasar, eri and munga. Sericulture with an agricultural base, industrial superstructure and an essentially labour intensive set up, is an effective tool for generating gainful employment for the backward sections of the society.

Tasar is one of the main forest produce and is mainly confined to tribal areas. Due to lack of food plants the tasar industry is facing a crisis. (Divakar and Yegneshwar, 1986).

The concept of tasar culture has undergone radical transformation with the recent introduction of tasar on oaks. Tasar culture has now two distinct entities *viz.*, tropical and temperate tasar.

Tropical tasar growing area forms a distinct belt of humid and dense forest sprawling over the central and southern plateau. It covers the principal states of Bihar, Madhya Pradesh, Orissa, West Bengal touching the fringes of Uttar Pradesh, Andhra Pradesh, Maharashtra and Karnataka. In these areas the tasar silkworm (*Antheraea mylitta* Drury) exists in the form of a member of nineteen ecoraces. However, Daba of Singhbhum is the most popular race, followed by Sukinda and Bogai of Orissa, Sarihan of Bihar and Bhandara of Maharashtra.

Temperate tasar extends from Jammu and Kashmir in the west to Manipur in the East including Himachal Pradesh, Uttar Pradesh, Assam, Mizoram, Arunachal Pradesh, Meghalaya and Nagaland. In these areas the tasar silkworm *Antheraea proylei* J. synthesised by interspecific hybridization (*A.* roylei X *A. pernyi*) is being used (Sengupta, 1986).

There is a vast scope for the expansion of tasar culture in our country. *Waste and marginal lands of farmers can be used for block plantation of arjun (**Terminalia arjuna**) and asan (**Terminalia tomentosa**). In agroforestry system farmers can plant tasar host plant at closer spacing under square, rectangular system or can be planted under alley cropping system in hilly area.*

*Farmers of Chotanagpur plateau cultivate brinjal, tomato, lady's finger, gora paddy, etc. in the interspace. After rearing of tasar silkworm the branches are pollarded. The branches are utilised as fuelwood by the farmers and intercrop get sufficient light for its growth* (Jha & Jha, 1989).

**Techniques of tasar culture**

* Super quality seeds of host plant must be collected at appropriate time and are to be preserved in a proper manner.
* The technology of raising *Terminalia arjuna* seedlings on a large scale has been successfully evolved by the Central Tasar Research and Training Institute, Ranchi. This technique of raising the arjun plants involves dipping the seeds in water for 96 hours. After treatment, soaked seeds are kept in heaps covered with moist gunny cloth in shade, till germination. This method gives a germination of about 85 to 90 per cent within a period of twenty days which permits to raise a stable planting stock material as against erratic and poor germination spreading nearly two months under the conventional methods.
* Suitable rooting media for raising *Terminalia arjuna* with seedlings in polythene tubes has also been found to be FYM, soil and sand in the ratio of 3:2:1 and the planting of three months old seedlings leads to about 90 per cent success (Sengupta, 1986).
* Pits must be prepared properly. 20 cm cube pits are generally preferred. Arjuna and asan are planted in close plantation at a distance of 1.5 x 1. 5m or more as per desire of the farmers. Before plantation, manure and insecticides must be put properly in each pit.

- Fencing is desirable before plantation. If it is not done before, it must be done immediately after plantation of seedlings.
- Regular removal of weeds are very advantageous for the growth of plants. Hoeing should be performed at regular intervals, to provide proper aeration and conserve moisture in the soil. For the luxuriant growth of plants fungicides, insecticides and manures should be put in proper quantity. Watering of plants during summer upto 3 years becomes necessary to check casualties.
- Proper pollarding of plants must be performed for keeping them at bushy stage.
- Supply of improved variety of silk "layings" is a basic necessity for scientific tasar culture. In India, the Central Silk Board in collaboration with Sericulture Department of different states is presently supplying disease free layings (D.F.L) and is thus, promoting the development of tasar culture.
- It is a notable fact that the tasar silk worm is semidomesticated in the sense that it plays 'outdoor and indoor games' while in process of breeding. It is found that for the rearing of larvae from hatching till cocoon is formed, it keeps out-doors, while the rest of activities take place indoors. When it is ready for harvest, the seed cocoons are selected and are left hanging in pest-proof-cages to look as if it were a garland in form.

  The cages have to be kept in a well ventilated and well lighted grainage room to facilitate its growth. The period of emergence of moth falls between mid-night and day-break. Its emergence from inside the cocoon shell is facilitated by secretion of enzyme which softens its exit.

Adult male and Female (pair) are kept in a sweet box for mating. Some of their matings do occur inside the cage itself. But, any and every weather is not akin to its mating tendency. Mating may take place, at a temperature between 26° to 30°C and in higher humidity 70-80% (RH). After mating, the sexes are kept apart and, as a consequence, the female alone is left in the sweet box to lay eggs. 80% eggs are laid within 70 hours. Presence of darkness is conducive to her egg laying. The average number of eggs laid by a female moth is estimably fixed in between 150-250. (See Fig. 68 and 69)

When egg-laying is over, the smears of the females moth are put up for examination. On examination, the eggs of the moths

Fig. 68 : Emergence of *Antheraea mylitta* moths for the second crop rearing. Moths are in search of partner for mating.

**Fig. 69 :** *Antheraea mylitta* male and female just before mating.

showing Nosema are set aside. Only the healthy eggs are taken to be dried in shade and then reserved for hatching. In suitable condition, it takes a week or so in hatching, the fit time for which is the early hours. On hatching, small twigs of host plants are fondly placed over the larvae. The twig together with the larvae are, later on fitted into the host plant in the manner of grafting. The larvae of *A. mylitta* is tetramoulter. The larval period in the first and second crops stretches over 25 to 45 days respectively. The physical colour of larvae is brownish yellow, while in hatching stage, but it gives place to green in the late first instar. It has gleaming spots on its dorsal surface, and on both sides of tubercles. In the late fifth instar, it falls to spinning cocoons. Away from the dim and bustle of human habitation, the larvae is apt enough to bind two or three leaves together with its silken thread to be shaped into cord (Hammock). The cocoon formation subject to atmospheric temperature is completed within 3 to 7 days. These cocoons are ready for harvesting on 6th day of spinning. Harvesting being done, the leaves are taken off from cocoon. The cocoons are grey yellow and white in colour (Jha, 1987). (See Figs. 70, 71, 72)

So long only two crops of tasar were being raised, first being the seed crop and the second one the commercial crop. The Central Tasar Research and Training Institute, Ranchi has isolated various stable trivoltine lines of Daba and Sukinda which enable the rearers to have a successful third crop rearing. Due to lower temperature and dry conditions prevailing during that period the third crop has been found to be a very successful one with higher effective rate of rearing.

During that period much of other agricultural activities not being there, the rearers have also been found to be very enthusiastic to take up the third crop. The same has already been tried very successfully in Orissa and Andhra Pradesh as also in some parts of Madhya Pradesh and Bihar. Introduction of the third crop has a potentiality to increase the tasar production substantially and also the earnings of the rearers (Sengupta, 1986).

**Gains of tasar culture (Jha and Sen, 1991)**

(1) Employment potential is generated in the collection and preservation of super quality seeds of plants, sowing of seeds and watering the seedlings.

(2) For the preparation of pits, for planting and watering of seedlings and fencing work, the poor tribals and other

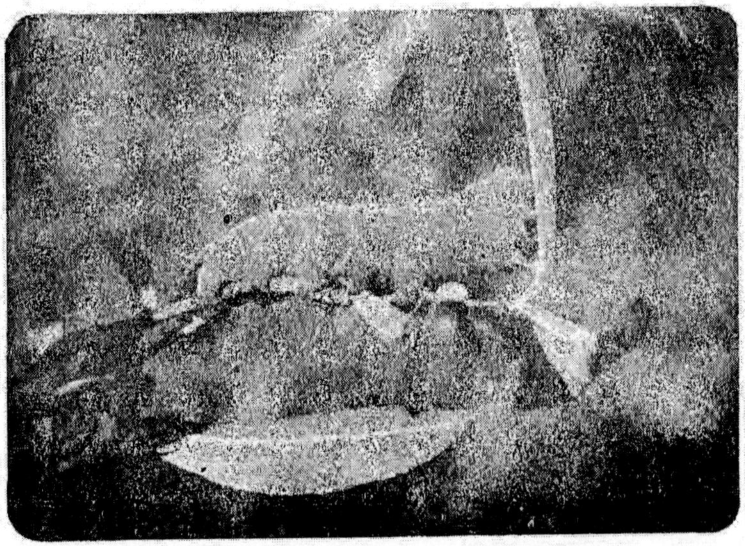

Fig. 70 : Larva of *A. mylitta* (Fifth instar).

Fig. 71 : Larva of *A. mylitta* before cocoon formation.

**Fig. 72 :** Cocoons of *Antheraea mylitta* ready for harvesting.

weaker sections of our rural society get employment. Pollarding of plants also provide job to them.

(3) Proper rearing of silk worms and collection of cocoons provide job to the people of weaker section of our society.

In brief, it may be said that tasar culture is a labour intensive programme and is helpful particularly to the weaker section of tribal area. According to Prasad (1984) the number of families associated with tasar culture in India were as follows:–

**Table 7**

| | State | No. of tribal families |
|---|---|---|
| 1. | Bihar | 60,000 |
| 2. | Madhya Pradesh | 20,000 |
| 3. | Orissa | 32,000 |
| 4. | West Bengal | 2,000 |
| 5. | Other states taken together | 12,000 |
| | Total : | 1, 26,000 |

Prasad and Jha (1985), has reported that it will be more profitable business if some spinning machines are provided to the tribals. Many tribal boys and girls are engaged already in this trade. These youngsters earn Rs. 20 to Rs. 25 per day and remain in their own villages.

In the year 1982-83 in Bihar, sixty thousand families were engaged in it and they earned Rs. 3.75 crores. Thus the income per family was Rs. 625.

It may be more advantageous for the inhabitants of tribal belt, if close plantation in waste land is done on a large scale. It has been proved by experts that after 3 to 4 years of close plantation of Arjun and Asan, in waste land under sunfed condition, may provide per ha. Rs. four to five thousands per rearing season. Such efforts will rapidly raise tasar production and income.

Many jobless tribals for their bare subsistence pilfer valuable woods from jungle, which is undoubtedly very harmful for the country. Such tribals after getting employment in tasar culture may stop pilfering of woods and thus the country will be benefited.

Tasar culture is a source of earning of foreign exchange and thus is advantageous for the developmental programmes of the country. The value of export of tasar was Rs. 60 lakhs in 1967. It rose to Rs. 6.12 crores in 1979. Scientific tasar culture may promote rapidly the earning of foreign exchange of India.

The gains of tasar culture are enormous. Needless to say that the growth rate of tasar production is not upto the mark. In 1967 we produced 300 m tonnes of tasar, which rose to 423 m tonnes in 1977. In the year 1983 the production was about 440 m tonnes. Slow growth of production is due to some hindrances. Following are the main hindrances for the rapid growth of tasar production (Prasad, 1984).

**Hindrances in tasar culture (Jha & Jha, 1985)**

1. Disease Free Layings (DFL) supplied in our country by the Government but still some tribals use personal layings, which contain disease like pebrin, etc. In Bihar Central Silk Board in collaboration of the state Sericulture Department has established a number of centres for supply of disease free layings of silkworms. As stated earlier, there was no systematic seed organisation, with the result poor quality seed used to come into market and that too in a irregular and insufficient quantity resulting in poor commercial crop (Prasad, 1984). Utmost care must be taken to keep D.F.L. without any deficiency.
2. Gall formation due to insects in the leaves of arjun and asan badly affects tasar culture. Hence by using insecticide at the time of infection, this melody must be controlled with due care.
3. Sporadic plantation of arjun and asan is not conducive for scientific tasar culture. Due to cutting of marginal forests for cultivation and for fuel-wood, the Adivasi are now forced, to go deeper in the forests in search of food plants, and as a result of this many tribals have abandoned tasar cultivation. For close plantation the tribals should be provided wasteland.
4. The plants of arjun and asan must remain at bushy stage for good tasar culture. Sometimes due to lack of nutrients and rainfed conditions, the plants are not bushy. Moreover, in the absence of proper pollarding by some cultivators, the

plants are not at bushy stage. Precaution is required to remove such shortcoming.

5. Tasar worms are devoured by birds before the formation of cocoons. Hence watch keepers must be appointed to remove this deficiency.
6. Central Silk Board and Sericulture Department of different states are providing farmers, scientific training in tasar culture. New methods of tasar silk production and their techniques should be taught to the new generation and at the same time the merits and returns from this industry should be highlighted (Divakar and Yegneshwar, 1986).
7. Subsidised seedlings, manure, insecticides, layings are provided by the government. The government also provide cheap institutional credit to them. But the poor farmers are at times forced to grease the palm of the government servants, which adversely affects the incentive for better tasar culture.
8. Generally rural and other weaker sections of interior area sell cocoons in village at a very cheaper rate. Thus they are exploited by the middle man.
9. Better marketing arrangement by the government is urgently required, so that the poor farmers may get their due shares. This will promote incentive for better tasar culture.
10. The employees of Central Silk Board and Sericulture Department of different states have to perform arduous task in the promotion of tasar culture. They should be provided better pay scale and promotional avenues. This will provide incentive to them for better participation. On the other hand the corrupt and inefficient staff must be dealt strictly. The above can be achieved only through a practical and realistic extension method. Another steps must be taken simultaneously along with motivation of farmers relates to upgradation of their skills. In other words, a farmer should not only be made mentally prepared to accept new technologies but he should be properly equipped to successfully use it in his farm land. This would involve both theoretical and practical training (Subarna, 1990).

*If above hindrances are removed and planting of tasar host plants in agroforestry system in the farm land may be started on a massive scale, the tasar culture may have bright future in uplifting the economic well-being of the weaker sections of our society.*

## (C) AGROFORESTRY AND LAC CULTIVATION

There is a vast scope for the expansion of lac cultivation in our country which can be achieved by planting lac host plants as a component of agroforestry system, on waste and marginal lands, degraded forest area, panchayat lands, field bunds, etc. The ideal host species should be such that it is fast growing, pollards well and can withstand heavy lac insect infestation (Jha and Jha, 1985).

Social land allocation projects will be much helpful to increase the productivity of lac cultivation. Under this programme, generally lands are allotted to the landless on a condition that they will raise trees on it. Thus, by allotting these land to the land-less these same are being brought under productive use and allottees get the benefit of a stable and sustained source of income.

To encourage the farmers various incentives like provision of water, health and educational services, co-operative loan facilities will go a long way in promoting productive work like cultivation of lac.

Under social forestry/agroforestry programmes large scale block plantations of lac host plants will increase the productivity of lac.

The Government of India has established during pre-independence era the Indian Lac Research Institute at Namkum, Ranchi to carry out research in different facets of lac cultivation and utilisation. Forest Departments of different lac producing states are also promoting lac cultivation.

### Technique of lac cultivation

*Host Plants*

Palas (*Butea monosperma*), Kusum (*Schleichera oleosa*), Ber (*Zizyphus mauritiana*), Babul (*Acacia nilotica*), Assamese-maj (*Albizia lucida*), Jalari, (*Shorea taluri*), Ghont (*Zizyphus xylopyra*), Acasi (*Acacia auriculiformis*), etc.

The lac host plants take few years to attain maturity. The farmers do not get any return in the initial years. Lac host plants may be planted in square, rectangle and triangular system. The design of planting of seedling varies according to topography and prospective use of the interspace.

In the square system, seedlings are planted at the corners of the square. The interspace can be utilised for growing agricultural crops, fodder and vegetables. In the rectangular system seedlings are planted at the corners of rectangle. Rectangular system helps in increasing the width between two rows and more number of plants are accommodated within rows. This design may be adopted in planting lac hosts under agroforestry system. The lac host plants may be planted under triangular system. This systems provides open space for intercrops as well as for the trees but intercrops cultivation and operations in the plot becomes difficult.

Under agroforestry system while selection of intercrops, farmers must take into consideration the nutritional requirements of intercrops and tree species. The intercrops should not exhaust the soil too much. In Chotanagpur plateau of Bihar, lac cultivators grow vegetable as intercrops (cabbage, potato, chillies, bean) and crops (gorra paddy), etc. (Jha and Jha, 1989).

The nursery techniques for different species of trees are different. But in general, to get healthy seedlings, super quality seeds of host plant should be collected and a mixture of sand, clay and farm yard manure in equal proportion is ideal for proper germination of seeds. If the seedlings are raised in polythene pots, it will facilitate easy carriage from nursery to the field. The cost of production of seedling varies from place to place. For planting, 30 $cm^3$ pits are generally preferred in village. The seedlings are planted at a distance of 3 m x 3 m. During summer, watering is necessary to prevent mortality of seedlings up to two years (Sen Sarma and Jha, 1989).

The possibility of shifting lac cultivation from the conventional tree hosts found scattered under the forest conditions to plantations of suitable quick growing host has been examined using bushy hosts and suitable tree hosts which can be trained into bushes. It has been established that *ber* and *galwang* can be trained into bushes suitable for lac cultivation which can be raised within four years of transplantation thereby reducing the time gap by 4 to 5 years. Lac cultivation trials on these bushes have shown that stick lac yields of 4.5-5.0 quintals per hectare can be obtained as compared to 2-2.5 quintals per hec:are on the conventional tree hosts. (Annon; 1985).

**Lac crops**

There are two strains of lac grown in India, the Rangeeni and Kusumi lac grows on kusum trees whereas Rangini crop grows on other host plants. The most important of them are palas (*Butea monosperma*) and Ber (*Zizyphus mauritiana*).

The crop produced are named after the Indian calendar and months in which these are harvested. Maturity period of the crops are :-

For Kusumi strain :
(1) JETHWI (January-February to June-July)
(2) AGHANI (June-July to January-February)

For Rangini :
(1) KATKI (June-July to October-November)
(2) BAISAKHI (October-November to June-July).

Both the strains are capable of producing two generations and hence two crops in a year (Jha, 1987).

**Preparation of host plants (Jha & Sen, 1991)**

The tiny lac larva settles and grows on young and succulent green shoots. In order to promote flush of green and succulent shoots, the host is pruned well in advance and as per requirements of the crop as follows :-

**Table 8** : Pruning time

| Sl.No. | Strain | Crop | Pruning time |
|---|---|---|---|
| 1. | Rangini | Katki<br>Baisakhi | February<br>April |
| 2. | Kusumi | Jethwi<br>Aghani | January-February<br>June-July |

**Broodlac :**

Broodlac is just like seeds for crops. This is piece of a branch of host plant carrying lac incrustation at egg laying stage, and this is used for the inoculation. Just before the emergence of the larvae, 15 cm to 23 cm long brood sticks are made and tied to the shoots of the host tree in bundles of two or more in one of the following three methods :-

(*a*) The longitudinal method

(*b*) The lateral method

(*c*) The interlacing method

The adult female lays about 200 eggs within the encrustation. The hatching time fluctuates between two to eight hours. After laying is done, the larvae emerge in great number and settle down on young succulent shoots. They usually settle on lowerside of branches to protect themselves from dessicating effect of sun. The larvae may crawl up to 7 metre distance and settle down in about 3 weeks time.

They suck the sap and from glands secrete lac which appears initially as a sticky shining liquid which form a uniform coating. It gradually increases in thickness and extends until the secretions of adjoining larvae eventually meet and coalesce to form a continuous encrustation. There are two kinds of adult males, winged and wingless. Both kinds are sexually potent and fertilize the females within the cells through the anal tubercular openings. The fertilized female continues to grow and secrete lac until the period of oviposition (Jha, 1987). (See Fig. 73)

**Lac harvesting**

Harvesting is done in two ways :

(a) Selection of broodlac for further inoculation is done when a tiny yellowish spot appears at the one end of encrustation indicating maturity of the broodlac.

(b) For collection of lac from host plant, harvesting is done when lac becomes mature. The encrusted shoots are cut and the lac is collected by scraping. The product is known as sticklac.

This lac has many impurities. These are separated and the product obtained from sticklac after cleaning, washing and drying is granular in form and is referred to as 'seedlac'. Seedlac after further purification either by heating process or by dissolution process is converted into shellac (Sen Sarma and Jha, 1987). (See Fig. 74)

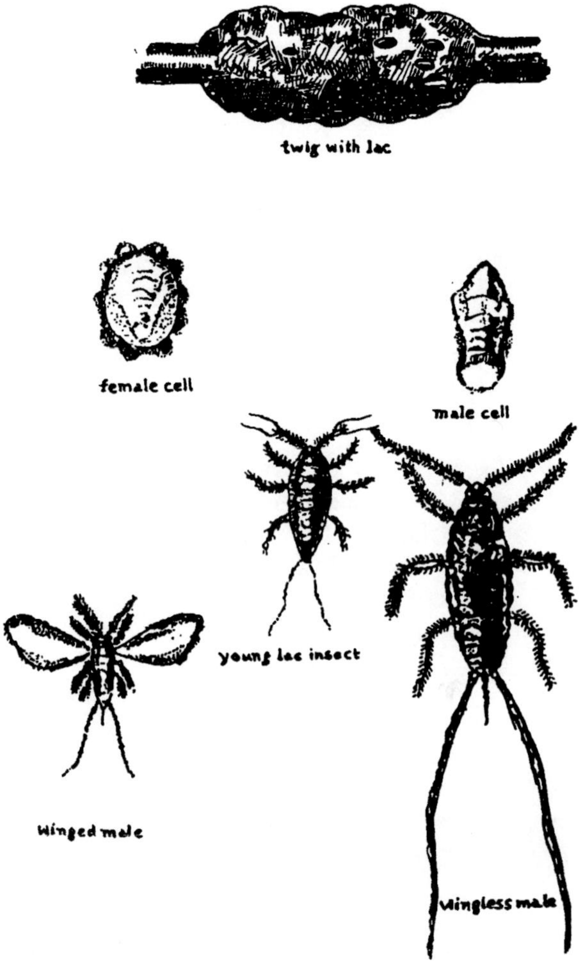


**Fig. 73 :** Life cycle of lac insect.

**Fig. 74 :** Twig with lac encrustation.

### Lac crop production

The gain from lac cultivation are enormous. Table 9 shows the quantity of sticklac production in different years. This table clearly shows a declining trend of production.

**Table 9 :** Sticklac production in India

| Year | Total sticklac production (tonnes) |
|---|---|
| 1983-84 | 11,605 |
| 1984-85 | 12,955 |
| 1985-86 | 18,175 |
| 1986-87 | 20,340 |

**Source :** 'Indian Shellac'.

### Constraints in lac cultivation

*Non availability of broodlac in time (Jha and Sen, 1991)*

Most of the trees are left un-inoculated due to non-availability of broodlac in time. This can be overcome by ensuring sustained and timely supply of broodlac.

### Shortage of host plant

Scarcity of the host plants at suitable places is one of the major constraints in lac production. Therefore, farmers should be encouraged to take up such plantation on wastelands etc. and this can be reinforced by adequate technology transfer in respect of pollarding and inoculation.

### Lack of marketing facilities and uncertainty in prices

Generally, the lac is reported to be harvested even at immature stage by cultivators due to urgency in getting returns. Steps should be taken to remove the financial constraints and also to ensure remunerative prices.

Besides, the cultivators in the interior areas are exploited by middlemen who purchase the product at cheaper rates. Therefore, due to lack of proper marketing facilities, lac production has suffered a tremendous set back. Better marketing arrangement should be made by the Government or Quasi Government agencies like BISCOLAMF in every areas where lac is cultivated, so that the cultivators may get their due share. This will act as an incentive for better and higher lac cultivation.

**Lack of knowledge**

Under the aegis of Lac Development Directorate cultivators are imparted training for adopting improved and scientific methods of lac cultivation. Needless to state that many cultivators still cannot participate in training and hence the production is adversely affected.

**Adverse climatic conditions**

Lac cultivation areas suffered most due to adverse climatic condition. Hot and dry weather are inimical to the lac production. Heat causes closing of the pores leading to suffocation of lac insects.

**Lac products (Annon, 1952)**

(a) *Insulating varnishes*

Shellac spirit varnishes are used for electrical insulation, such as for finishing coils of small motors, transformers, etc. The heat resistance of such varnishes could be improved by the addition of urea and their use extended for impregnating built-up coils of large motors.

Shellac based oil-varnishes have been found to be suitable as baking varnishes for insulating armature coils.

(b) *Insulating cloth*

Cambric cloth impregnated with a modified shellac oil-varnish makes an insulating material similar to 'Empire Cloth' widely used in the construction of electrical equipment.

(c) *Laminated boards*

Unsized paper can be impregnated with solutions of lac in methylated spirit or ammonia and pressed together to form tough sheets, boards or tubes. The addition of urea and formaldehyde or casein confers improved strength and adhesion between laminae. Plywood made with a shellac adhesive is more resistant to the action of water.

(d) *Jute-lac boards*

Lac-bonded hessian boards can be made either by melting powdered lac between successive layers or by previously treating the fabric with lac solutions. The boards can be used for the construction of chests, containers, panelling, etc.

(e) *Oil cloth*

When lac is modified with drying oils, it yields tough glossy, water-proof films. By suitable admixture with mineral fillers and pigments,'oil-cloth' can be manufactured in various colours and attractive finishes.

(f) *Plastic-moulded articles*

Shellac is ideally suited for the plastic moulding of gramophone records. Chemical modifications of lac through the use of urea or urea and formaldehyde can be used for plastic moulding to obtain products with greater heat-resistance and strength.

(g) *Commercial enterprises*

Among the article mentioned above, insulating varnishes, oil-cloth, laminated boards, water proof abrasive papers, flexible micanite and plastic moulding have been manufactured and marketed in India and others are in the process of large-scale development.

**Benefits from lac cultivation**

In lac cultivation poor people of tribal areas get employment in the following ways :

Employment potential is generated in pit digging, planting, watering of seedlings and fencing work. Proper rearing of lac insects, collection of sticklac and broodlac and their processing provide job to the people of weaker section.

Many tribal men and women are already engaged in this trade. They earn Rs. 30 to 40 per day and remain in their own village.

Many jobless tribals for their bare subsistence pilfer valuable woods from forest, which is undoubtedly one of the causes for forest degradation. These tribals when employed in lac cultivation may cease pilfering and illicit cutting from the forest for mere subsistence. Beside this, they can solve their problem of fuelwood from pruned branches.

Scattered lac host is not conducive for scientific lac cultivation this has been found uneconomical. For close block plantation the farmers should be provided wastelands, marginal land, etc.

It was observed that due to various reasons lac production has declined considerably during the last few years. In Bihar BISCOLAMF ensures adequate price to the cultivators and also for supply to broodlac to needy growers. It is apparent that lac cultivation in India can be boosted up by providing proper incentives and adequate supply of seedling, manures, insecticides, broodlac at subsidised rate by various agencies. Many tribal families are still not getting these inputs and adversely affects production of lac.

Better marketing arrangement by the Government is urgently required in all the lac growing states, so that the distress selling should be avoided resulting in better and higher lac cultivation (Sen Sarma and Jha, 1989).

## (D) SISAL PLANTATION IN AGROFORESTRY SYSTEM

Sisal (*Agave sisalana*) plantation will utilise unproductive soil, conserve soil from erosion and may fetch considerable amount of revenue within a short span. Thus it is worthwhile encouraging its plantation on a large scale all over India specially in degraded jhum land and marginal wastelands under agroforestry system.

Farmers will get work throughout the year, specially in those areas where decortication plants are established. The farmers can also grow sisal as a fillars in between the tree species under alley cropping system and supply the harvested ones to the factory. Sisal fibre obtained from these plants is used for various purposes such as marine cordages, different types of brushes, carpets, bags, etc. There may be a good market of this in future, for manufacturing hard board also. Infact sisal has multiple uses. The pulp residue left after decortication of leaves contains 11 per cent of fermentable alcohol and also some acids from which some spirit, liquor and vinegar can be prepared. A kind of wax (sisal wax) is obtained as a byproduct from the pulp residue, which is used for the manufacture of shoe, car and floor polishes.

Dry sisal waste can be utilised for production of sodium pectate, a gelatinous material, which is of special interest to pharmaceutical and cosmetic industries.

Sisal can also be used as raw material for the manufacture of high quality paper. *It is worthwhile trying agave under agroforestry system to conserve soil and moisture of the land and improve the productivity of site. Agave may be intercropped with perennial tree species. Tree species fixing atmospheric nitrogen are preferred. Leucaena is best suited as an associated tree species with agave (Itnal,*

*1992). It may be an important substitute of other plants, planted for soil conservation measures in hilly area.*

The requirement of soil and moisture of this plant is very negligible. The important point in its favour is that the plantation may be carried out for utilising the soil which is neither fit for agricultural crop not for tree plantation.

It can flourish on many types of soil. It is drought resistant and frost hardy. A rainfall of 750 mm to 1000 mm is sufficient for its growth.

**Technique of planting (Sen and Jha, 1988)**

The plants are raised both by root suckers and bulbils. Root suckers appear almost every year till its maturity but bulbils occur only once in its entire life cycle. These bulbils are generally of small sizes and are taken to the mother bed filled with sand, watered both in the morning and evening and when it sprouts it is transferred in the nursery beds for about twelve to fifteen months and transplanted at a spacing of 10 cm x 10 cm. These bulbils are transplanted in the fields during the plantation season. In the third year, suckers grow from the roots which can be replanted from the second week of June to July end. These suckers do not require any type of nursing in the nursery. Suckers are directly planted in the plantation areas.

**Fencing**

Cattles do not browse but may trample the young transplants. Therefore, some fencing is needed.

**Soil work**

Soil work is done one year before and usually after the monsoon. Hoeing is done in strips (15 cm to 20 cm deep) by pick axes. 1 to 1.5 metres space between rows is sufficient. The suckers are planted at .5 x .5m spacing.

**Treatment before planting**

Before planting the uprooted root suckers or nursery raised bulbils are subjected to pre-treatment root trimming with a view to initiate root growth. Adventitious roots are trimmed by a sharp scythe. The exterior layers of leaves are also removed exposing the cream colour bulb. Root trimmed plants may be transplanted

Fig. 75 : Sisal (*Agava sisalana*) plantation.

immediately. The entire operation from uprooting to transplanting may be completed within two days.

**Transplanting**

Root suckers and the bulbils are transplanted preferably with the onset of monsoon. This operation should be completed at any cost by the end of July. Since it is drought resistant plants, premonsoon transplanting is also possible.

**Cultural operation**

Contrary to the general belief sisal plants require attention throughout its entire life cycle, normally ten years. It is also desirable for economic return as well. After production of bulbils the entire plants die and in its place next plantation is done.

**Weeding and hoeing**

The soil is loosened up to the depth of 20 cm in 150 cm wide strip except around the plants. This is very important for the healthy growth of plants and production of root suckers.

**Manuring**

The plants are *calciphilous*. Hence calcium fertilizers should be applied to these plants besides application of urea, single super phosphate and potash. Green manure and FYM are commonly used by farmers.

The following dose per plant has been found to be satisfactory for the successful growth : Urea-14 gms, single superphosphate-18 gms, potash-4 gms, calcium may be added at the rate of 4 kg per hectare.

The mixture is added to each plant after two weeks of planting. The application of the fertilizer must be completed by the second week of August.

**Harvesting of leaves**

Preliminary harvesting is done from the third year of planting. During harvesting all the peripherial leaves are cut, leaving behind 5 leaves in the centre. The regular harvesting is done from the third or fourth year.

The yield of harvested leaves is about 4665 kgs per hectare in the first year of harvesting. In the subsequent operation only 5-10 leaves are left excluding central column of unfolded leaves. 10-15 leaves may be obtained from each plant. The leaves are cut by a curved scythe having long handle as close to the base as possible. Care should be taken not injured the trunk. The leaves vary from 45 cm to 105 cm in length (Sen and Jha, 1991).

**Extraction of fibres**

Sisal leaves contain only 4 per cent of fibre. These leaves are decorticated on decortication machine run with the help of electric current. The fibre is washed at least three times and dried in sun for three days. It is spread over a towel rack, underneath of which sand is spread so that the colour of the fibre may not be lost in the soil. In the beginning the fibre generally has the greenish colour but gradually it changes into white colour after drying in the open for three days.

The length of the fibre is generally between 45 cm to 105 cm. The longer the fibre the greater is the demand in the market.

The leaves are harvested from the month of October and this operation lasts upto April. The leaves must be decorticated within twenty four hours from the time of harvesting.

The residue left out after decortication of fibre is also of great use. It is used in a compost pit and also applied directly around the plants in place of fertilizer. It has given a marvellous effect on the growth of agricultural crops and trees.

Normally the decortication machine may run for eight hours at one stretch.

100 kg of leaves yield 10 kg of fibre with moisture and ultimately 4 kg of dry fibre is obtained. Thus, 100 kg of green leaves in weight will yield 4 kg of marketable fibre (Kanan, 1973).

**Baling of fibre**

Dry fibre is finally baled out in baling machine made of wood and each baled out bundle of fibres is of 50 kg to 100 kg in weights. These bundles are stored in shed on a wooden platform. The stored fibres are well protected by rain water and other injurious insects.

**Yield**

In the 3rd and 4th year dry fibre yield will be 150 kg/ha/yr. From the 5th year to 15th year the yield will be about 300 kg/ha/yr.

During 15th year period total expenditure will be Rs. 37,000 per ha and likely income Rs. 53,000 (Sen and Jha, 1989).

## (E) BAMBOOS IN AGROFORESTRY SYSTEM

Bamboo plays an important role in the economy of the country and is used for a variety of rural purposes. Bamboo is associated with mankind since ancient times and is a natural gift to the tropical people. Every part of the babmoo is utilized in one way or the other and it has a high calorific of 4600 to 5400 cal/kg which makes it an energy crop also.

India has perhaps the world's richest resources of bamboos, having about 130 species occurring over an area of 9.57 million ha. This is about 12.8 per cent of the total forest area of the country and represents 20 per cent of India's total forest produce. Bamboos are widely distributed in Andhra Pradesh, Arunachal Pradesh, Karnataka, Kerala, Manipur, Meghalaya, Mizorum, Nagaland, Sikkim, Tripura, Orissa, West Bengal and Madhya Pradesh (Maithani, 1988).

The bamboos are an important woody perennial. Bamboo as a component of agroforestry system would help in minimisation of soil erosion and maintain the productivity.

Several species of bamboos have been introduced into agricultural lands in several part of India. In the eastern parts of India, *Bambusa tulda*, *Bambusa giagantea*, *Melocanna bambusoids* are commonly planted. *Bambusa vulgaris* is widely planted in Orissa, Bihar and eastern parts of Uttar Pradesh, *Oxytenanthera stocksii* in the coastal areas of Maharashtra, *Dendrocalamus strictus* and *Bamnusa arudinacea* over extensive areas in the semi-arid tracts of Andhra Pradesh, Uttar Pradesh, Madhya Pradesh, Maharashtra and Gujarat and *Dendrocalamus hamiltonii* (also for fodder) in the hills of Himachal Pradesh. Bamboos are generally planted in homesteads or on the boundaries of agricultural lands. Mature culms are harvested annually (Chaturvedi, 1991).

Extension worker should demonstrate the farmers that specific type of bamboo can be profitably grown on the marginal or sub-marginal land under agroforestry system. Bamboo growing in hill sides can be more profitable. Bamboo rhizomes keep soil in place and conserve soil moisture. The bamboo species, in general, are light demander but bamboo seedlings are not suppressed by shade of other trees planted in agroforestry system. (See Fig. 76)

**Area under Bamboo cultivation**

Bamboo covers an area of about 0.1 million sq. km. and its annual potential yield is about 4.5 million tonnes. The statewise bamboo area and the estimated potential availability of bamboo are as follows (Tiwari, 1981) :

**Table 10**

| Name of State | Bamboo Area sq.km | Potential annual cut (Lakh Tonnes) |
|---|---|---|
| Andhra Pradesh | 19,790 | 2.55 |
| Arunachal Pradesh | 7,779 | 2.00 |
| Assam | 10,000 | 12.10 |
| Bihar | 5,294 | 2.00 |
| Gujarat | 1,936 | 0.46 |
| Himachal Pradesh | 104 | 0.03 |
| Jammu & Kashmir | Negligible | — |
| Karnataka | 6,000 | 4.75 |
| Kerala | 631 | 1.08 |
| Madhya Pradesh | 14,864 | 8.00 |
| Maharashtra | 8,500 | 3.00 |
| Manipur | 2,500 | 2.00 |
| Orissa | 10,500 | 4.89 |
| Punjab | N.A. | 0.09 |
| Tamil Nadu | 5,388 | N.A. |
| Tripura | 2,849 | 2.15 |
| Uttar Pradesh | 4,000 | 0.41 |
| West Bengal | 164 | 0.08 |
| Total | 1,00,299 | 45.59 |

N.A.= Not available.

Amongst the many bamboo species found in India the ones that are commercial importance are as follows : *Bambusa arundinacea, Bambusa vulgaris, Bambusa tulda, Dendrocalamus strictus, Melocanna baccifera, Oxytenanthera stocksii, Ochlandra travancorica*, and *Thyrsostachys oliveri*. The flowering cycles of these bamboos vary from 7 years to 60 years except for *Bambusa vulgaris* which has not gregariously flowered in India for the past 90 years. Most bamboo species, however, have a more or less sharply defined

Fig. 76 : Bamboo plantation.

flowering cycle of 3, 7, 11, 15, 30, 48, 60 or 120 years (Chaturvedi, 1988).

Bamboo flowerings are of three kinds :

(*i*) *Gregarious flowering* : It is the general flowering within one or a few years, and cover considerable areas, of all or most of the individuals of certain species that do not flower annually; in some cases followed by the death of the plant.

(*ii*) *Sporadic flowering* : It is the flowering of one or a few culm in clump or a few clumps in a locality.

(*iii*) *Annual flowering* : In addition to the above two kinds of flowering, annual flowering is also met to certain species of bamboos (Khanna, 1984).

**Regeneration of Bamboo**

(a) *Artificial Regeneration*

Suri and Chauhan (1984) have stated that the bamboo can be regenerated artificially by showing seeds, planting nursery raised seedlings and by vegetative methods. The bamboo seeds look like paddy grain with a small horizontal clip in the middle. A seed generally measures about 6mm long and 1.95 mm wide.

For planting the land is prepared by ploughing and cross-ploughing. Sowing is done in September and October. Germination takes place in 10 to 20 days. Three month old germinated seedlings are picked out and transplanted in polythene bags, filled with well-powdered soil mixed with manure and insecticides. Chemical fertilizers are used to boost seedling growth. The seedlings are regularly watered. Polythene bag seedlings by September and October become fit for out planting during July next.

Preparation of proper pits promotes the growth of plants. Generally 30 $cm^3$ pits are preferred in planting seedlings at 5m x 5m spacing. Proper spacement promotes rapid growth. Before planting, the dug out pits and excavated soil is allowed to weather for 4 months and are refilled with weathered excavated soil after the seedlings are lodges in the pits. Regular weeding is very essential for the growth of plants. Hoeing should be performed at regular intervals to conserve moisture in the soil. Watering of plants during the summer up to period of three years is necessary to prevent mortality due to desiccation. (See Fig. 77)

It is desirable to construct fencing before plantation. If it cannot done before planting of seedlings it must be done immediately after. Live hedge fencing is also a popular and cheap. Resource poor cultivators can adopt this method and avail of the benefit of fencing.

(b) *Vegetative Reproduction by Rhizome*

The rhizome is a thickened, underground stem provided with distinct nodes and internodes, scaly leaves at the nodes with a bud in the axil of each such leaf. Some adventitious roots grow from its lower side. In the vegetative reproduction, one year old culm with its rhizome and root system is first dug up, the culm is then cut to about a meter high and the whole thing is planted during the season of rest, so that during the season of active vegetation which usually begins with the rains, these may be well in position and are capable of taking root early (Suri and Chauhan 1984).

Other vegetative methods of reproduction adopted in bamboo plantation are culm cutting, nodal cutting and layering. (See Fig. 78)

**Management of bamboo in agroforestry system**

* All bamboo culms older than two years should be harvested before April and after October.
* Harvesting should be carried out each year and intensive cultural operations should be done to facilitate the growth and establishment of bamboo and intercrops.
* If bamboos are planted along with tree species. Trees providing light shade to bamboos should not be removed as bamboos grow better under the shade of some tree like, neem, siris, amla, and such light crowned species (Chaturvedi, 1988).

**Insect pests of bamboo and their control**

There are large number of species of insects that attack bamboos belonging to the orders Coleoptera, Lepidoptera, Orthoptera and Hemiptera. Of the defoliator greater leaf roller *Pyrausta coclesalis* is the most important occurring in epidemics in post-monsoon season. Other defoliators are *P. bambucivora*, *Pionia flavofimbriata, Massepha absolutalis, Heiroglyphus banian* and *Poecilocerus pictus*. The defoliators can be controlled by spraying/ dusting insecticides. The bamboo weevils *Cyrtotrachelus dux* and *Cyrtotrachelus longimanus* are important shoot borers. The hispine

Figs. 77a&b : Bamboo seedlings and new bamboo culm.

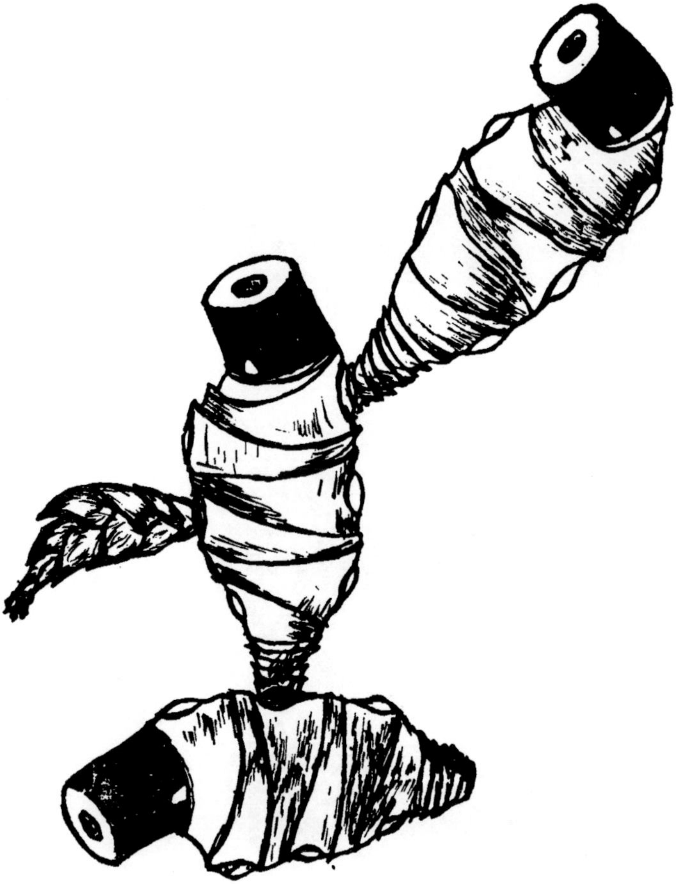

**Fig. 78 :** Rhizome of bamboo.

beetle *Estigmena chinensis* inflicts heavy damage to *Dendrocalamus strictus* culms. The borers of standing bamboos are controlled by silvicultural thinnings. The other shoot boring insects *Argyroploce paragramma, Chelyophoro ceratatina, C. striata* are not important. There are many species of scale insects belonging to the genera, *Chionaspis*, *Odonaspis* and *Asterolecanium*, which attack both wild and cultivated species but are not important. Of the sap suckers bamboo aphid *Oregma bambusae* and seed bug *Ochrophara* occurs in epidemic form. The sap-suckers can be controlled by chemical sprays. Ghoona beetles, *Dinoderus brevis*, *D. minutus* and *D. ocellaris* that attack dried bamboos are most destructive. Prophylactic and preservative treatments are recommended for their control. Detailed studies are carried out on the bioecology and population dynamics of bamboo pests, and their impact on bamboo growth. Integrated pest management practices laying emphasis on cultural, biological and genetic control should be evolved for natural and man-made bamboo forests. Safer, cheaper and effective methods are also required to be evolved for control or borers of felled bamboos (Singh and Bhandari, 1985). (See Fig. 79)

**Uses**

* Bamboo leaves are used as fodder. If bamboos are planted along with grasses, it produce maximum amount of fodder per hectare. Forage from bamboo is intended as a supplement to existing fodder supplies.
* Bamboo as a tree component of agroforestry would help in minimisation of soil erosion. Agricultural crops occupy the fields for part of the year and in part the field remain vacant. Introduction of bamboo in agroforestry system would reduce the soil loss and maintain the soil productivity on sustained basis.
* Seeds are eaten by the poor people.
* Tender shoots are used for pickles and curries.
* It is extensively used in place of timber, *e.g.* agricultural implements, bridges, fences, buildings, combs, fans, utensils, furniture, handicrafts, umbrellas, musical instruments, walking sticks, thatching, tobacco pipe, sports goods, water vessels, boxes, fuel, etc.

Fig. 79 : The borer of standing green bamboo.

* It provides raw material for paper industry. According to Tewari (1988) more than half of bamboos produced in India is used in paper making. The paper industry in India has steadily increased its capacity from 0.14 million tonnes in 1951 to 2.16 million tonnes at the beginning of 1983. The paper industry claims that today it consumes about 2 per cent of the country's annual consumption of wood and 51 per cent of bamboo. Natural forests are unable to meet their demands of bamboo, therefore, production of bamboo has to be increased outside forest areas. There are over 7500 units in the handmade sector of match industry, employing half million people directly and indirectly. Bamboo is one of the major sources for making match sticks.

It is apparent that the economy of a large part of both rural and urban areas depends upon forest based industries directly and indirectly. The native dwellers of forest areas, who entirely depend upon the forest produce are migrating to urban areas in search of employment.

It should be the foremost duty of the concern departments of government of India to tap the available village resources so that the villagers may be engaged in occupation in their own villages so that they do not have to migrate to urban areas in search of employment. *As bamboo is a fast growing species, and it attains maturity within a short period, planting of bamboo in agroforestry system in the farm land may be started on a massive scale and rural people may be engaged in making baskets, mats, cycle carriers etc., which are in great demand.*

## REFERENCES

Annon, (1952). *Lac and Lac products*, Indian Lac Research Institute, Namkum, Ranchi, pp. 1-8.

Annon, (1985). *Salient research findings*, Indian Lac Research Institute, Namkum, Ranchi, pp. 1-4.

Baumer, M. (1989). Statements related to apiculture, (Cited from article by Pawlick, T., The unexploited bee ), *Agroforestry Today*, Vol. I, No. 2, pp. 9-10.

Chaturvedi, A.N. (1988). Management of Bamboo forests Bamboo special-I, *Indian Forestor*, Vol. 114, No. 9, pp. 489-495.

Chaturvedi, A.N. (1991). Agroforestry in India with particular reference to economic factor 8, In: *Agroforestry in Asia and the Pacific*, Mellink, W., Rao, Y.S. and Mac Dicken, K.G., (eds.), Rapa Publications, pp. 278-287.

Divakar, P. and Yegneshwar, V. (1986). Tasar culture and tribal development, *Kurukshetra*, Vol. XXXIV, No. 4, New Delhi, pp. 19-20.

Editorial Board, F.R.I. (1986). Lac and Manufacture of Shellac, *Indian Forest Utilization*, Vol.-II, Govt. of India. Pub., Delhi.

Itnal, C.J. (1992). Agave for marginal and submarginal lands, In : *Agroforestry Principles and practices*, Nadagoudar, B.S. and Madiwalar, S.L. and Mutanal, S.M., eds., NWDB, Forestry Deptt., Univ. of Agril. Sciences, Dharwad, pp. 92-93.

Jha, L.K. (1987) *Apiculture; Applied agricultural entomology*, New Central Book Agencies, Calcutta, pp. 215-221.

Jha, L.K. and Jha, K.N. (1985). Tasar culture and its impact on generation of income and employment, *My forest*, Vol. 21(4), pp. 289-294.

Jha, L.K. (1989). Survey of existing agroforestry practices in south Bihar, *Annual Report*, I.C.A.R. Regional Centre on Agroforestry, B.A.U., Ranchi, pp. 5-6.

Jha, L.K. and Jha, K.N. (1985). Social Forestry Programme in ensuring better quality of life for the poor, *My forest*, Vol. (21), No. 4, pp. 317-325.

Jha, L.K. (1987). *Applied Agril. Entomology*, New Central Book Agency, Calcutta, pp. 1-275.

Jha, L.K. and G. Jha (1989). Survey of existing agroforestry practices in south Bihar, *Annual Report*, I.C.A.R. Regional Centre of Agroforestry, B.A.U., Ranchi, pp. 5-6.

Jha, L.K. and Sen, P.P. (1991). *Social Forestry*, Himalayan Publication, Bombay pp. 195-200.

Jha, L.K. and Sen, P.P. (1991). *Social Forestry*, Himalayan Publishing House, Bombay, pp. 63-75.

Kanan, (1975). Sisal and other leaf fibres in India, J.A.C.L., *Journal of Forestry*, Vol. III(2), (C.R).

Khanna, L.S. (1984). *Principles and Practice of silviculture*, Khanna Bandhu, Dehra Dun.

Maithani, G.P. (1988). Editorial, Bamboo special-I *Indian Forestor*, Vol. 114, No. 9 pp. 487-488.

Pawlick, T. (1989). The unexploited bee, Agroforestry Today, Vol. I, No. 2, pp. 8-10.

Prasad, N.M. (1984). Tasar culture and tribal economy in Bihar, 38-44, *Training Programme on Social and Agroforestry*, B.A.U., Ranchi (India).

Prasad, V.N. and L.K. Jha (1985). Creating employment through Social Forestry, *Yojna*, 29(19),11.

Rajan, B.K.C. (1986). Apiculture and farm forestry in semi-arid tracts of Karnataka, My forest, Vol. 22 (1), pp. 41-49.

Sengupta, K. (1986). Tasar Silk Industry in India, C.T.R and T.I; Central Silk Board, Govt. of India, Ranchi, pp. 1-40.

Sen Sarma, P.K. and Jha, L.K. (1989). Lac cultivation and its impact on generation of income and employment, *Souvenior, BISCOLAMP*, Ranchi, pp. 32-34.

Sen, P.P. and Jha, L.K. (1988). Technique of agave plantation and its importance as a boon to tribal economy and as a soil conservation measure in degraded forests and wastelands, *My forest*, Vol. 24, No. 4, pp. 295-300.

Sen Sarma, P.K. and Jha, L.K. (1989). Lac cultivation and its impact on generation of income and employment, *Souvenior, BISCOLAMP*, Ranchi, pp. 32-34.

Sen, P.P. and Jha, L.K. (1988). Technique of agave plantation and its importance as a boon to tribal economy and as a soil conservation measure in degraded forests and wastelands, *My forest*, Vol. 24, No. 4, pp. 295-300.

Sen, P.P. and Jha, L.K. (1989). Cultivation of Agave, *Vanbithi*, West Bengal, pp. 31-32.

Shravan, K. (1990). *An improved baking type insulating varnish*, Indian Lac Research Institute Namkum, Ranchi, pp. 1-3.

Singh, P. and Bhandari, R.S. (1988). Insect Pests of bamboo and their control, *Indian Forestor*, Vol. 114, No. 10, pp. 680.

Subarna, M.C. (1990). Importance of Socio-economic and Cultural Attitude in Planning for Agriculture Development of the tribal area, B.A.U. Ranchi.

Suri, S.K. and Chauhan, K.S. (1984). *Indian Timbers Bamboo*, F.R.I. Publication, Dehra Dun.

Tiwari, D.N. (1988). Bamboo as poverty alleviator, *Indian Forestor*, Vol. 114, No. 10, pp. 610-612.

14

# Pest Management in Agroforestry System

Pest plays a very important role in the productivity of crops. The evolution of phytophagous insects has been dependent upon the types of plant communities that provide food and shelter to them. Insect pests, cause extensive damage to trees and agricultural crops. The usefulness of study of insect communities with reference to sylvan and agro-ecosystems is ultimately measured in terms of its contribution towards solving actual pest problems.

The pest population depends upon the nature of sylvan and agro-ecosystem. Agroforestry system is not helpful in multiplication of pest population due to insufficient food availability on account of polyculture. However, tree pest may change its food habit and becomes the pest of crops and vice-versa rather infrequently.

Management of pest population in this system is different from the suppression of pest population in pure agriculture crops or in pure tree stands. Thus, there is need to study in depth about the nature of damage biology, population dynamics etc., under agroforestry system and development of appropriate technology to combat 'population influx' of pests.

The study of pest management in agroforestry system would be interesting and quite important in order to understand the ecology of insect communities to forest and agro-ecosystem. Only few reports are available with regard to pest management in agroforestry system. This chapter deals with the pest problem in nursery, standing trees and crops or *vice-versa*.

**Insect pest host interaction in agroforestry system**

The multiplication of pest depends upon the type of plant communities that provides food and shelter. A system may be congenial for the pest multiplication due to regular supply of food and shelter throughout the year. A tree pest may develop taste for agricultural crops or pest of agricultural crops may develop a fancy for the tree species during shortage of food. This system may lead to change in the food habit of pests from mono or oligophagous to polyphagous and may lead to great loss to both tree species as well as crops or fodder. There are many examples when pests of agricultural crops develop a fancy for tree species and vice-versa.

*Helopeltis antonii*
*Helopeltis theivora*
(Tea Mosquito Bug)
Order—Hemiptera
Family—Capsidae.

The two stages (nymphs and adults) suck the sap from the tender shoots and leaves of tea and neem (*Azadirachta indica*) mahogony (*Swietenia macrophylla*) and cashew (*Anacardium occidentale*). The females lay eggs on tender shoots. Fecundity of a female is 12 to about 500 egg. The nymphs continue to suck the sap until they become adult. Nymphal period is 30 days. There may be several generations in a year. Adults hibernate during winter. If tree species neem, mahogony and cashew are planted near the tea garden, there may be problem in keeping the pest population at endemic level.

*Stephanitis typicus*
(Banana lace-wing bug)
Order—Hemiptera.
Family—Tingidae.

In Assam and West Bengal, farmers grow banana, turmeric along with coconut trees. The Banana lace-wing bug cause whitish or yellowish spots on the leaves or leaflets of above crops and coconut palm. It is suspected to be vector of coconut virus. Under this cropping system, pest gets sufficient food and shelter throughout the year. Due to their fancy for tree species as well as agricultural crops.

*Chrysocoris stollii*
(Litchi Bug)
Order—Hemiptera
Family—Scutelleridae.

In North Bihar this pest causes damage to litchi and pulses. It also feeds on *Croton sparsiflorus*. In the litchi orchard, pulses and other crops/grasses are grown in the interspace. This pest attacks pulses, litchi and croton. Thus it may not be checked naturally as sufficient food supply is available throughout the year.

The adult is small and greenish in colour, with 6 spots on the thorax and 7 spots on the scutellum. A female lays up to 30 eggs. Its population can be reduced by uprooting Croton sparsiflorus an alternate host plant.

*Tricentrus bicolor*
(The tree hopper)
Order—Hemiptera
Family—Membracidae.

This is a common pest of berseem (*Trifolium alexandrinum*), Arhar (*cajanus cajan*), wheat, pea, acacia and cassia in Rajasthan and other arid parts of the country. Both tree species acacia and cassia are planted in the agroforestry system. Thus, farmers should be careful in selecting the tree species and crops to prevent the attack by the tree hopper or suitable control measure should be adopted. Both the adult and the nymphs are devastating by nature. They extract sap by sucking foliage. The females generally lay eggs on the lower surface of the leaves.

*Oxyrhachis tarandus*
Order—Hemiptera
Family—Membricidae.

This is common pest on many crops; maize, rice, wheat, berseem, pea, mung, arhar, cabbage, chillies and tree species; *Grevillea robusta, Albizia lebbeck* etc.

The eggs are laid on the bark in two rows which are glued. It undergoes five moults within 33 to 35 days. This can be controlled under agroforestry system by using lindane, phosphamidon or malathion.

*Nephotettix virescens*
(Leaf hopper)
Order—Hemiptera
Family—Cicadellidae.

Roy Chaudhuri (1983) reported that Sandal (*Santalum album* L.) ; known for its fragrant oil and heartwood, is affected by a devastating desease, causing reduction in size of leaves and shortening of internodes resulting in a 'spikey' appearance of branches. Shivaramakrishnan and Sen Sarma (1975) was the first to test *Nephotettix virescens* for transmitting sandal spike and 12 out of 15 test plants got infected by using 60 infective leaf-hopper per test plant. They transmitted the disease from these diseased test plants to healthy sandle wood plants which was confirmed by bud grafting (Shivaramakrishnan and Sen Sarma, 1977 and 1978). Roy Chaudhuri reported that rice yellow draft is caused by MLO which is also transmitted by *Nephotettix virescens*. Rice is grown in close proximity of sandal tree plantations in Karnataka, Tamil Nadu and Kerala. It will be worthwhile testing (i) Is *N. virescens* the natural vector of sandal spike? (ii) Whether the disease can be transmitted by the vector from the spiked sandal trees to healthy rice developing yellow dwarf disease and *vice-versa* ? Further research in this direction is essential to understand the role of insect vector in transmitting disease in sandal tree and rice. If it is established, then in order to protect costly sandle wood stand and to maintain the productivity of rice, we have to find out suitable insect suppression technique.

*Eurybrachys tomentosa*
(Common hopper)
Order—Hemiptera
Family—Ricaniidae.

This is a pest on tree species, *Santalum album*, *Albizia lebbeck*, *Cassia fistula*, *Terminalia tomentosa* and crops like cotton, arhar etc.

The male is smaller than female. The nymphs suck sap from the foliage, new shoot and flowers of trees and agricultural crops. *E. tomentosa* may become a key pest in agroforestry practice when above tree species are grown along with agricultural crops like arhar and cash crop cotton. The female lay eggs in cluster on the leaves. The eggs are covered with a thick mat of white flocculent efflorescence. Incubation period lasts for about 20 days. Life cycle is completed from 109 to 132 days.

*Pyrilla perpusilla*
(Sugarcane leaf hopper)
Order—Hemiptera
Family—Lophopidae.

This is common pest of sugarcane, wheat, maize, napier grass and bamboo is sugarcane growing state, like, Bihar and U.P. The bamboos are an important woody perennial component of agroforestry system in Bihar. The damage intensity of the pest is the

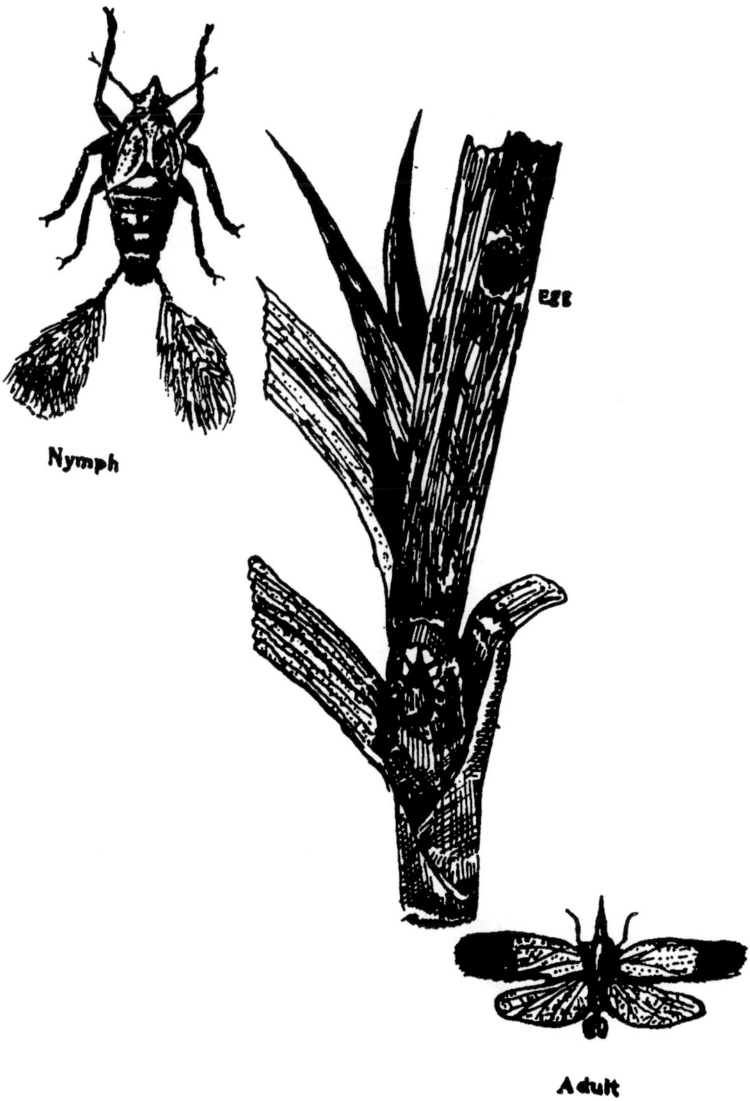


Fig. 80: *Pyrilla perpusilla*

highest from April to October. Nymphs and adults suck the sap from leaves. The leaves attacked by nymphs, have, in the middle, black fungus which develops due to the honey dew. It is detrimental to the growth of plants.

The straw coloured adult insect has two pairs of wings which are folded on the abdomen in shape of a roof. They have long pointed snout. The female has a pair of pods on the abdominal end of the body. The female lays egg on the underside of the leaves. Number of eggs varies from 10 to 65. The eggs take a week to hatch into young nymphs. Fecundity of a female is high (@ 720 eggs). Soon after hatching, the young nymphs become brown. There are five instars. Nymphs suck the sap till it becomes adult. It takes nearly 65 days to become an adult. There are three to four generations in a year. During the scarcity of the principal food plants, they start feeding on other host plants like grass species or bamboo. In North Bihar and U.P., farmers usually grow bamboo on field bunds or in one corner of the field. They also cultivate napier grass in the litchi and mango orchard or even on the farm bund. Thus, the pest does not experience shortage of food. This pest may become serious when sugarcane crops and bamboo or grasses are grown together. During summer season, pyrilla may be held in check by dusting BHC, 5 per cent, at the rate of 25 kg per hectare. In case of developed crops, malathion 50 per cent E C, 1250 ml or endosulphan 35 per cent EC, 800ml per hectare may sprayed over the standing crops/grass and bamboo culms.

*Aphis gossypii*
(The cotton aphid)
Order—Hemiptera
Family—Aphididae.

It is a polyphagous pest, commonly called cotton or melon aphid. The pest is active throughout the year, maximum activity being during cloudy but not rainy days when humidity is high and rainfall low. The winged and wingless forms reproduce parthenogenetically and Viviparously. A single female produces 12 or more nymphs per day and newly born nymphs take 7 to 9 days to mature. There is sexual reproduction also specially during cold weather.

The adults and nymphs damage foliage by sucking the sap. Under agroforestry system, when agricultural crops on cash crops cotton etc., are grown along with tree species like *Tectona grandis*,

*Morus alba* etc. *Aphis gossypii* nymphs feed on the foliage of above tree species also. For management of the pest in the agroforestry system, the tree species as well as crops should be treated with thiometon 0.08 per cent, fenitrothion 0.05 per cent, phosphamidon 0.1 per cent, parathion 0.025 per cent etc.

*Aphis craccivora*
(Bean aphid)
Order—Hemiptera
Family—Aphididae.

This is commonest pest infesting a variety of plants. Jha and Singh (1990) reported that under alley cropping when *Leucaena leucocephala*, red gram (arhar) were grown *A. cracivora* caused damage to pulse and leucaena foliage. This is considered as serious pest of mung, glyricidia, sesbania etc.

Clusters of nymphs and adults may be seen in the undersurface of tender leaves, pods and flower stalks, sucking the sap and giving out honeydew. The infested leaves turn yellowish pale. In case of severe infestation the pods are malformed. The black ant, *Camponotus compressus* is associated with this aphid.

Reproduction is usually viviparous and parthenogenetic. A single female produces 10 or more nymphs per day. The nymphs undergo four moults before reaching the adult stage. In sexual reproduction the life cycle from egg to adult occupies 8-9 days.

*Pulvinaria psidii*
(scale insects)
Order—Hemiptera
Family—Coccidae.

This is commonest pest in South India and infest tea, coffee, *Ficus retusa*, guava and many other trees.

*Ceroplastodes cajani*
(Scale insect)
Order—Hemiptera
*Family—Coccidae.*

This is a pest on *Zizyphus* species, *Cajanus cajan* and on other pulses. When pulses are grown along with Zizyphus species, its rate of multiplication becomes very high and require proper management to suppress its population.

*Coccus viridis*
(Coffee green scale)
Order—Hemiptera
Family—Coccidae.

Recently it has been reported that *C. viridis* is a serious pest on coffee as well as Ficus species, Citrus species, *Alstonia scholaris* and *Aegle marmelos*.

The bugs infest the undersurface of leaves and midrib. In case of severe attack, tender shoots droop. Therefore, it is considered as an important pest, when coffee is grown in close proximity of above tree species. It is essential to do further research in this direction for proper understanding of insect-host interactions in coffee and other host tree species.

*Dichocrocis punctiferalis*
(Castor shoot and capsule borer)
Order—Lepidoptera
Family—Pyralidae.

The damage is caused by caterpillars, which attack the crops like castor, ginger, turmeric, sunflower at the flowering stage and tree species, *Tectona grandis*, *Psidium guajava*. Singh and Singh (1978) reported this insect causing epidemic defoliation of khasi pines. It is some time serious pest of teak as well. Beeson (1941) reported that *Dichocrocis punctiferalis* is a widespread general feeder that is frequently a pod borer and stem borer attacking pods and seeds and also bores the shoot and stems. Caster seed borer is not a regular pest of teak plantation. Under agroforestry system when teak is planted along with agriculture crops like caster, ginger, turmeric, sunflower etc., this system may be most congenial for the multiplication of caster shoot borer due to abundance of food and shelter throughout the year.

The moth is medium sized with white fore wings of 4 cm expanse. The female lays eggs in groups of 2 to 3 mostly on flower buds and tender capsule. A female lays about 76 eggs. The eggs hatch in about a week. In two or three weeks the caterpillar become full grown. Pupation occurs inside the webbed cocoon of the capsule. The pupal period lasts 7 to 25 days. The life cycle is completed in 4 to 5 weeks. There are 3-4 generations in a year.

To suppress the pest population, infested shoots and capsule containing larvae inside must be destroyed. This pest can also be suppressed under agroforestry system by spraying 0.05 per cent DDT or 0.05 per cent BHC at monthly intervals.

*Trabala vishnou*
(Eggar moth)
Order—Lepidoptera
Family—Lasiocampidae.

It is distributed throughout India and is a major pest of caster. Under agroforestry system, (taungya), when *Shorea robusta, Terminalia tomentosa, Syzygium jambolanum* is planted along with caster, damage to tree species is most frequent.

The male moth is pale green in colour with faint dark markings and reddish brown patch along with a dark spot on the fore wing. Eggs are laid in straight double rows on the foliage. The larva takes shelter during the day and feeds mainly during the night. The larva is pinkish-brown with tufts of dark hairs arising from blue spots along the sides and black and an intervening row of tufts of pale pink hairs and projecting tufts of longer black hairs on each side of the head (Sevastopulo, 1939). Pupation occurs in the cocoon, interwoven with larval hairs. Beeson (1934) stated that there were 4 or more generations in a year. In the North, larvae and pupae may appear at any time in January-April. The pupal period in November to January is about 50 days falling to about 35 days in February-March and to 15 days in April. There may thus be 1 or 2 complete life-cycles in the hot weather. The egg-stage last 11 days in April. In July the pupal period is 17 days. The rains generation lasts 7 to 8 weeks (August to October) with the egg 10 days larva 25 days and pupa 16 days. The pupal period of later broods in October and November is about 22 days. The coldest season may be passed as caterpillar or pupa. There are 6 larval stages.

When caster is grown along with tree species, then proper management to suppress the pest population is essential.

*Metanastria hyrtaca*
Order—Lepidoptera
Family—Lasiocampidae.

The main host plants are *Acacia nilotica, Albizia* species, *Eucalyptus* species and *Terminalia* species. When agricultural crops are grown along with above tree species, the pest shifts its interest on a number of agricultural crops. This insect is considered as an occasional pest of crops. But this may pose threat to crops as well as tree species under agroforestry system. Thus, judicious management to control the pest is essential.

The forewing of male has a chocolate brown spot with a white centre. In the female the forewings are marked with transverse bands of light and dark brown. Eggs are laid in group on the underside of the young leaves. The larva is covered with a close grey pubescence and a fringe of longer hairs at the sides and tufts on the mesonotum projecting forward on either side of the head; between the thorax and abdomen is a velvety black band bording a patch of dark reddish hairs (Subramaniam and Anantanarayanan, 1938). Two or more generations occur in a year. The incubation, larval and pupal periods last 9-12, 45-100 and 9-18 days respectively.

*Stauropus alternus*
(Crab caterpillar)
Order—Lepidoptera
Family—Notodontidae.

In South India, the *Stauropus alternus* is a pest on pigeon pea, *Acacia*, *Tamarindus indica*, *Cassia fistula*, *Grevillea robusta*, etc., Pigeon pea, arhar crops are grown along with above tree species. There may be much fluctuations in population of *S. alternus* due to sufficient food availability throughout the year. They may feed on the arhar, and in the absence of arhar crops they will continue their life cycle on the alternate host tree species and *vice-versa*.

The damage is caused by caterpillars which feed on leaves. The adult moth is pale greyish brown, suffusued with darker shades and pale spots on both wings. The eggs are flattened and are deposited in clusters on the leaves. The egg hatch in 4 to 20 days and when full-grown, the caterpillar is brownish black to greyish black in colour. The length is 40 to 45 mm. Larval period varies 21 to 42 days. Pupation takes place in loosely woven cocoons of yellow fibrous silk. The pupae vary from 18 to 20 mm in length and are dark brown. Pupal period varies from 10 to 23 days.

Spraying with 0.2 per cent carbaryl has been proved effective (Jha, 1987).

*Pericallia ricini*
(woolly Bear)
Order—Lepidoptera
Family—Arctiidae.

This species is a sporadic pest found all over Indian sub-continent. Its main host plant is banana; besides, it has also been observed feeding on brinjal, caster, cucurbits, sweet potato, moringa, sesbania and various ornamental plants.

Farmers generally plant moringa and sesbania on the bund of agricultural field. Under this condition insect may attain status of pest due to congenial atmosphere of multiplication in the presence of seasonal and perennial food crops. Caterpillars are nocturnal in habit and feed voraciously at night (Jha, 1987). In case of severe infestation, the entire plants are defoliated affecting adversely the yield under agroforestry system. Adults are of medium size, having grey head and thorax and crimson abdomen with black strips on it. Forewings are ornamental with pale black pattern while the hindwings are crimson pink with dark spot. A female lays 20 to 150 eggs. Egg, larval and pupal periods are 4 to 6, 26 to 32 and 10 to 12 days respectively while entire life cycle is completed in about 4 weeks during April-May. Eggs are laid in clusters on the lower surface of leaves. Caterpillars are blackish brown in colour, enclosed in a very thin white cocoon. The moth is grey with dark spots on the pink hindwings.

To suppress, hand-picking of the egg masses and caterpillars and their prompt destruction is suggested in the initial stage of attack. In case of severe infestation, dusting with 5 per cent carbaryl or 10 per cent BHC is recommended. In nature, these larvae are parasitised by *Apanteles ricini*, *Meteorus* species, *Thelarra nigripa*, etc., but none of these parasites has been able to check the pest population.

*Achaea janata*
(The caster semilooper)
Order—Lepidoptera
Family—Noctuidae.

This is one of the important pest of castor and many tree species (*Acacia nilotica*, *Shorea robusta*, *Zizyphus jujuba*, *Albizzia* species etc.) suitable for agroforestry system.

The caterpillar feeds voraciously on leaves, tender petioles, young capsules etc. Pest is sporadic and is abundant in the rainy season. The adult is greyish brown, with a wing expanse of 6-7 cm. The forewings have wavy markings. There are prominent black and white patches on the hindwings. These moths appear in May and June and lay eggs on early sown caster crops or any alternate host plants. A female lays 318 to 648 eggs on leaves either singly or in cluster. It is blue green in colour, 1 mm in diameter and bears ridges and furrows. Incubation period is 3 to 4 days. On hatching the young larvae feed on the same leaf on which eggs were laid and later move to other leaves. The larva is a semi-looper, slender, measuring about 5.5 cm when full grown. The larval period lasts 11-17 days. The larva undergoes 5 moults. When about to pupate, the larva descends to the ground and constructs a loose cocoon of coarse silk with particles of earth webbed together. The length of pupal stage varies from 11 to 27 days. The life cycle is completed in 28 to 47 days. There are 5-6 generations in a year (Jha, 1987).

The pest can be suppressed by spraying with 0.05 per cent parathion or 0.1 per cent carbaryl. Dusting the crop with 5 per cent DDT or BHC has also proved effective.

*Euproctis fraterna*
(Tussock moth)
Order—Lepidoptera
Family—Lymantriidae.

The larvae feeds on castor, cotton, red-gram and many tree species like, *Tectona*, *grandis*, *Terminalia tomentosa*, *Zizyphus jujuba*, *Eucalyptus* species etc.

It is recognised as a sporadic pest of tree species and agricultural crops. The newly hatched larva feeds by scraping the epidermal leaf tissues. At a later stage, skeletonization of leaf in patches is observed.

The moth is yellow with pale transverse lines on the forewings. The eggs are minute, flat and yellow and are laid in masses on shoot, small leaf, flower, bud and plant tops. The early instar larvae have whitish hairs and feed gregariously. The full grown larva has red head, reddish brown body with white hairs surrounding the head with a single long tufts in the head and another in the anal region. It pupates in a cocoon of hairs on the leaves or on branches. The life cycle lasts 45 to 57 days, the egg, larval and pupal periods respectively being 5 to 9, 29 to 35 and 10 to 12 days (Nayar, *et al.*, 1984).

*Euproctis scintillans*
Order—Lepidoptera.
Family—Lymantriidae

It is a polyphagous species and feeds on many agricultural crops and defoliates tree species, like, *Aesculus indica*, *Acer* species, *Pyrus* species, *Salix* species in Kashmir and the North-West Himalayas.

This may be a serious pest under agroforestry system in North-West Himalayas. The moth is snow white with tuft of anal brush.

*Porthesia scientillans*
Order—Lepidoptera
Family—Lymantriidae.

*Porthesia scientillans* is a pest on crops like, castor, cotton, pigeon pea, linseed, sunhemp, pumpkin and tree species like *Acacia arabica*, *Cassia* species.

Acacia and cassia are considered suitable for agroforestry system. When these tree species are grown along with above crops, P. scientillans may attain pest status. The moth is yellow with a reddish line on the edge of the wings. The life cycle completed within 60 days.

*Lymantria ampla*
Order—Lepidoptera
Family—Lymantriidae.

The larvae feed on agricultural crops as well as few important tree species suitable in agroforestry system like, *Tectona grandis*, *Terminalia* species etc. The eggs are laid on the food plants in masses covered with hairs. The full grown larva is coffee brown, a black velvety band between pro and mesothorax, each segment bearing a transverse row of papules with tufts of long hairs, the tufts on each side of the prothorax and a group of tufts on the last abdominal segment being more conspicuous. The pupa is chocolate brown 1.5 cm long, with tufts of short hairs on the abdomen and thorax. No cocoon is formed. There are probably 4 broods in a year with moths in January, April, June and October. The life cycle is from 7 to 10 weeks (Beeson, 1934).

*Holotricha insularis*
(The white-grub)
Order—Coleoptera
Family—Melolonthidae.

The grub feeds on the roots and other parts of the seedlings and beetles feed on the leaves and tender parts. The tree species attacked by the adults are *Acacia arabica*, *Grewia* species, *Azadirachta indica*, *Psidium guajava*, etc. Grubs are destructive to crops like maize, sugarcane, sorghum, cucurbit, cowpea, brinjal etc.

The beetles are thick bodied, and moderate sized, with copper coloured flat head. The prothorax is large. The eggs are laid two to four in number up to a depth of 10 cm in moist or sandy loam soil. The female may lay as many as 30 eggs. Eggs hatch in 8 to 12 days. The full grown grub is dirty white, fleshy and curved. The full grown grubs migrate in the soil for pupation. The pupal period lasts 15 to 22 days. The adults as well as pupae hibernate from November to June. There is only one generation in a year.

This pest may be controlled by mixing thoroughly with the soil 5 per cent heptachlor dust at the time of sowing at the rate of 30 to 35 kg per hectare. Collection and destruction of beetles with the help of light traps will also keep in check their number.

*Holotrichi serrata*
(Palas white grub)
Order—Coleoptera
Family—Melolonthidae.

It is one of the destructive pests to palas (*Butea monosperma*), kusum (*Scheleichera olcosa*), teak (*Tectona grandis*), sugarcane and tobacco. This pest is very common in north as well as South Bihar.

In South Bihar, when teak or palas are planted and agricultural crops are grown in the interspace, it defoliate the tree species as well as agricultural crops and grub feeds on the roots of both the plants. The beetle feed on leaves in the morning and evening hour. Beetles are active during July to September. The female lays eggs in the soil at the depth of 10 to 12 cm. Single eggs are laid in each earthen cell. They hatch within two weeks. They become full grown by September. The pupal period lasts 10 to 15 days. There is only one generation in a year. The life cycle is completed in 150 to 180 days (Jha & Singh , 1990).

*Oryctes rhinoceros*
(Rhinoceros beetle)
Order—Coleoptera
Family—Dynastidae.

The beetle is serious pest of palms of various kinds, casuarina, sugarcane and pineapple. In North Eastern states, farmers cultivate pineapple in a large scale in the interspace of palm trees (betel nut and coconut).

The pest is abundant during March to June and bores in to the top of the tree (soft parenchyma) and burrows in downward direction, killing the bud, ultimately the coconut palm dies. Presence of fibrous chewed material near the holes indicate its attack. The beetle is black or reddish black, with a prominent curved horn which projecting backwards on the head. Females lays up to 100 eggs at a depth of 5 to 15 cm in the soil or other refuse present in the farm. The incubation period varies from 8 to 18 days. The grub feeds on the decaying trees and also consume the roots of intercrops like pineapple. The grubs move from one plant to another and one grubs may destroy number of pineapple plants by devouring roots and seedlings. The grub period lasts for 90 to 200 days. The pupal stage ends within 25 days. The life cycle is completed within 200 to 240 days.

Cleaning and destruction of all breeding place are important remedy to control the pest population. In order to save the field from grub, spraying of BHC, carbryl and aldrin at 0.1 per cent concentration is effective.

*Ceuthorrhynchus asperulus*
(Red gram bud weevil)
Order—Coleoptera
Family—Curculionidae.

The weevil feeds on the pigeon pea in South India ana also attacks the buds of *Gliricidia maculata*.

If pigeon pea is grown in the interspace of *Glyricidia maculata*, farmers must be careful about this pest. Its population may increase beyond tolerable level due to supply of food and shelter throughout the year. Although, this pest is considered as of minor importance.

*Myllocerus maculosus*
(Cotton ash weevil)
Order—Coleoptera
Family—Curculionidae.

*Myllocerus maculosus* infest cotton, lady's finger, maize, ragi, sugarcane, *Dalbergia sissoo* , etc.(Nayar *et al.*, 1984), Stebbing (1914) reported *Myllocerus* species as a pest of *Tectona grandis* in taungya cultivation. *Dalbergia sissoo* and *Tectona grandis* are considered good tree species for agroforestry system.

Both adults and grubs cause serious damage by feeding on leaves and on roots of host plants respectively. The adult is black or chest-nut brown colour, with brown and green scaling, the sides and under parts pale metallic green; the prothorax dark brown above, with two narrow green stripes, the elytra indistinctly striped, the intervals 1, 3, and 5 being brown, the suture and alternate intervals rather paler and more or less irrorated with green scales. Rostrum about as long as broad, longer than the head, sometime dilated at the apex, the apical margin deeply and acutely incised, the raised dorsal area depressed in the middle without any central carina, the lateral carinae elevated far above the upper margin of the eyes and forming an angular prominense above the insertion of the antennae (Stebbing, 1914). Female lays up to 361 eggs at a depth of 8 cm. The incubation period varies from 3 to 11 days. The grub is light yellow coloured, stout and fleshy. The life cycle is completed within 42 to 54 days.

The pest population can be managed by using dieldrin 1.5 per cent dust in to the soil before sowing @ 2 kg active ingredient / ha is most effective.

**Insect Pest of Tree Species Suitable for Agroforestry System**

(A) SAL ***(Shorea robusta)***

*Hoplocerambx spinicornis* New.
(Sal heart wood borer)
Order—Coleoptera
Family—Cerambycidae

It is one of the most serious and widely distributed pest of *Shorea robusta*. It has been reported from Madhya Pradesh, Chotanagpur, Western Bengal, Assam and Himachal Pradesh. The grubs that hatch out of the egg feed under the bark, and then in the sapwood, and finally bore into the heartwood.

The adult is about 6.5 cm long. The elytra varies in colour from black to reddish brown. The beetles emerge during the first few weeks of monsoon. Eggs are laid on the bark of sal tree. A single female can lay as much as 500 eggs. Incubation period is 3 to 7 days. The grub feeds on the inner bark and sapwood of the tree and constructs irregular galleries. The grubs attain maturity from March to November and spend the cold weather hibernating in the pupal chamber constructed in the heart of the tree. The grubs transform into beetles in May-June and wait until the monsoon arrives to emerge. Trap-tree operation carried out regularly to prevent/ minimise the borer attack.

*Drosicha stebbingii*
Order—Hemiptera
Family—Margarodidae.

The mealy bug occurs in North India where it is periodically a pest of *Shorea robusta*.

According to Stebbing the damage caused by sal mealy bug is manifolded *e.g.* (a) the clogging-up of the stomata and pores of the leaves and twigs by the sugary secretion, which envelops them in an impermeable varnish. In wet seasons, the rain washes the film off, but in a dry season it remains for several months. It also interferes with photosynthesis, (b) Twigs and branches dry up and die, (c) The leading shoot young sapling are killed and the laterals replacing it may also be killed so that growth is deformed. (See Fig. 81)

This pest can, to some extent, be kept under control by some natural insect enemies as well. These enemies are *Coccinella* species, *Rodolia fumida* etc. Apart from these, there are predatory birds also, which feeds on this harmful insect pest.

*Paectes subapicalis* walk.
Order—Lepidoptera
Family—Noctuidae.

The young caterpillars feed upon the tender epidermis, thus partly skeletonise or perforate young leaves. As they develop in size they devour the entire leaf structure with the exception of the midribs. When they are sufficiently plentiful the host may be defoliated severely. Mature leaves are eaten sparingly (Mathur, 1961).

The adults are dark fuscous greyish brown. The eggs are circular, dome shaped in outline, light greenish white, and about 1 mm in diameter. The eggs hatch in about 2 to 3 days in summer, 4 to 7 days in October and 7 to 10 days in November. Full grown larva is pale green, having two pairs of sub-dorsal white lines and two faint white lines on each side. The duration of the larval period ranged from 8 to 13 days in June, 10 to 14 days in July, 12 to 16 days in August, 13 to 22 days in September and 20 to 30 days in October. The full grown larva pupate in the soil at the depth of four inches. The pupal period lasts 7 to 28 days, depending up on season. This pest is suppressed by birds and bacterial disease in the nature.

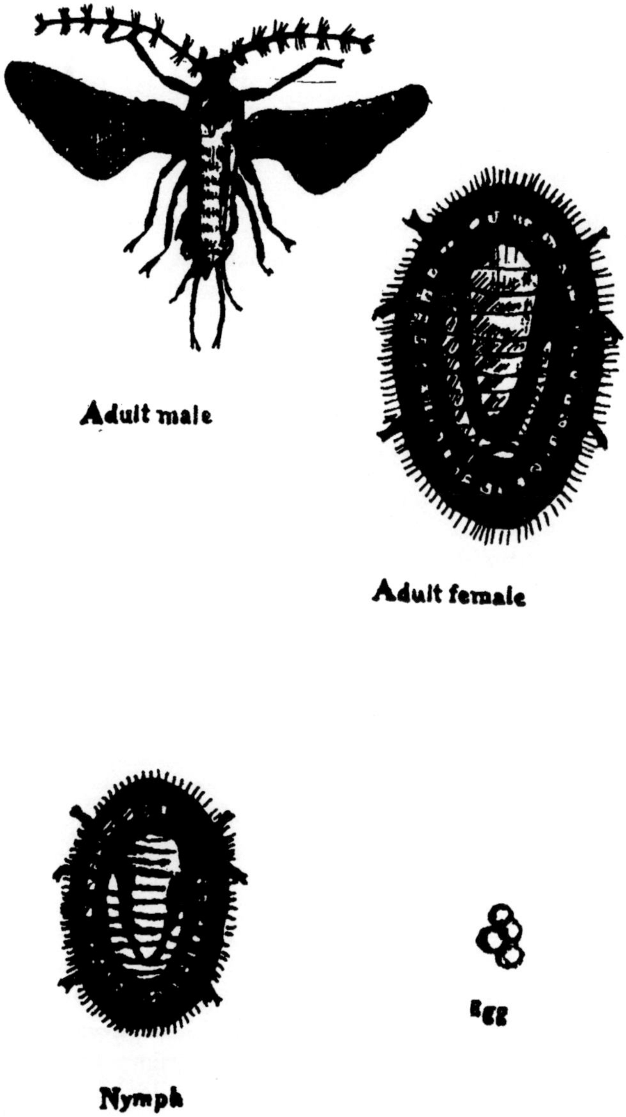


Fig. 81 : Drosicha stebbingi.

(B) TEAK *(Tectona grandis)*

*Paectes subapicalis* Walk.
(See pests of *Shorea robusta*)
*Hyblaén puera* (Teak defoliator)
Order—Lepidoptera
Family—Hyblaeidae.

The caterpillars feed on the leaves and do extensive defoliation. The newly hatched larva must have young soft leaf tissue for its food. The leaves are entirely stripped off, nothing being left except the stronger ribs in severe epidemic.

The adult is a small moth having head and thorax greyish to reddish-brown; the abdomen dark brown with orange segmental bands. The female lay eggs on tender parts of plants. Single female can lay as much as 1,000 eggs during her life time. The period of incubation varies throughout the year from 2-4 days. Moulting takes place in leaf-folds. Larval period varies from 8-26 days. Pupation takes place in a triangular leaf fold cut specially and strongly spun together. The pupal period varies from 4 to 25 days.

The following plants are not desirable in and near, teak plantation because they are the food plants of the teak defoliators.

(*a*) *Anogeissus latifolia*, *Cassia fistula*, *Shorea robusta*, *Garuga pinnata*, etc.

(*b*) Aerial application of sevin, carbaryl have proved successful, but it expensive and is not commensurate with the expected value of the timber after harvesting.

*Amsacta lactinea*
Order—Lepidoptera
Family—Arctiidae.

In initial stage the caterpillar remains gregarious and skeletonise leaves and cause defoliation. After feeding for about a week the larvae abandon *Tectona grandis* as a food and remaining instars feed on other plants.

The female lays about 400 to 500 eggs on the leaves of *Tectona grandis*. The dark larvae feed gregariously on the leaves. In about a week they slowly spread from plant to plant feeding voraciously. The mature larvae are very hairy and black with tufts of reddish-brown hairs. Mature larva measures about 1.5 inches in length.

**Fig. 82 :** Teak defoliator (*Hyblaca puera*).

Pupation takes place in the soil in a cocoon of cast-off larval hairs. Pupal period takes about 20 days. Aerial application of carbaryl, sevin have proved successful but the cost is prohibitive.

*Dihammus cervinus*
(Canker-Grub of Teak)
Order—Coleoptera.
Family—Cerambycidae

Attack is indicated by globular cankers that are occasionally formed as high as three feet above ground. In teak, canker is formed due to partial girdling of and injury to the cambium. The stem gets too weakened due to extensive tunnelling by the larvae that it breaks off (See Fig. 83).

Adult beetles are about 15-22 mm long, they are light or greyish brown in hue. Eggs are inserted in the cambium. A single female can lay up to 60 eggs during an oviposition period @ 2 to 7 eggs per day. Incubation period lasts 5 days in May, and 5-7 days in August. The young larvae bore just under the bark of the tree downwards from the point of insertion of the eggs. In 2 or 3 weeks the larva enters the sap wood. The mature grubs measure up to 2.5 cm. The grubs pupate inside pupal chamber, which may be up to 5 cm long. It is plugged up with coarse fibres. The pupal period is 14 to 19 days. *D.cervinus* primarily develops in the roots of *Clerodendron infortunatum*. Hence its removal in the plantations is recommended to prevent spread of *D. cervinus*.

Infested teak should be cut below the level of infestation to obtain new sound stem.

*Cossus cadambae*
Order—Lepidoptera
Family—Cossidae.

The larvae tunnel through the sapwood, they make small tunnels in the wood similar to the bee holes of the *Xyleutes ceramica*.

The female beetle lays the eggs in the wounds caused by the dying back of branches. Full grown larva is about 3.7 cm long and bright red in colour. It bores into the sound wood a tunnel of irregular shape and a few inches long. Larval period varies from ten to eleven months. The larvae pupate in the pupal chamber. Emergence takes place in the spring. Partial reduction of its potential damage can be effected by silvicultural measures only.

**Fig. 83 :** Teak canker grub beetle (*Dihamus cervinus*).

(C) **SISOO** ***(Dalbergia sissoo)***

*Plecoptera reflexa* Guen
Order—Lepidoptera
Family—Noctuidae.

The larvae which destroy the foliage of sisoo, feed on the tender epidermis, thus skeletonizing the leaves of leading shoots which later on wither and dry up. As larvae increase in size, they devour the entire leaf structure. Young foliage is essential to the young larva, whereas on old foliage they develop very slowly.

Adult moths have greyish-brown forewings with oblique irregular bands and wing span of 1.25 inches. A single female can lay 400 eggs which are laid during the night on young leaves, usually one egg only per leaf. The incubation period is 1 to 5 days. Full grown larva is pinkish with 8 white lines down the back. Larval period lasts for 10 days in hot weather; 13 days in spring and 20 days or more in autumn. Most caterpillars fall to ground and pupate among leaves or in loose soil; a pupal cell consisting of silk and particles of soil etc., is usually constructed. Pupal period varies from 5 to 7 days in hot weather and 8 to 11 days in autumn. Some times pupa remains inside up to 80 to 100 days. Life cycle is completed. within a month and duration of the hibernating life-cycle is about 140 days. At Dehra Dun 10 complete generations were observed in a year. Partial reduction of damage can be obtained by silvicultural measures.

*Apoderus sissu*
(Leaf-roller)
Order—Coleoptera
Family—Curculionidae.

The pest starts its activity right from March and continues up to August. The beetles cut leaf across near its base or through the midrib. They rolle up the partially detached part of the leaf so as to form a cylindrical roll. The larva feeds on the inner rolls leaving the outer sheath untouched.

The adult beetle is golden yellow with dark suffusion and three black spots on the elytra. The beetle lays 20 eggs. Eggs is creamy white, oblong , 1 mm. in diameter. Incubation period is 3 to 6 days during March-September and in 5 to 8 days in October. The larva attains full growth in 10-16 days during March-April, 4-7 days during May-October, and 20-30 days in November-December. Its body is wrinkled curved, white or yellowish. It measures 5 mm long. It pupates in a cavity within the roll. Pupal period extends from 3 to 6 days during March-October and 5 to 9 days in November-December. There are eight generations in a year in North India. To suppress the pest population, trimmings of affected trees should be burned.

*Cladobrostis melitricha*
(Borer of living twigs)
Order—Lepidoptera
Family—Blastobasidae.

Beeson (1941) stated that the attack of this species amounted to a heavy pruning of the crowns of shisham trees. The branches break off at the point of girdling, leaving a clean-cut conical depression at the end. Two or more larvae work in one shoot with the result that several metre of growth may be destroyed.

Female deposites sub-cylindrical, brownish eggs on the young shoots of *Dalbergia sissoo*. Full size larva is white, 18 mm long. The caterpillar bores into the pith and excavates a tunnel several inches long. It pupates in tunnel below the girdled portion opposite to previously bored exist hole. Pupa is yellowish brown 11 mm long. Pupal period takes about three weeks. To suppress the pest population infested plants should be cut below the level of infestation to get a new stem. Biological insect pest suppression techniques may be useful in suppression of pest population.

**(D) POPLARS**

Like other trees, poplars are also vulnerable to the attack of Insect pests and severe damage to poplars is being caused by insects.

*Apriona cinerea*
(Stem and root borer)
Order—Coleoptera
Family—Lamiidae.

Damage is caused by the beetles as well as by their grubs. Beetles feed on bark of living shoots. The grubs enter into central pith region, where they feed, making a tunnel downwards. This insect keeps its tunnels comparatively free of frass by pushing this waste material out of opening through the bark.

The adults are dark ashy-grey and have irregular patterns of black punctures. Beetles are 55 mm in body length. The mature grubs measure up to 50 mm. The head bears conspicuous jaws which are rounded at the tip. The grubs pupate inside pupal chamber, which may be up to 20 mm wide and 110 mm long, is plugged up at either end with wood fibres and dust. (See Figs. 84 and 85).

For control dimecron spraying on plants or furadan mixing in soil gave good results.

(2) *Aeolesthes sarta* Sol.
(The Quetta borer)
Order—Coleoptera
Family—Cerambycidae.

This pest attacks poplar in Kashmir Valley.

The young grubs feeds in the soft sappy bark. The older grubs make tunnels, oval in cross section and about 22 mm, at their widest point. As a result of their heavy infestation entire tree withers and dries up in 3-4 years.

The adults are black or dark brown with a fine satiny grey pubescence. The adult measures from 33 mm. to 44 mm in length and 9 mm to 13 mm breadth. The thorax is covered with dense brown puctures. The female lays eggs in groups of 50-75 in batches of 10. The egg is milky white in colour, elliptical in shape with pointed ends. Incubation period is five to ten days. The young larva on hatching out feeds in the soft bark until its mouth parts are strong enough. The larva pupates in the long axis of the tree. The larval period lasts 7 to 10 days, depending upon the season. In pupal stage grub changes into a white or whitish yellow object bearing some resemblance to the future beetle. Adult cuts on exist hole to the surface by making a short horizontal tunnel. This pest can be checked by fumigating the infested stem with paradichlorobenzene in kerosene oil (saturated solution).

*Zeuzera coffeae* Nietn.
(The red borer)
Order—Lepidoptera
Family—Cossidae.

Almost all the clones and species of poplars are attacked by this pest. The larva tunnels along branches and in trunks. The larval tunnels do not readily heal.

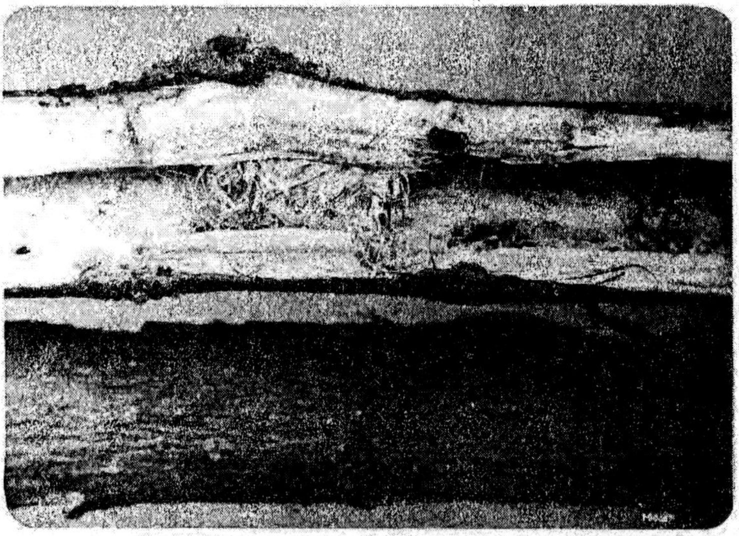

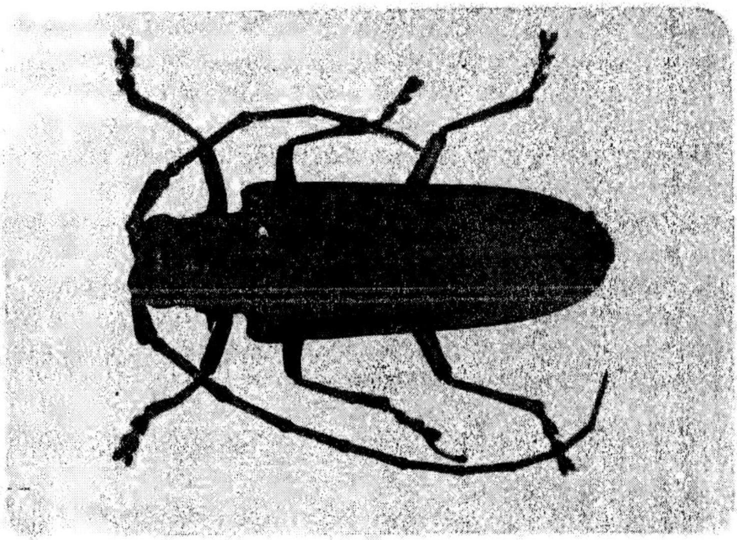

**Figs. 84 and 85 :** Damage to poplar stem, caused by *Apriona cinerea*. Poplar stem borer (*Apriona cinera*).

The adult is white, with a wing expanse 35 to 45 mm. The forewings are marked with small black spots and streaks. There are few black spots on the hind wings.

The egg period lasts about 8 to 10 days; and there may be 500-1000 reddish yellow eggs in 1-2 weeks on the barks of small stems or branches. The eggs hatch in about 10 days. The full grown larva is about 40 mm long. The larvae are stout bodied and dark reddish in colour with a black head. Pupation takes place just underneath the bark, a pupal cell of silk and wood particles is usually constructed. The pupal period lasts for 21 to 30 days.

*Eucosma* species
Order—Lepidoptera
Family—Eucosmidae.

The larvae bore 7-shaped holes with vertical shafts in the growing stems and feed on the callus tissue at the opening of the hole.

The eggs are flat, circular and white. Fully grown caterpillar is about 20 mm long and 5 mm in diameter which are purplish-white with a harden dark brown head capsule. The mature caterpillar pupates in the vertical shaft after closing the top with a fibrous disc. Larvae may be killed in individual plant with carbon disulphide+fuel oil @ 5 ml. per hole and plugging the hole after treatment.

*Agonoscelis nubila* Fabr.
(Sap sucking bug)
Order—Hemiptera
Family—Pentatomidae.

The adults suck the sap from the twigs and branches. Heavily infested plants may show light defoliation and stunting but serious growth retardation seldom occurs.

Adults are yellowish bodied, only about 10 to 11 mm. long. There are 5 generations annually. Life-cycle is completed in 40 to 63 days during March (1st generation), June-July (2nd generation), July-September (3rd generation), while 50 to 70 days in October-December (5th generation).

Spraying of dimecron 20 E.C. 0.2 per cent is effective.

*Tetranychus urticae*, Koch
(Leaf curl mite)
Order—Acarina
Family—Tetranychidae.

Numerous tiny mites can be seen on the undersurface of the leaf. They feed on the terminal foliage and tender stem. The infested leaves may then dry up.

The full grown female is red or greenish, measuring about 0.4-0.5 mm long. The male is slightly smaller. A single female lay up to 50 eggs. Female deposits the eggs on the undersurface of leaves. The eggs are spherical whitish, about 0.1 mm. in diameter. Incubation period is 12 to 15 days. The larvae is six-legged, pinkish and slightly larger than the egg. The larval period lasts 3-5 days. Later, they enter into nymphal stage. Nymphs are greenish or reddish coloured. Nymphal period is completed in an average of 8 days. This may be controlled by spraying DINOCAP 0.06 per cent, and by using predaceous mite, *Phytoseilus riegeli* (Phytoseidae).

*Pygaera fulgurita* Walk
(The poplar defoliator)
Order—Lepidoptera
Family—Notodontidae.

The larvae feed on leaves and in years of serious outbreaks, nurseries and young plantation of poplar may be completely defoliated.

The forewings are yellowish brown with conspicuous dark brown transverse line in the middle bends sharply towards the outside, where as hind wings are cream colour. The moth has a wing span of 30-38 mm. Females lay their eggs in group on the foliage of host plant. A single female can lay 650-700 eggs. Incubation period lasts 4-7 days. Larval development takes about 40 days, at which time the caterpillar measures about 40 mm long. The colouring is very variable, ranging from yellowish red, through reddish-orange, to pale-green. Caterpillars are frequently striped with dark and light longitudinal lines. The larvae tie two or three leaves with the help of silk thread and construct yellowish-brown loosely woven cocoons. The pupal stage lasts 8 days. The life-cycle is completed in 25 to 30 days in North India.

Aerial application of 0.25 per cent sevin or 0.2 per cent carbaryl have proved to be successful method of control. Recently

Fig. 86 : Poplar and Salix defoliator (*Pygaera fulgurita*).

Ahmad and Sen Sarma has used crude suspension of virus infected cadavars in distilled water after centrifuging the disintegrated cadavar larvae in double distilled water. The results are highly encouraging in the dissemination and spreading of the viral disease among the healthy caterpillars.

*Lymantria obfuscata* Walk.
(The India Gypsy moth)
Order—Lepidoptera
Family—Lymantridae.

The larvae feed on foliage of poplars and in the case of severe infestation, they may cause complete defoliation often killing the trees.

Adult body is bulky and has feebly developed wings. Females lay eggs in cluster on the trunks of the tree during June-August. The young caterpillars feed for about six weeks. The full grown larva measures about 45 mm in length. Pupation takes place during June and adult moth emerges in the middle of July. Pest population can be checked by scrapping the eggs mass from the tree trunk or spraying the egg mass with an ovicide affords much protection. In severe defoliation aerial application of carbaryl 0.25 per cent is recommended.

(E) **Gmelina arborea** ***(Gamhar)***

*Calopepla leayana*
Order—Coleoptera
Family—Chrysomelidae.

Defoliation of *Gmelina arborea* occurs at the beginning of rains which continues till October. The species is a serious pest of plantations. The beetle feeds on the leaf and also eats the young buds and shoot. The larva in its early stage eats the leaf parenchyma preferably on the underside. From the beginning of the 4th instar the larva eats more voraciously. Complete skeletonisation leaving only the mid rib and main veins intact is characteristics of the 5th instar larva.

Adults are metallic colour. The elytra coarsely wringkled and are deep bluish-green in the young and deep violet-blue to black in colour in the old beetle. Eggs are laid in groups of 10 to 100. The eggs are embedded in a frothy secretion which hardens to form domed brownish oothecae. Oothecae are laid on the under surface of leaves or on shoots. There are five larval instars. The length of life of the beetle is about 50 days (maximum 120 days) in the first generation and six to eight month in the over wintering generation (Chaudhuri, 1925, Garthwaite, 1939 and Beeson, 1931). Larvae pupate among fallen leaves, cracks or barks, buncle of straws etc. (See Fig. 87).

Population of this pest can be kept with in control by using white light traps just after the rains when sunshines. This pest can, to some extend, be kept under control by pupal parasites; *Brachymeria excarinata* and a fungus; *Beauveria bassiana*, etc. Pupae can be destroyed by providing artificial pupation sites which may be physically destroyed later on.

*Tingis bessoni*
(Lace bugs)
Order—Hemiptera
Family—Tingidae.

Generally one or two years old plantations are attacked by the lace bugs. The nymphs and adults feed at the base of leaf blade. The leaf becomes patched at the base. The sapless leaf start defoliating. A sooty mould develops on the shoots.

Bhandari and Kumar (1992) stated that this pest may be controlled by appling systematic insecticides viz. monocrotophos (0.03%) or synthetic pyrethroids like cypermethrin or fenvalerate (0.03%).

(F) ***Leucaena leucocephala*** (Subabool)

*Heteropsylla cubana*
Order—Hemiptera
Family—Psyllidae.

*Leucaena leucocephala* is a widely accepted tree species in agroforestry systems. The psyllids has become a threat to subabool in India. The insect is native to Central America but in the 1980s it spread in Asia. Insect is very small, 1-2mm in size. Waage and Huxley have stated that psyllids lay clusters of minute orange eggs between the leaflets of very young subabool leaves. The nymphal stages are orange or green, wingless and crawling.

The nymphs attack on young leaflets. Due to heavy drain of cell sap the leaves get malformed turn brown and often drop. This pest is now assuming serious pest status. As the pest is an exotic, its

**Fig. 87 :** Gamhar defoliator (*Calopepla leayana*).

natural enemy complex is either known or non-existing in India (Sen Sarma, 1992).

(G) ***Albizia lebbeck*** **(L.) Benth.**

*Heteropsylla* species
Order—Hemiptera
Family—Psyllidae

Hegde & Relwani (1985) reported that during last two year severe attack of psyllids was observed on 1-2 year old saplings in Maharashtra and Karnataka states. Large number of nymphs accumulate on the growing shoots and suck sap from the shaded portion of the young leaves and tender stems. This results in a decrease in stem growth and leaf expansion. Leaflets remain small in size, turn yellow and become cup-shaped. The nymphs are pale to light green in colour, flat and wide in shape, and can be distinguished from aphids by their fast movement. Adult psyllids have a green tapering abdomen, with a pair of light brown transparent wings. This psyllid seems to be different from the species which attacks *Leucaena leucocephala* because the leucaena plants grown around infested albizia were not affected. This pest may be control by spraying twice with Nuvacron (0.05%) at an interval of one week.

*Eurema blanda silhetana*
Order—Lepidoptera
Family—Pieridae.

The larvae are gregarious feaders, eating, away the leaflets and sometimes striping leaves of big tree.

The butterfly is bright yellow with a wing expanse of 3.7 to 5.0 cm. The forewings are with a black border. A female lays 20 to 50 eggs singly or in clusters on the undersurface of new leaves. Incubation period is 3 to 6 days. The young caterpillar is about 1 inch long, is green with a pale lateral stripes and a black head. The duration of larval period ranges from 14 to 16 days. The pupae is black, triangular shaped. It takes 30 days at low levels at 40 to 50 days above 4,000 feet to complete its total life cycle. In order to put this pest under control dust the seedlings or saplings with 10 per cent BHC dust.

## PEST OF NURSERY

Like other crops, forest nurseries are also vulnerable to the attack by insect pests. The loss caused by insect pests can be considerably reduced by proper pest management. Many species, of insects and a few mites have been reported damaging forest nurseries in India. Most of these cause little damage and only a few are of major importance. These pests includes, cockchafers, cut worm (*Agrotis ipsilon*), *Elater* species, *Laphygama exiqua*, *Diacrisia obliqua*, Termites, etc.

Cockchafers (White Grubs)
Order—Coleoptera
Family—Scarabaeidae
Sub-Family—Melonthinae

Cockchafers have attained importance during the last two decades. The white grubs are usually present in all types of soil. They feed on the small roots of plants; at times most injurious in seed beds, seedlings are killed. Generally damages is caused during wet and cool weather.

The beetles are yellow or dark brown black in colour. Eggs are deposited in the soil and on hatching the grubs measure about 1 cm. in length. The full grown grub is about 6 cm. long. Grubs are white in colour. The grubs ar large. thick, curved 'C' shaped having three pairs of well developed legs. Grubs feed on the roots up to winter. Larval development is completed within 8-10 months.

**Control**

- The grub population can be minimized by preparation of the seed beds a month earlier to the swarming of the beetles *i.e.,* in autumn.
- Damage is also prevented by keeping the nursery field free from weeds.
- At the time of seed beds' preparation, collection and destruction of grubs also plays important role in keeping pest population under check.
- In seed beds and nurseries insecticides should be mixed with soil. Soil application of aldrin or heptachlor in dust formulation at the rate of 2 kg. a.i/ha is very effective in keeping the pest under control. Adults are attracted to neem twigs which may be used to trap the adults.

**Fig. 88 :** White grub (*Holotricha serrata*) adult.

*Elaterid* species
(Wire-worm)
Order—Coleoptera
Family—Elateridae.

The grub cuts the root beneath the surface of the soil and numerous, may cause considerable damage to seedlings in nurser It remains active throughout the year.

Adults head being sunk in the large prothorax, the latter bein squarish in out line, with the posterior angles prolonged into points behind. Full grown larva is cylindrical shining reddish brown in colour with chitinous external coverings. Life-cycle of elater species is still not known.

**Control :** See cockchafers.

*Agrotis ipsilon* Rott.
(cut worm)
Order—Lepidoptera
Family—Noctuidae.

The larvae cut young plants at the base just above the soil surface and either consume portions of the plants on the bed or drag them to their holes to feed on.

The eggs are laid in clusters on the lower surface of leaves. A single female may lay as many as 2,000 eggs. The eggs are greenish white in colour. The larva is dark brown, smooth and flattened. When fully fed the larva is 2.5 to 3.7 cm long. The larval duration depends upon the types of food available and prevailing climate condition of the regions. Pupation occurs in the earthen cocoon in the soil, the incubation, larval and pupal period lasts 2-13, 28-35 and 10-30 days respectively.

**Control**

1. Damage is prevented by keeping the nursery free from weeds.
2. Flood the area and force the larvae out of their tunnels in soil and kill as they come to soil surface.
3. Spraying with 0.075 per cent chlorfenvinphos, 0.05 per cent fenitrothion and 0.15 per cent carbaryl has proved effective in killing the larvae.

4. Dusting with 10 per cent chlordane or 5 per cent heptachlor at the rate of 25 kg. per hectare proved effective.

*Diacrisia oblique* Wlk.
(Bihar hairy caterpillar)
Order—Lepidoptera
Family—Noctuidae.

In the nursery, in initial stage the larva remains gregarious and skeletonise the leaves. Later they feed on the leaf tissue and cause defoliation.

The moth has pale buff body. The moth bears black spots on the wings. Each female may lay eggs up to 1000 alone or in group in three or four days. Incubation period lasts 3-4 days. The caterpillar takes three week to grow in full. The caterpillars defoliate the nursery plants and move from one bed to another. They pupate in the soil. The pupal stage ends in 10 days. It takes four to six weeks to complete its lifecycle.

**Control**

1. Hand picking in the initial stage.
2. Apply Dichlorvos (300 ml) when the caterpillars are small (1 cm. long)
3. Spray endosulfan or quinalphos (1000 ml) or chlorpyriphos (500 ml). Add 450 ml of sandovit or teepot for every hectare of spray liquid.

*Laphygma exiqua* Hub.
Order—Lepidoptera
Family—Noctuidae.

*Laphygma exiqua* has a wide distribution and is a destructive sporadic pest of agricultural and forest plants at seedling stage specially in Bihar.

The eggs are laid in cluster having 10 to 200 or more eggs, on the leaf. The eggs are pearly white, spherical and covered with short whitish hair. In the nursery two or three days old larvae devour steadily the leaves during morning and evening. The incubation, larval and pupal period last 1-2, 9-10 and 5-7 days, respectively.

**Fig. 89 :** *Diacrisia obliqua*

**Control :** See Bihar hairy caterpillar.

Termite
(White ants or Dimak)
Order—Isoptera
Family—Termitidae.

Termites are widely distributed. In Nursery, this pest attacks the plant from the time of seed germination and transplanting. Much of the subsequent termite attack on transplants is due to injury to the roots at the time of lifting. Termites first attack the dead roots and can continue to feed on the living tissues when the resistance of the plant is lowered. Attacked young plant show the outer leaves drying up, which could be easily removed at this stage.

The most striking features of the termite is the great development of the social system. The nest is in mates are mostly two sterile castes and three fertile reproductive castes. The queen is a helpless individuals, whose sole function is production of eggs. Each female may lay eggs up to 15000-(30,000 in *Macrotermes*) per day for several years. The eggs are oval shaped and is yellowish in hue. The egg hatching takes place within 10 days. The newly hatched nymphs develop into workers and soldiers in about six months. The nymphs resembles the adults except in size and colour, has not traces of wings and is sexually immature. In summer the eggs may give rise of reproductive forms. These then emerge and fly out to start new colonies.

**Control**

1. Nursery snould first be cleared. The termite nest should be located and destroyed by mechanically or blowing into them 5 per cent BHC dust or spraying into them 0.5 per cent BHC.
2. To prevent termite attack in the seedling beds, 5 per cent BHC dust at the rate of 25 kg. per hectare should be used at the time of nursery preparations.

**Minor pest**

In addition to major pests there are some minor pest. Among them a cricket (Orthopterans) was reported as a pest of nursery plant. It attacks and defoliate them.

The Periodical cicada, *Magicicada* species attacks the plant by sucking the sap. They inflict severe injury to forest nurseries. They live for about 17 years in the nymphal stage, the longest nymphal stage for any sap sucking insect.

Many aphid species injure nursery plants by sucking the sap from the young leaves. The infested leaves turn yellowish pale affecting the growth. They also secrete, the honey dew, which help in the infestation of fungus and pathogen.

The plant bug, Lygus species and lace bug, Corythuca species also suck the vital sap of the seedlings. In case of Lygus there may be several generations in a year while in Corythuca species there are only 2 generations in a year.

In case of infestation of periodical cicada, aphids, plants bug and lace bug, malathion and endosulfan may be sprayed at weekly intervals.

## BIRDS AS PEST

Certain activities of birds are responsible for economic loss. Graminivorous birds, get nourishment from plant. We have been observing for a decade how much loss is caused by them to agricultural output. Hence it has been practically realised that birds play serious roles in settings as well as up-setting our life. The concept of pest management of birds in India dates back to the early studies of Mason and Lefroy (1912). In latter years, De Abreu (1920) made comprehensive study on food habit of birds and shown in recent years importance of birds as pests. They are Bindra and Tur (1972), Bhatnagar (1976a, b) Bhalla and Pawar (1977), *et al.*

*Various species of birds have been observed depredating almost all crops, even in fields and orchard. The quantity of food intake of birds is conditioned by various factors like ecology, biology and behaviour. In agroforestry system damaged caused by birds may be more in comparison to pure agricultural crops. This system provides shelter as well as food in the same site throughout the year.*

Reports given by Van Tyne and Berger (1959) show that food consumption and body weight are correlated and the capacity for food consumption is also regulated by the temperature and other physiological conditions. According to the reports of Mehrotra *et al.* (1967) in *Passer domesticus,* food intake rate amounts to 20 to 21 gram of millet per day. Beri *et al.*, (1972) reported that adults of common sparrow feed on vegetable portion and seed during nine months in a year (Contd. Page 401).

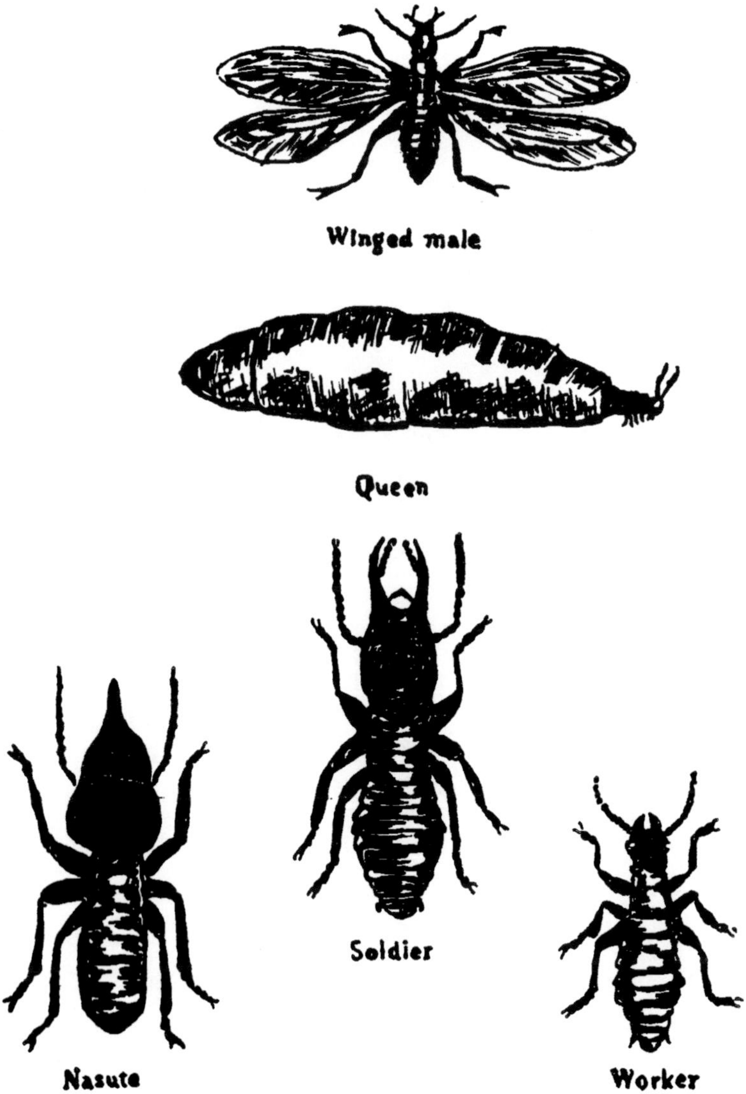


**Fig. 90 :** Different castes of termite.

**Major and minor pest of trees, crops and vegetables**

| Sl. No. | Tree species | Insect pests | Order/ family | Damaging stage | Nature of damage |
|---|---|---|---|---|---|
| 1 | 2 | 3 | 4 | 5 | 6 |
| 1. | Sissoo (*Dalbergia sissoo*) | (*a*) *Perissus dalbergiae* | Coleoptera Cerambycidae | Grub | Shoot borers. |
| | | (*b*) *Apoderus sissu* | Curculionidae | ,, | Leaf-roller. |
| | | (*c*) *Cladobrostis melitricha* | Lepidoptera Blastobasidae | ,, | Borer of the living twings. |
| | | (*d*) *Dichomeris eridantis* | Gelechiidae | ,, | Leaf defoliator |
| | | (*e*) *Dasychira dalbergiae* | Lymantriidae | ,, | ,, |
| | | (*f*) *Plecoptera reflexa* | Noctuidae | ,, | ,, |
| | | (*g*) *Phytometra orichalcea* | ,, | ,, | ,, |
| | | (*h*) *Aspidiotus orientalis* | Rhynchota Cocidae | Nymph & adult | Sheding of the foliage and drying of the Branches. |
| | | (*i*) *Icerya purchasi* | ,, | ,, | ,, |
| 2. | Neem (*Azadirachta indica*) | (*a*) *Laspeyresisa koenigana* | Lepidoptera Eucosmidae | Larva | Bores in the superimposed shoot. |
| | | (*b*) *Boarmia varigata* | Geometridae | ,, | Leaf defoliator. |
| | | (*c*) *Helopeltis antonii* | Hemiptera Miridae | Nymph & adult | Suck the sap from leaves and tender shoots. |

(*Contd.*)

| 1 | 2 | 3 | 4 | 5 | 6 |
|---|---|---|---|---|---|
| | | (*d*) *Pulvinaria maxima* | Coccidae | Nymph & adult | Suck the sap from leaves and tender shoots. |
| | | (*e*) *Aspidiotus orientalis* | ,, | ,, | Sap sucker. |
| | | (*f*) *Holotrichia serrata* | Coleoptera Scarabaeidae | Grub & adult | Feeds on foliage (in the beetle stage) and injurious to rootlets (in grub stage). |
| | | (*g*) *H. consanguinea* | ,, | ,, | ,, |
| | | (*h*) *Adoretus decanus* | ,, | ,, | ,, |
| 3. | Poplar (*Populus spp.*) | (*a*) *Chrysomela populi* | Coleoptera chrysomelidae | Larva & adult | Leaf defoliator. |
| | | (*b*) *Pygaera fulgurita* | *Lepidoptera* Notodontidae | Larva | Leaf defoliator |
| | | (*c*) *Pygaera cupreata* | ,, | ,, | ,, |
| | | (*d*) *Malacosoma idicum* | Lasiocampidae | ,, | Feeds on bud and leaf lamina. |
| | | (*e*) *Lymantria obfuscata* | Lymantriidae | ,, | Leaf defoliator. |
| | | (*f*) *Eucosma glaciata* | Eucosmidae | ,, | Shoot borer. |
| | | (*g*) *Apriona cinerea* | *Coleoptera* Cerambycidae | Grub | Stem borer. |
| | | (*h*) *Aeolesthes sarta* | ,, | ,, | ,, |
| 4. | Gambar (*Gmelina arborea*) | (*a*) *Dihammus cervinus* | Coleoptera Cerambycidae | ,, | Shoot borer. |
| | | (*b*) *Calopepla leayana* | chrysomelidae | Grub & adult | Feeds on leaf. |

(*Contd.*)

| 1 | 2 | 3 | 4 | 5 | 6 |
|---|---|---|---|---|---|
| | | (*c*) *Alcides gmeli nae* | Curulionidae | Grub | Shoot borer. |
| | | (*d*) *Tingis beesoni* | Rhynchota/ Tingidae | Nymph & adults | Feed at the base of leaf-blade on the under-surface. |
| 5. | *Acacia* spp. | (*a*) *Psiloptera cupreo-spledeas* | Coleoptera/ Buprestidae | Adult | Adult gnaws the bark of thin shoot. |
| | | (*b*) *Celosterna scabrator* | Lamiidae | Grub | Stem borer. |
| | | (*c*) *Psiloptera fastuosa* | Buprestidae | ,, | ,, |
| | | (*d*) *Hyposidra successaria* | Lepidoptera/ Geometridae | Larva | Leaf defoliator. |
| | | (*e*) *Stauropus alternus* | Notodontidae | ,, | ,, |
| 6. | *Albizia lebbeck* | (*a*) *Derolus volvulus* | Coleoptera/ Cerambycidae | Larva | Eats bark. |
| | | (*b*) *Rhesala imparata* | Noctuidae Lepidoptera/ | ,, | Leaf defoliator. |
| | | (*c*) *Eurema blanda silhe-tana* | Pieridae | ,, | ,, |
| 7. | Siris (kala) (*Albizia procera*) | (*a*) *Ascotis selenaria imparata* | Lepidoptera/ Geometridae | ,, | Leaf defoliator. |
| | | (*b*) *Indarbela quadrinotata* | Indarbelidae | ,, | Eats bark. |
| | | (*c*) *Rhesala imparata* | Noctuidae/ | ,, | Leaf defoliator. |
| 8. | Bakain (*Melia azedarach*) | (*a*) *Zeuzera coffeae* | Lepidoptera/ Cossidae | ,, | Stem borer. |

(*Contd.*)

| 1 | 2 | 3 | 4 | 5 | 6 |
|---|---|---|---|---|---|
| | | (*b*) *Ascotis selenaria imparata* | Geometridae | Larva | Leaf defoliator. |
| 9. | Teak (*Tectona grandis*) | (*a*) *Dihammus cervinus* | Coleoptera/ Cerambycidae | Grub & adult | Canker produced in the stem. |
| | | (*b*) *Alcides ludificator* | Curculionidae | ,, | Shoot borers. |
| | | (*c*) *Amsacta lactinea* | Lepidoptera/ Arctiidae | Larva | Leaf Defoliator. |
| | | (*d*) *Hyblaea puera* | Hyblaeidae | ,, | ,, |
| | | (*e*) *Dichocrocis puneti-feralis* | Pyralidae | ,, | Pod and stem borer. |
| | | (*f*) *Pyrausta macheralis* | ,, | ,, | Teak skeletonizer. |
| | | (*g*) *Cargara carinata* | Hemiptera/ Membracidae | Nymph & adult | Damage the inflorescence. |
| 10. | Eucaluptus (*Eucalyptus tereticornis*) | (*a*) *Celosterna scabrator* | Coleoptera Lamiidae | Grub | Stem-borer. |
| | | (*b*) *Odontotermes spp.* | Isoptera/Termitidae | Adult | Damage to seedlings. |
| | | (*c*) *Microcerotermes minor* | ,, | ,, | Damage to young plants. |
| 11. | Khezri (*Prosopis cineraria*) | (*a*) *Celosterna scabrator* | Coleoptera/ Cerambycidae | Grub & adult | Stem borer. |
| | | (*b*) *Indarbela quadrinotata* | Lepidoptera/ Indarbelidae | Larva | Bark eater. |
| | | (*c*) *I. tetraonis* | ,, | ,, | ,, |
| | | (*d*) *Oxyrachis tarandus* | Rhynchota/ Membracidae | Adult | Sap-sucker |
| | | (*e*) *Drosicha magiferae* | Coccidae | Nymph & adult | ,, |

(*Contd.*)

| 1 | 2 | 3 | 4 | 5 | 6 |
|---|---|---|---|---|---|
| 12. | *Mangifera indica* | (*a*) *Idiocerus atkinsoni* | Hemiptera Cicadellidae | Nymph & adult | Sap-sucker. |
| | | (*b*) *I. clypealis* | ,, | ,, | ,, |
| | | (*c*) *Drosicha stebbingi* | Margarodidae | ,, | ,, |
| | | (*d*) *D. mengiferae* | ,, | ,, | ,, |
| | | (*e*) *Apsylla cistellata* | Psyllidae | ,, | ,, |
| | | (*f*) *Pulvinaria polygonata* | Coccidae | ,, | ,, |
| | | (*g*) *Aleurocanthus mangiferae* | Aleyrodidae | ,, | ,, |
| | | (*h*) *Indarbela quadrinotata* | Lepidoptera Metarbelidae | Larva | Feeds on the bark. |
| | | (*i*) *Phycita umbratelis* | Phycitidae | ,, | Bore in to fruits and feed on interior soft part. |
| | | (*j*) *Amradiplosis echinogalliperda* | Diptera Caeidomylidae | Maggot | Bores into mid rib and forms gall. |
| | | (*k*) *Erosomia indica* | ,, | ,, | Growth of inflorescence stunted. |
| | | (*l*) *Dasineura amaramanjarae* | ,, | ,, | Feed on internal portion of buds. |
| | | (*m*) *Batocera rufomaculata* | Coleoptera Cerambycidae | Grub | Stem borer. |
| 13. | Litchi (*Litchi chinensis*) | (*a*) *Chrysocoris stollzi* | Hemiptera Lacciferidae | Adult | Sap sucker. |

(*Contd.*)

| 1 | 2 | 3 | 4 | 5 | 6 |
|---|---|---|---|---|---|
| | | (*b*) *Toxoptera aurantii* | Aphididae | Nymph & adult | Suck sap from the inflorescence. |
| | | (*c*) *Drosicha stebbingi* | Margorodidae | ,, | Suck sap from the leaves and flowers. |
| | | (*d*) *Indarbela tetraonis* | Lepidoptera Metarbelidae | Larva | Stem borer. |
| | | (*e*) *Argyroploce aprobola* | Eucosmidae | ,, | Feed on fruits. |
| | | (*f*) *Virachola isocrates* | Lycaenidae | ,, | Feed on seeds. |
| 14. | **Guava** (***Psidium guajava***) | (*a*) *Drosicha stebbingi* | Hemiptera Margorodidae | Nymph & adult | Sap sucker. |
| | | (*b*) *Coccus viridis* | Coccidae | ,, | Drain sap from the leaves and honey dew secretion favours development of sooty mould. |
| | | (*c*) *Indarbela quadrionata* | Lepidoptera Metarbelidae | Larva | Stem borer. |
| | | (*d*) *Rapala varuna* | Lycaenidae | ,, | Feed on the fruits. |
| | | (*e*) *Virachola isocrates* | Lycaenidae | Larva | Feeds on the soft portion of fruit. |
| | | (*f*) *Strumeta ferrugineus* | Diptera Trypetidae | Maggot | Damage fruits. |

(*Contd.*)

| 1 | 2 | 3 | 4 | 5 | 6 |
|---|---|---|---|---|---|
| | | (*g*) *Dacus dorsalis* | Tephritidae | Maggot | Damage fruits |
| 15. | Ber (*Zizyphus jujuba*) | (*a*) *Tricentrus bicolor* | Hemiptera Membracidae | Nymph & adult | Feed on leaves. |
| | | (*b*) *Indarbela tetraonis* | Lepidoptera Metarbelidae | Larva | Shoot borer. |
| | | (*c*) *Meridarchis scyrodes* | Carposinidae | ,, | Feed on the pulp of fruits. |
| | | (*d*) *Dacus dorsalis* | Diptera Tephritidae | Maggot | ,, |
| | | (*e*) *Holotricha insularis* | Coleoptera Melolonthidae | Grub | Leaf defoliator |
| | | (*f*) *Anoplophora versteegi* | Cerambycidae | Adult | Bark feeder. |
| 16. | *Citrus* spp. | (*a*) *Alcurocanthus spiniferus* | Hemiptera Aleyrodidae | Nymph & adult | Sap sucker |
| | | (*b*) *A. woglumi* | ,, | ,, | ,, |
| | | (*c*) *Diaphorina citri* | Psyllidae | ,, | ,, |
| | | (*d*) *Icerya purchasi* | Margarodidae | ,, | ,, |
| | | (*e*) *Othreis conjuncta* | Lepidoptera Noctuidae | ,, | Pierce and suck sap of ripening fruits. |
| | | (*f*) *O. ancilla* | ,, | ,, | ,, |
| | | (*g*) *O. materna* | ,, | ,, | ,, |
| | | (*h*) *Papilio spp.* | Papillionidae | ,, | Leaf defoliator, |
| | | (*i*) *Phyllocnistis citrella* | Phyllocnistidae | ,, | Feed on the upper and lower epidermis of leaves in zigzag manner. |

(*Contd.*)

| 1 | 2 | 3 | 4 | 5 | 6 |
|---|---|---|---|---|---|
| | | (*j*) *Anoplophora versteegi* | Coleoptera Cerambyciae | Grub | Wood borer. |
| 17. | Apple (*Pyrus malus*) | (*a*) *Quadraspidiotus perniciosus* | *Hemiptera* Diaspididae | Nymph & adult | Suck sap from the leaves & fruits. |
| | | (*b*) *Eriosoma lanigerum* | Aphididae | ,, | Suck sap from the leaves and forms galls. |
| | | (*c*) *Icerya purchasi* | Magrarodidae | ,, | Suck sap from the leaves and calyx. |
| | | (*d*) *Psylla mali* | Psyllidae | ,, | Suck sap from the leaves. |
| | | (*e*) *Malacosoma indicum* | Lepidoptera Lasiocampidae | Larva | Leaf defoliator. |
| | | (*f*) *Virachola isocrates* | Lycaenidae | ,, | ,, |
| | | (*g*) *Aeolesthes holosericia* | Coleoptera Cerambycidae | Grub & adult | Shoot borer, also feeds on bark. |
| | | (*h*) *Apriona cinerea* | ,, | ,, | ,, |
| | | (*i*) *Myllocerus maculosus* | Curculionidae | ,, | Leaf defoliator. |
| 18. | Dhan (Paddy) (*Oryza sativa*) | (*a*) *Hieroglyphus banion*, Fabr. | Orthoptera Acrididae | Nymph & adult | Leaves and ears are eaten. |
| | | (*b*) *H.oryzaevorus*, Carl. | Orthoptera Acrididae | Nymph & adult | Leaves and ears are eaten. |
| | | (*c*) *H. nigrorepletus*, Bol. | Orthoptera Acrididae | Nymph & adult | Leaves and ears are eaten. |

(*Contd.*)

| 1 | 2 | 3 | 4 | 5 | 6 |
|---|---|---|---|---|---|
| | | (*d*) *Patanga succinata*, L. | Orthoptera Acrididae | Nymph & adult | Leaves and ears are eaten. |
| | | (*e*) *Schistocerca gregaria* Forsk. | Orthoptera Acrididae | Nymph & adult | Leaves and ears are eaten. |
| | | (*f*) *Oxya velox*, fabr. | Orthoptera Acrididae | Nymph & adult | Leaves and ears are eaten. |
| | | (*g*) *Leptocorisa varicornis*, Fbr. | Hemiptera coreidae | Nymph & adult | Suck sap from the milky grain. |
| | | (*h*) *Nephotettix nigropicta*, stal. | Jassidae | Nymph & adult | Suck sap from the leaf sheaths and transmits the viral diseases. |
| | | (*i*) N. virescens Dist. | Jassidae | Nymph & adult | Transmit tungro of rice. |
| | | (*j*) Sogatella furcifera, Horv. | Araeopidae | Nymph & adult | Infested leaves show greenish yellow colouration. |
| | | (*k*) *Nil aparvata lugens*, stall. | Araeopidae | Nymph & adult | Chaffy ears appear after heavy infestation. |
| | | (*l*) *Peregrinus maidis*, Ashm. | Araeopidae | Nymph & adult | Emergence of ears is prevented due to twisting of top leaves. |
| | | (*m*) *Pyrilla perpusilla*, walk | Lophopidae | Nymph & adult | Sucks the sap from leaf sheath. |

| 1 | 2 | 3 | 4 | 5 | 6 |
|---|---|---|---|---|---|
| | | (*n*) *Heterococcus rehi*, L. | Pseudococcidae | Nymph & adult | Sucks the sap from leaf sheath. |
| | | (*o*) *Tryporyza incertulas*, wsk. | Lepidoptera Phyrastidae | Larva | Bore into, the stem near the root. |
| | | (*p*) *Nymphula depunctalis*, Guen. | Pyrastidae | Larva | Feed on leaves and form tubular cases which are seen floating on water generally. |
| | | (*q*) *Spodoptera mauritia*, Boised. | Noctuide | Larva | Larva appear in swarms and feed on the foliage of young seedling. |
| | | (*r*) *Pseudoletia separata*, walk | Noctuide | Larva | Half ripe paddy ears are cut off by larva at night. |
| | | (*s*) *Mythimna separata*, Wlk. | Noctuide | Larva | Feed on milky ears. |
| | | (*t*) *Sesamia inferens*. walk | Noctuide | Larva | Feeds on central delicate portion of stem. |
| | | (*u*) *Cirphis unipuncta* Moore | Noctuide | Larva | Feed on the stem |
| | | (*v*) *Pachydiplosis oryzae* | Diptera Cecidomyuiidae | Maggot | Forms stem galls. |
| | | (*w*) *Atherigona oryzae*, Mal. | Anthomyiidae | Maggot | Generally infest in seedling stage. |

(*Contd.*)

| 1 | 2 | 3 | 4 | 5 | 6 |
|---|---|---|---|---|---|
| | | (*x*) Hydrellia philippina. Fer. | Ephydridae | Maggot | Mine into leaves of rice and inflict damage to crop. |
| | | (*y*) Echinocnemus oryzae, ms. | Coleoptera curculionidae | Grub. | Feeds on the root of paddy. |
| | | (*z*) *Sitophilus oryzae*, Linn. | Coleoptera curculionidae | Grub. | Infest the grain both in the field and in storage. |
| 19. | Gehun (Wheat) (*Triticum* sp.) | (*a*) *Patanga succinata* Linn. | Orthoptera Acrididae | Nymph & adult | In the early stage the leaves are eaten. |
| | | (*b*) Chrotogonus brachyp-terus, B. | Orthoptera Acrididae | Nymph & adult | In the early stage the leaves are eaten. |
| | | (*c*) *Odontotermes obesus*, Ramb. | Isoptera Termitidae | Worker | Attack on wheat roots, feed on leaves, earhead and on their decaying matter. |
| | | (*d*) *Mi[illegible]termes obesi*, Hol. | Isoptera/Termitidae | Worker | Attack on wheat roots, feed on leaves, earhead and on their decaying matter. |
| | | (*e*) *Aphis maidis*, Fitsch | Hemiptera Aphididae. | Nymph & adult | Suck sap from the tender leaves and shoots. |
| | | (*f*) *Rhopalosiphum maidis*, Fitsch. | Hemiptera Aphididac | Nymph & adult | Suck sap from the leaves and tender shoots. |

(*Contd.*)

| 1 | 2 | 3 | 4 | 5 | 6 |
|---|---|---|---|---|---|
| | | (*g*) *Macrosiphon miscanthi*, Tak. | Aphididae | Nymph & adult | Suck the sap from leaves and tender shoots in early stage and later get confined to the ears. |
| | | (*h*) *Schizaphis granarium* Rond. | Aphididae | Nymph & adult | Suck the sap from leaves and tender shoots in early stage and later get confined to the ears. |
| | | (*i*) *Agrotis flammatra*, schiff. | Lepidoptera Noctuidae | Larva | Leaf eater |
| | | (*j*) *A ypsilon*, Rott | Lepidoptera Noctuidae | Larva | Leaf eater |
| | | (*k*) A.biconica, Koll. | Lepidoptera Noctuidae | Larva | Leaf eater |
| | | (*l*) *Atherigona simplex*, thom. | Diptera Anthomyiidae | Maggot | Sub-sucker |
| | | (*m*) *Tanymecus indicus*, Fst. | Coleoptera Curculionidae | Grub & adult | Feeds on young seedling. |
| 20. | Jau (Barley) (*Hordeum vulgare*) | (*a*) *Hieroglyphus nigrorepletus*, Bol. | Orthoptera Acrididae | Nymph & adult | Feeds on the foliage and on the young grain. |
| | | (*b*) *Microtermes Obesi*, Hol. | Isoptera Termitidae | Worker | Feeds on the foliage and on their decaying matter. |

(*Contd.*)

| 1 | 2 | 3 | 4 | 5 | 6 |
|---|---|---|---|---|---|
| | | (*c*) *Rhopalosiphum maidis*. Fitsch. | Hemiptera Aphididae | Nymph & adult | Suck sap from leaves and tender shoots. |
| | | (*d*) *Pyrilla perpusilla*. wlk. | Lophopidae | Nymph & adult | Suck the sap of the leaves. As result of it leaves dry up. |
| | | (*e*) *Agrotis ypsilon*. Rott | Lepidoptera Noctuidae | Larva | Some times new leaves and flowers may be cut off. |
| 21. | Jai (oats) (*Avenus sativa*) | (*a*) *Agrotis ypsilon*, R | Lepidoptera Noctuidae | Larva | Caterpillar feeds on the foliage of young seedling. |
| | | (*b*) *Heliothis armigera*, Hubn. | Noctridae | Larva | Feed on leaves and, thus destroy the seedlings in the early stages. |
| | | (*c*) *Mythimna separata*, wlk. | Noctridae | Larva | Feed on the leaves and immature ears. |
| 22. | Marua (Fingermillet) (*Eleusine coracana*) | (*a*) *Chrotogonus* spp. | Orthoptera Acrididae | Nymph & adult | Cut the seedling. |
| | | (*b*) *Acrotylus* spp. | Orthoptera Acrididae | Nymph & adult | Cut the leaves and some time immature ears. |
| | | (*c*) *Orthacris simulans* | Orthoptera Acrididae | Nymph & adult | Leaf feeder. |
| | | (*d*) *Tetraneura nighriab-dominalis* Sasoki. | Hemiptera Aphididae | Nymph & adult | Some time attack leaves. |

(*Contd.*)

| 1 | 2 | 3 | 4 | 5 | 6 |
|---|---|---|---|---|---|
| | | (*e*) *Schizaphis graminum*, Rond. | Hemiptera Aphididae | Nymph & adult | Its infestation results in yellowish patches on and ultimate drying of plant. |
| | | (*f*) *Rhopalosiphum maidis*, Fitsch. | Hemiptera Aphididae | Nymph & adult | Its infestation results in yellowish patches on and ultimate drying of plant. |
| | | (*g*) *Leptocorisa acuta*, thunb. | Coreidae | Nymph & adult | Suck sap from the leaves and unripe grains. |
| | | (*h*) *Nezara viridula*, Linn. | Pentatomidae | Nymph & adult | Suck sap from the leaves and unripe grains. |
| | | (*i*) *Dolycoris indicus*, Sasaki. | Pentatomidae | Nymph & adult | Suck sap from the leaves and unripe grains. |
| | | (*j*) *Sitotroga cerealella*, oliv. | Lepidoptera Gelechiidae | Larva | Damage grains. |
| | | (*k*) *Heliothis armigera*, Hb. | Noctuidae | Larva | Feed on leaves |
| | | (*l*) *Mythimna separata*, wlk. | Noctuidae | Larva | Feed on leaves and immature grains. |
| | | (*m*) *Spodoptera exigua*, Hubn. | Noctuidae | Larva | Feed on leaves and young seedlings. |

(*Contd.*)

| 1 | 2 | 3 | 4 | 5 | 6 |
|---|---|---|---|---|---|
| | | (*n*) *Sesamia inferens*, wlk. | Noctuidae | Larva | Feed on leaves and young seedlings. |
| | | (*f*) *Simplicia robustalis*, G. | Noctuidae | Larva | Feed on leaves. |
| | | (*g*) *Amsacta albistriga*, Mat | Arctiidae | Larva | Feed gregariously on the lower surface of leaves. |
| | | (*h*) *A.moorei*, Butt. | Arctiidae | Larva | Feed gregariously on the lower surface of leaves. |
| | | (*i*) *Myllocerus maculosus*, Desb. | Coleoptera Curculionidae | Grub & adult | Adult feeds on leaves while grub feeds on roots of host plants resulting in drying of plants. |
| | | (*j*) *Myllocerus discolor*, F. | Coleoptera Curculionidae | Grub & adult | Adult feeds on leaves while grub feeds on roots of host plants resulting in drying of plants. |
| | | (*k*) *Holotrichia consanguinea*, Bl. | Melolonthidae | Adult | Adult beetle feeds on the leaves. |
| 23. | Jowar (sorghum) (*Sorghum vulgare*) | (*a*) *Patanga succinata*, Linn. | Orthoptera Acrididae | Nymph & adult | In early stage feeds on leaves and later gets confined to the ears. |

| 1 | 2 | 3 | 4 | 5 | 6 |
|---|---|---|---|---|---|
| | | (*b*) *Chrotogonus saussurei*, Bol. | Acrididac | Nymph & adult | Feed and cut the leaves. |
| | | (*c*) *Hieroglyphus nigrorepletus*, Bol. | Acrididac | Nymph & adult | Feed on the foliage and on the immature grains. |
| | | (*d*) *Peregrinus maidis* | Hemiptera Delphacidae | Nymph & adult | Emergence of ears is prevented due to twisting of top leaves. In case of severe attack, the plant will show yellow appearance. |
| | | (*e*) *Neomaskellia* bergii, Sign. | Aleyrodidae | Nymph | In the later stage ears are formed with ill-filled grain and drying of leaves are observed |
| | | (*f*) *Tetraneura nigriabdominalis*, Sac. | Aphididae | Nymph & adult | Suck sap from the roots. |
| | | (*g*) *Rhopalosiphum maidis*, Fitch. | Aphididae | Nymph & adult | Suck sap from the tender shoots and green leaves. |
| | | (*h*) *Saccharicoccus sacchari*, Ckll. | Pseudococeidae | Adult | Sucks sap. Its heavy infestation results in loss of vigour. |

(*Contd.*)

| 1 | 2 | 3 | 4 | 5 | 6 |
|---|---|---|---|---|---|
| | | (*i*) Heterococcus rehi, Lind. | Pseudococcidae | Adult | Suck the sap. |
| | | (*j*) *Leptocorisa acuta*, Thumb. | Coreidae | Nymph & adult | Suck sap from the leaves and immature grains. |
| | | (*k*) *Thosea aperiens*. wlk. | Lepidoptera Cochlidiidae | Larva | Feed by scraping the under-surface of leaves and cause drying of the leaves. |
| | | (*l*) *Chilo partellus*, Sin. | Crambidae | Larva | The newly hatched larva bores into stem. |
| | | (*m*) *Chilo auricilia*, Dudg. | Crambidae | Larva | Feed on leaves and than bores into stem for pupation. |
| | | (*n*) *Cryptoblabes angustipennella*, Hemps. | Pyraustidae | Larva | Feed on leaves and than bores into stem for population |
| | | (*o*) *Dichocrocis punctiferalis*, Guen | Pyraustidae | Larva | Feed on the ears of sorghum |
| | | (*p*) *Amsacta albistriga*, M | Arctiidae | Larva | Feed on lower surface of leaves and move from field to field. |
| | | (*q*) *A.moorei*, M. | Arctiidae | Larva | Feed on lower surface of leaves and move from field to field. |

(*Contd.*)

| 1 | 2 | 3 | 4 | 5 | 6 |
|---|---|---|---|---|---|
| | | (*r*) *Heliothis armigera*, Hb | Noctuidae | Larva | Feed on the ears of sorghum. |
| | | (*s*) *Agrotis ypsilon*, Rott, | Noctuidae | Larva | Feeds on leaves blade and some time on the ears. |
| | | (*t*) *Mythimna separata*, wlk. | Noctuidae | Larva | Feed on the leaves |
| | | (*u*) *Holotrichia consanguinea*, Blanch | Coleoptera Melolothidae | Grub | Feed on the leaves and cut the foliage and ears |
| 24. | Makka (Maize) (*Zea mays*) | (*a*) *Hieroglyphus banian*, Fabr. | Orthoptera Acrididae | Nymph & adult | Feed on leaves. |
| | | (*b*) *H.nigrorepletus* Bol. | Orthoptera Acrididae | Nymph & adult | Feed on leaves. |
| | | (*c*) *Rhopalosiphum maidis*, Fitsch. | Hemiptera Aphididae | Nymph & adult | Suck sap from the tender shoot and green leaves |
| | | (*d*) *Hysteroneura setariae*, Thomas. | Hemiptera Aphididae | Nymph & adult | Suck sap from the tender shoot and green leaves |
| | | (*e*) *Heliothis armigera*, Hb. | Lepidoptera Noctuidae | Larva | Feed on leaves and fruiting bodies. |
| | | (*f*) *Sesamia inferens*, wlk. | Lepidoptera Noctuidae | Larva | Bore into stem and feed on the central portion of stem. |
| | | (*g*) *Agrotis flammatra*, Schiff. | Lepidoptera Noctuidae | Larva | Feed on leaves. |

(*Contd.*)

| 1 | 2 | 3 | 4 | 5 | 6 |
|---|---|---|---|---|---|
| | | (*h*) *A ypsilon*, Rott. | Lepidoptera Noctuidae | Larva | Feed on leaves. |
| | | (*i*) *Mythimna separata*, walk. | Lepidoptera Noctuidae | Larva | Feed on leaves. |
| | | (*j*) *Amsacta moorei*, Butt. | Arctiidae | Larva | Feed on the lower surface of leaves and sometimes feed on the ears. |
| | | (*k*) *Diacrisia obliqua*, wlk. | Arctiidae | Larva | Feed on the lower surface of leaves and sometimes feed on the ears. |
| | | (*l*) *Holotrichia consanguinea*, Balanch. | Colcoptera Melolonthidae | Grub & adult | Feed on roots and adults devour leaves. |
| | | (*m*) *Sitophilus oryzae*, Linn. | Curculionidae | Grub & adult | Feed voraciously on the kernel. |
| | | (*n*) *Rhizopertha dominica*, Fab. | Bostrichidae | Grub & adult | The adult appears to chew a great deal more grain than in needs for food. In both the adult and larval stages, it feeds on the interior of the grains. |
| 25. | Ganna (sugarcane) | (*a*) *Hieroglyphus nigrorepletus*, Bol. | Orthoptera Acrididae | Nymph & adult. | Feed on the leaves |

(*Contd.*)

| 1 | 2 | 3 | 4 | 5 | 6 |
|---|---|---|---|---|---|
| | | (*b*) *H. banian*, Fabr. | Orthoptera Acrididae | Nymph & adult. | Feed on the leaves |
| | | (*c*) *Oxya hyla hyla*, servile | Orthoptera Acrididae | Nymph & adult. | Feed on the leaves |
| | | (*d*) *Colemania sphe-narioids*, Bol. | Orthoptera Acrididae | Nymph & adult. | Feed on the leaves |
| | | (*e*) *Odontotermes obsesus*, Rheonb. | Isoptera Temitidae | Worker | The infested root and stalk of mature cane becomes almost hollow. As a result, the plants dry up. |
| | | (*f*) *O.assumthi*, Holm. | Isoptera Temitidae | Worker | The infested root and stalk of mature cane becomes almost hollow. As a result, the plants dry up. |
| | | (*g*) *Microtermes obesi*, Holm. | Isoptera Temitidae | Worker | The infested root and stalk of mature cane becomes almost hollow. As a result, the plants dry up. |
| | | (*h*) *Pyrilla perpusilla*, wlk. | Hemiptera Lophopidae | Nymph & adult. | Suck the sap of the sugarcane leaves. |

(*Contd.*)

| 1 | 2 | 3 | 4 | 5 | 6 |
|---|---|---|---|---|---|
| | | (*i*) *Peregrinus midis*, Ashm. | Delphaaclae | Nymph & adult. | Sap feeers of sugarcane leaves. |
| | | (*j*) *Aleurolobus barodensis*, Mask. | Aleyt odidae | Nymph | Suck the sap from the leaves and in a badly infested field, growth of sugarcane plant retarded. |
| | | (*k*) *Neomaskellia bergii*, sign. | Aleyt odidae | Nymph | Suck the sap from the leaves and in a badly infested field, growth of sugarcane plant retarded. |
| | | (*l*) *Bemisia tabaci*, Genn. | Aleyt odidae | Nymph | Suck the sap from the leaves and in a badly infested field, growth of sugarcane plant retarded. |
| | | (*m*) *Saccharicoccus sacchari*, cki. | Pseudococcidae | Nymph & adult female | Suck the sap from the leaves. |
| | | (*n*) Pseudococcus saccharicola, Taka. | Pseudococcidae | Nymph & adult remale | Suck the sap from the leaves. |

(*Contd.*)

| 1 | 2 | 3 | 4 | 5 | 6 |
|---|---|---|---|---|---|
| | | (*o*) *Melanaspis glomerata*, Green. | Dispididae | Adult | Infests stalk & leaf sheath. |
| | | (*p*) *Chilosacchari-phagus indicus*, Kapur | Lepidoptera Crambidae | Larva | Infests in the first five internodes. |
| | | (*q*) *Chilotraea infus-catellus*, snell. | Lepidoptera Crambidae | Larva | Larva causes dead heart |
| | | (*r*) *Scirpophaga nivella*, Fabr. | Pyraustidae | Larva | Bores into the growing shoot from the top and actively feed on the soft portion and cause dead heart. |
| | | (*s*) *Emmalocera depres-sella*, Swinh. | Pyraustidae | Larva | Bore into the base of the shoot and feed inside. |
| | | (*t*) Pelopidas mathias, Fb. | Hesperiidae | Larva | Feed on the green matter of leaves. |
| | | (*u*) *Mythimna separata*, walk | Noctuidae | Larva | Feed on the leaves. |
| | | (*v*) *Lepidiota mansueta*, Fb. | Coleoptera Melolonthidae | Gurb | Feed on the underground portion of stem. |
| | | (*w*) *Asmangalia cuspidata*, Mau. | Hispidae | Gurb | Feed on the leaves. |

(*Contd.*)

| 1 | 2 | 3 | 4 | 5 | 6 |
|---|---|---|---|---|---|
| 26. | Cotton (*Gossypium sp.*) | (*a*) *Poecilocerus pictus*, Fb. | Orthoptera Acrididae | Nymph & adult. | Feed on the lower surface and blade of leaves. When the flower appear the nymphs and adults also damage them. |
| | | (*b*) *Schistocerca gregaria*, Fork | Orthoptera Acrididae | Nymph & adult. | Feed on the lower surface and blade of leaves. When the flower appear the nymphs and adults also damage them. |
| | | (*c*) *Chrotogonus brachypterus*, Bl. | Orthoptera Acrididae | Nymph & adult. | Feed on the lower surface and blade of leaves. When the flower appear the nymphs and adults also damage them. |
| | | (*d*) *Atractomorpha crenulata*, Bl. | Orthoptera Acrididae | Nymph & adult. | Feed on the lower surface and blade of leaves. When the flower appear the nymphs and adults also damage them. |

*(Contd.)*

| 1 | 2 | 3 | 4 | 5 | 6 |
|---|---|---|---|---|---|
| | | (*e*) *Cyrtacanthacris ranacea*, Stoll. | Orthoptera Acrididae | Nymph & adult. | Feed on the lower surface and blade of leaves. When the flower appear the nymphs and adults also damage them. |
| | | (*f*) *Odontotermes obesus*, Rhamb. | Isoptera Termitidae | Worker | Destroy root, stalk, The infested portion of plant becomes almost hollow. |
| | | (*g*) Microtermes obesi, Holm. | Isoptera Termitidae | Worker | Destroy root, stalk. The infested portion of plant becomes almost hollow. |
| | | (*h*) *Dysdercus cingulatus*, Fb. | Hemiptera Pyrrhocoridae | Nymph & adult. | Suck plant sap as well as that of bolls. Attacked seeds become worthless for sowing and oil extraction. |
| | | (*i*) *Nezara viridula*, Linn. | Pentatomidae | Nymph & adult. | Suck sap from the green bolls. |
| | | (*j*) *Oxycarenus hyalinipennis*, costa | Lygaeidae | Nymph & adult. | Suck sap from immature seeds and stain the lint, |

(*Contd.*)

| 1 | 2 | 3 | 4 | 5 | 6 |
|---|---|---|---|---|---|
| | | (*k*) *Amrasca biguttula*, Ishida | Jassidae | Nymph & adult. | Suck sap from the under-surface of leaves. |
| | | (*l*) *Empoasca flavescens*, Dist. | Jassidae | Nymph & adult. | Suck sap from the under-surface of leaves. |
| | | (*m*) Bemisia tabaci, Genn. | Aleyrodidae | Adult | Attack on premature leaves. In the case of severe infestation, it results in complete drying of leaves and development of sooty mould. |
| | | (*n*) *Aphis gossypii*, Glover | Aphididae | Nymph & adult. | In case of severe infestation, black sooty mould develops. Suck sap from the immature leaves. |
| | | (*o*) Thrips tabaci, Lind. | Thysanoptera Thripidae | Nymph & adult. | Feed on green bolls. |
| | | (*p*) *Scirtothrips oligochaetus*, Karny. | Thysanoptera Thripidae | Nymph & adult. | Feed on green bolls. |
| | | (*q*) *Sylepta derogata*, Fb. | Lepidoptera Pyraustidae | Larva | Roll the leaves and feed on them. |

(*Contd.*)

| 1 | 2 | 3 | 4 | 5 | 6 |
|---|---|---|---|---|---|
| | | (*r*) *Phycita infusella*, Mcyr. | Phycitidae | Larva | Larva web the leaves together into small compact masses, which turn dark brown. |
| | | (*s*) *Platyedra gossypiella*, Saund. | Gelechiidae | Larva | Bores inside the boll. They feed some times on young buds. |
| | | (*t*) *Earias fabia*, stoll. | Arctiidae | Larva | Bores into the tender portion of growing shoot and also feeds on green bolls. |
| | | (*u*) *E. insulana*, Boisd. | Arctiidae | Larva | Bores into the tender portion of growing shoot and also feed on green bolls. |
| | | (*v*) *Amsacta moorei*, Butt. | Arctiidae | Larva | Bores into the tender portion of growing shoot and also feed on green bolls. |
| | | (*w*) *Spodoptera exique*, Hubn. | Noctuidae | Larva | Feed on leaves and young buds. |
| | | (*x*) *Euproctis fraterna*, Moore | Noctuidae | Larva | Leaf defoliator. |
| | | (*y*) *Astycus lateralis*, Fb. | Coleoptera curculionidae | Grub and adult | Feed on cotton leaves |

(*Contd.*)

| 1 | 2 | 3 | 4 | 5 | 6 |
|---|---|---|---|---|---|
| 27. | Tori (Mustard) (*Brassica spp.*) | (*a*) *Siphonocoryne indo-trassico*, Das. | Hemiptera Aphididae | Nymph and adult | Suck sap from the leaves and tender shoots and later get confined to the flowers and fruits. |
| | | (*b*) *Lipaphis erysimik*, Kalt. | Hemiptera Aphididae | Nymph and adult | Suck sap from the leaves and tender shoots and later get confined to the flowers and fruits. |
| | | (*c*) *Brevicoryne brassicae*, Linn. | Hemiptera Aphididae | Nymph and adult | Suck sap from the leaves and tender shoots and later get con-fined to the flo-wers and fruits. |
| | | (*d*) *Bagrada cruciferarum*, Kirk. | Pentatomidae | Nymph and adult | Suck sap from the leaves and buds. |
| | | (*e*) Cruciferous leaf webber *Crocidololmia binotalis*, Z | Lepidoptera Pyraustidae | Larva | Larvae web the leaves and feed by scraping the green tissues of leaf. They also feed on flowers buds and bore into pods. |

(*Contd.*)

| 1 | 2 | 3 | 4 | 5 | 6 |
|---|---|---|---|---|---|
| | | (*f*) Cabbage butter fly *Pieris brassicae*, Linn. | Pieridae | Larva | Feed on leaves. |
| | | (*g*) Bihar hairy caterpillar, *Dicrisia oblique*, wlk. | Arctiidae | Larva | Feed on leaves and pods. |
| | | (*h*) Pea leaf winer *Phytomyza atricornis*, Meign. | Diptera Agromyzidae | Maggot | Mine into the leaves and eat the green matters. |
| | | (*i*) Mustard sawfly *Athalia lugens proxima*, K. | Hymenoptera Tenthredinidac | Grub | In case of heavey attack, all the leaves are eaten and the planty dry up. |
| 28. | Linseed (*Linum usitatisinum*) | (*a*) Leaf caterpillar *Heliothis armigera*, Hb. | Lepidoptera Noctuidac | Larva | Feed on the leaves and, thus destroy the seedling in its early stage. |
| | | (*b*) *Plusia orichalcea*, | Lepidoptera Noctuidac | Larva | Feed on the leaves. |
| | | (*c*) Cut worm *Agrotis ypsilon*, Hufn. | Lepidoptera Noctuidac | Larva | Larvae cut new leaves and pods. |
| | | (*d*) *Porthesia scintillans*, W. | Lymantriidae | Larva | Feed on leaves. |
| | | (*e*) Linseed midge *Dasineura lini, barn*. | Diptera cecidomyiidae | Maggot | Maggot tunnels into bud and feeds on them. |
| | | (*f*) Pea-leaf-miner *Phytomyza atricornis*, Meign. | Agromyzidae | Maggot | Maggot mines in to leaves and eats mesophyll layer. |

(*Contd.*)

| 1 | 2 | 3 | 4 | 5 | 6 |
|---|---|---|---|---|---|
| 29. | Coconut *Cocus nucifera* | (*a*) White ant *Odontotermes obesus*, Ramb. | Isoptera Termitidae | Worker | Feed on roots, shoots and leaves. |
| | | (*b*) *Microtermes obesi*, Holm. | Isoptera Termitidae | Worker | Feed on roots, shoots and leaves. |
| | | (*c*) Black headed caterpillar, *Nephantis serinopa*, Meyt. | Lepidoptera cryptophaisdae | Larva | Damage is caused to young coconut fronds. |
| | | (*d*) *Contheyla rotunda*,H | Cochlidiidae | Larva | Feed on leaves and make tunnel into coconut tree. |
| | | (*e*) *Macroplectra nararia*, M. | Cochlidiidae | Larva | Feed on leaves and make tunnel into coconut tree. |
| | | (*f*) *Gangara thyrsis*, M. | Hesperiidae | Larva | Feed on leaves and make tunnel into coconut tree. |
| | | (*g*) *Suastus gremius*, F. | Hesperiidae | Larva | Feed on leaves and make tunnel into coconut tree. |
| | | (*h*) *Oecophylla smaragdina*, F. | Hymenoptera Formicidae | Larva | Some times feed voraciously on leaves. |

(*Contd.*)

| 1 | 2 | 3 | 4 | 5 | 6 |
|---|---|---|---|---|---|
| | | (*i*) *Oryctes rhinoceros*, Linn. | Coleoptera Dynastidae | Grub | It causes damage by boring into young fronds. |
| | | (*j*) *Rhynchophorus ferrugineus*. F. | Curculionidae | Grub | Damage is caused to young fronds. |
| 30. | Till (Gingelly) (*Sesamum indicum*) | (*a*) *Orosius albicinctus*, Dist. | Hemiptera Cicadellidae | Adult | Vector of the *sesamum phylody* virus. |
| | | (*b*) *Aphis gossypii*, G. | Aphididae | Nymph and adult | Suck sap from the leaves. In case of severe attack, the leaves curl up and finally dry up. |
| | | (*c*) *Nysius inconspicuous*, Fb. | Lygacidac | Nymph and adult | Suck sap from the immature leaves. |
| | | (*d*) *Antigastra catalaunalis* Dup. | Lepidoptera Pyraustidae | Larva | Bore into the shoots, flowers. Young caterpillar feeds on epidermal layer. |
| | | (*e*) *Diacrisia obliqua*. wlk. | Arctiidae | Larva | Feed on the leaves and thus destroy the seedling in the early stage. |

(*Contd.*)

| 1 | 2 | 3 | 4 | 5 | 6 |
|---|---|---|---|---|---|
| | | (*f*) *Acherontia lachesis*, Westor. | Sphingidae | Larva | Feed on leaves. |
| | | (*g*) Heliothes armigera, Hb. | Noctuidae | Larva | Feed on leaves and immature pods. |
| | | (*h*) *Asphondylia sesami*, Felt. | Diptera Cecidomyiidae | Maggot | Eat flowers and buds and form galls. |
| | | (*i*) *Dasyneura sesami*, Grover. | Diptera Cecidomyiidae | Maggot | Feeds on the internal content of bud. |
| | | (*j*) *Oxycetonia albo-punctata*, Fb. | Coleoptera Cetoniidae | Grub | Infest roots and tender shoots. |
| | | (*k*) *Trogoderma granarium*, Evert. | Dermestidae | Grub and adult | Infest seeds. |
| 31. | Ground nut (*Arachis hypogaea*) | (*a*) *Chrotogonus brachypterus*. Blan. | Orthoptera Acrididae | Nymph and adult | Feed the leaves. |
| | | (*b*) *Euborellia annulipes*, Luca. | Dermaptera Labiduridae | Adult | Bore into tender pods and feed on the kernels. |
| | | (*c*) *Aphanus sordidus*, Fb. | Hemiptera Lygaeidae | Nymph and adult | Suck the sap from the leaves and cause the leaves to shrivel. |
| | | (*d*) Caliothrips indicus, Bagn. | Thysanoptera Thripidae | Adult | Suck the sap from the leaves and make the tender foliage spotted. |

(*Contd.*)

| 1 | 2 | 3 | 4 | 5 | 6 |
|---|---|---|---|---|---|
| | | (*e*) *Stomopteryx* spp. | Lepidoptera Gelechiidae | Larva | Mines and eats the leaflets. |
| | | (*f*) *Amsacta moorei*, Bull | Arctiidae | Larva | Voraciously feed on plant leaves. |
| | | (*g*) *Diacrisia obliqua*, Wlk. | Arctiidae | Larva | In the early stage, feed on the lower surface of leaves and when grown up, eat the leaves and top shoot. |
| | | (*h*) *Heliothis armigera*, | Noctuidae | Larva | Feed on the leaves and destroy the seedling in the early stage. |
| | | (*i*) *Sphenoptera perotetti* Guer. | Coleoptera Buprestidac | Grub | Bore into stems and top roots. |
| | | (*j*) *Holotrichia consanguinea*, Blan. | Melononthidae | Grub | Bore into roots, stems and defoliate plants. |
| 32. | Caster (*Ricinus communis*) | (*a*) *Empoasca flavescens*, F | Hemiptera Cicadellidae | Nymph and adult | It forms yellowish patch on the leaf margin. |
| | | (*b*) *Trialeurodes ricini*, Mis. | Aleyrodidae | Nymph | Suck sap from the under-surface of leaves. |

(*Contd.*)

| 1 | 2 | 3 | 4 | 5 | 6 |
|---|---|---|---|---|---|
| | | (*c*) *Nezara viridula*, Linn. | Pentatomidae | Nymph & adult | Suck sap from the tender shoots and leaves. |
| | | (*d*) *Achaea janata*, Linn. | Lepidoptera Noctuidae | Larva | Feed voraciously on leaves. |
| | | (*e*) *A melicerta*, Linn. | Lepidoptera Noctuidae | Larva | Feed voraciously on leaves. |
| | | (*f*) *Heliothis armigera* Hb | Lepidoptera Noctuidae | Larva | Feed on leaves and fruiting bodies. |
| | | (*g*) *Spodoptera litura*, F. | Lepidoptera Noctuidae | Larva | Feed on the leaves. |
| | | (*h*) *Dichocrocis puncti-feralis*, Guen. | Pyraustidae | Larva | Bore into the shoots and later on into the seeds and capsule. |
| | | (*i*) *Ergolis merione*, cam. | Nymphalidae | Larva | Feed on the leaves and destroy the seedling in the early stage. |
| | | (*j*) *Asphondylia ricini*, Mani. | Diptera Cecidomyiidae | Maggot | Infests flower, bud and immature fruit. |
| 33. | Gram (*Cicer arictinum*) | (*a*) *Agrotis ypsilon*, Rott | Lepidoptera Noctuidae | Larva | Cut new leaves and flowers. |
| | | (*b*) *A.flamatra*, Schiff | Lepidoptera Noctuidae | Larva | Cut new leaves and flowers. |

(*Contd.*)

| 1 | 2 | 3 | 4 | 5 | 6 |
|---|---|---|---|---|---|
| | | (*c*) *A.spinifera*, Hb. | Lepidoptera Noctuidae | Larva | Cut new leaves and flowers. |
| | | (*d*) *Heliothis armigera*, Hb. | Lepidoptera Noctuidae | Larva | Feed on the leaves and destroy the seedling in the early stage. |
| | | (*e*) *Spodoptera exigua*, Hb. | Lepidoptera Noctuidae | Larva | Feed on the foliage and young seedling. |
| | | (*f*) *Plusia nigrisigna*, wlk. | Lepidoptera Noctuidae | Larva | Feed on the leaves. |
| | | (*g*) *Tanymecus indicus*, Fst. | Coleoptera Curculionidae | Grub and adult | Destroy and feed the leaves. |
| | | (*h*) *Melanagromyza obtusa*, Mall. | Diptera Agromyzidae | Maggot | Feed on tender shoots. |
| 34. Tomato | (*Lycopersicon esculentum*). | (*a*) *Myzus persicae*, Sulz. | Hemiptera Aphididae | Nymph and adult | Suck sap from the tender parts of plants. |
| | | (*b*) *Bemisia tabaci*, Genn. | Aleyrodidae | Adult | Attack premature leaf. |
| | | (*c*) *Cryptopeltis crassicornis*, Dist. | Miridae | Nymph adult | Suck sap from the green leaves, shoots and buds. |
| | | (*d*) *Scirtothrips dorsalis*, Hood. | Thysanoptera Thripidae | Adult | Suck sap from the leaves and buds. |
| | | (*e*) *Caliothrips indicus*, Bag. | Thysanoptera Thripidae | Adult | Suck sap from the leaves, and buds; |

(*Contd.*)

| 1 | 2 | 3 | 4 | 5 | 6 |
|---|---|---|---|---|---|
| | | (*f*) *Trichoplusia ni*, Hb. | Lepidoptera Noctuidae | Larva | Eat away the leaves of growing plant. |
| | | (*g*) *Heliothis armigera*, Hb. | Lepidoptera Noctuidae | Larva | Feed on green leaves and bore into fruits. |
| | | (*h*) *Spodoptera litura*, Fb. | Lepidoptera Noctuidae | Larva | Bore into fruits and spoil them. |
| | | (*i*) *Strumeta cucurbitae* Coque. | Diptera Trypetidae | Maggot | Bore into fruits and make them unfit for consumption. |
| 35. | Lady's finger (*Abelmoschus esculentus*) | (*a*) Aphis gossypii, Genn. | Hemiptera Aphididae | Nymph and adult | Suck sap from the leaves and buds. |
| | | (*b*) *Bemisia tabaci*, Genn. | Aleyrodidae | Adult | Feed on immature leaves and buds. |
| | | (*c*) *Tricentrus bicolor*, Dist. | Membracidae | Nymph & adult | Suck sap from the leaves and tender fruits. |
| | | (*d*) *Saissetia hemisphaerica*, Targ. | Coccidae | Adult | Infest under-surface of leaves and tender shoots |
| | | (*e*) *Agrotis* spp. | Lepidoptera Noctuidae | Larva | Feed and cut leaves and tender shoots. |

(*Contd.*)

| 1 | 2 | 3 | 4 | 5 | 6 |
|---|---|---|---|---|---|
| | | (*f*) *Earias insulana*, Boisd. | Lepidoptera Noctuidae | Larva | Bore into fruits and feed on them. |
| | | (*g*) *Anomis flava*, Fb. | Lepidoptera Noctuidae | Larva | Larva attack on apical buds and top tender leaves. |
| | | (*h*) *Heliothes armigera*, Fb. | Lepidoptera Noctuidae | Larva | Larva attack on apical buds and top tender leaves. |
| | | (*i*) *Mealanagromyza obtusa*, Mall. | Diptera Agromyzidae | Maggot | Feed on tender leaves and tender shoots. |
| 36. | Brinjal (*Solanum melongena*) | (*a*) *Aphis gossypii*, G. | Hemiptera Aphididae | Adult | Suck sap from the leaves and buds |
| | | (*b*) *Tricentrus bicolor*, Dist. | Membracidae | Adult | Suck sap from the buds and tender fruits |
| | | (*c*) *Bemisia tabaci*, Genn. | Aleyrodidae | Adult | Feed on leaves |
| | | (*d*) *Urentius hystricellus*, Rich | Tingidae | Nymph and adult | Feed on leaves |
| | | (*e*) *Caliothrips indicus*, Bog. | Thysanoptera Thripidae | Nymph and adult | Suck sap from the leaves |
| | | (*f*) *Scirtothrips dorsalis*, Hood. | Thysanoptera Thripidae | Nymph and adult | Suck sap from the leaves |
| | | (*g*) *Leucinodes arbonelis*, Guen. | Lepidoptera Pyraustidae | Larva | Bore into young shoots and young fruits |

(*Contd.*)

| 1 | 2 | 3 | 4 | 5 | 6 |
|---|---|---|---|---|---|
| | | (*h*) *Eublemma* spp. Walk. | Noctuidae | Larva | Feed inside folded leaves |
| | | (*i*) *Agrotis ypsilon*, Rott. | Noctuidae | Larva | Feed on leaves and tender shoots. |
| | | (*j*) *Spodoptera exigua*, Hb. | Noctuidae | Larva | Feed on green leaves |
| | | (*k*) Epilachna *vigintiocto-punctata*, F. | Coleoptera Coccinellidae | Grub & adult | Feed voraciously on the leaves and tender part of the shoots. |
| | | (*l*) *Myllocerus subfasciatus*, G. | Curculionidae | Grub & adult | Grubs feed on the roots while adults on the leaf blade in characteristic manner. |
| 37. | Cabbage (*Brassica* spp.) | (*a*) *Lipaphis erysimi*, Kalt. | Hemiptera Aphididae | Nymph and adult | Suck sap from the leaves and tender shoots. |
| | | (*b*) Bagrada cruc-iferarum, F. | Pentatomidae | Nymph and adult | Suck the sap from the tender parts of the plant. The affected plants are deprived of their vitality. |
| | | (*c*) *Plusia orichalcea*, Fb | Lepidoptera Noctuidae | Larva | Feeds on the foliage. |

| 1 | 2 | 3 | 4 | 5 | 6 |
|---|---|---|---|---|---|
| | | (*d*) *Agrotis ypsilon*, Rott. | Lepidoptera Noctuidae | Larva | Feed on new leaves and flowers. |
| | | (*f*) *A.segetum*, schiff. | Lepidoptera Noctuidae | Larva | Feed on new leaves and flowers. |
| | | (*g*) *Diacrisia obligua*, Wlk. | Arctiidae | Larva | Feed on the leaves. |
| | | (*h*) *Hellula undalis*, Fb. | Pyraustidae | Larva | Mines the leaf along the side of vein |
| | | (*i*) *Plutella xylostella*, | Plutellidae | Larva | Feed on the lower surface of leaves. |
| | | (*j*) *Pieris brassicae*, Linn | Pieridae | Larva | Feed on immature leaves. |
| | | (*k*) *Phytomyza atricornis*, Meig. | Diptera Agromyzidae | Maggot | Mine into leaves mesophyll layer. |
| | | (*l*) *Athalia lugens proxima*, Klug. | Hymenoptera Tenthredinidae | Larva | Eat up the edge of leaves |
| 38. | Radish (*Raphanus sativus*) | (*a*) *Chrotogonus oxypterus*. Bl. | Orthoptera Acrididae | Nymph & adult | Feed on the leaves. |
| | | (*b*) *Lipaphis erysimi*, Kalt. | Hemiptera Aphididae | Nymph & adult | Suck sap from the leaves |
| | | (*c*) *Bagrada cruciferum*, F. | Pantatomidac | Nymph & adult | Suck sap from the leaves |
| | | (*d*) *Phytomyza atricornis*, Meig. | Diptera Agromyzidae | Maggot | Feeds on the mesodermal layer. |

(*Contd.*)

| 1 | 2 | 3 | 4 | 5 | 6 |
|---|---|---|---|---|---|
| | | (*e*) *Spodoptera litura*, F. | Lepidoptera Noctuidae | Larva | Feed on the leaves |
| | | (*f*) *Athalia lugens proxima*, Klug. | Hymenoptera Tenthredinidae | Larva | Feed on the leaves. |
| 39. | Carrot (*Dacus carota*) | (*a*) *Phytomyza atricornis*, Meign. | Diptera Agromyzidae | Maggot | Feed on the mesodermal layer of leaves. |
| | | (*b*) *Chaetochnema basalis*, Bally | Coleoptera Alticidae | Grub | Feed on interior portion of roots. |
| 40. | Potato (*Solanum nigrum.*) | (*a*) *Myzus persicae*, Sulz. | Hemiptera Aphididae | Nymph & adult | Suck sap from the leaves. |
| | | (*b*) *Aphis gossypii*,G. | Hemiptera Aphididae | Nymph & adult | Suck sap from the leaves. |
| | | (*c*) *Bemisia tabaci*, Genn. | Aleyrodidae | Adult | Feed on leaves. |
| | | (*d*) *Nezara viridula*, Linn. | Pentatomidae | Nymph & adult | Suck sap from the leaves and tender shoot. |
| | | (*e*) *Phthorimaca operculella*, Z. | Lepidoptera Gelechiidae | Larva | Feed on leaves. |
| | | (*f*) *Agrotis* spp. | Noctuidae | Larva | Feed on stems and tubers. |
| | | (*g*) *Heliothis armigera*, Hb. | Noctuidae | Larva | Feed on the leaves. Sometimes defoliate entire crop. |
| | | (*h*) *Dasychira mendosa*, Wlk. | Arctiidae | Larva | Attack premature leaves. |

(*Contd.*)

| 1 | 2 | 3 | 4 | 5 | 6 |
|---|---|---|---|---|---|
| | | (*i*) *Phytomyza articornis*, Meign. | Diptera Agromyzidae | Larva | Mine and feed on the meso-dermal layer of leaves. |
| | | (*j*) *Epilachna vigintictopunctata*, Fab. | Coleptera Coccinellidae | Grub & adult | Grub as well as adult feed on leaves by scrapping in a characteristic manner. |
| | | (*k*) *E.odecastigma, Mule.* | Coleptera Coccinellidae | Grub & adult | Grub as well as adult feed on leaves by scrapping in a characteristic manner. |
| | | (*l*) *Gonocephalum hoffmanseggi*, St | Tenebrionidae | Grub & adult | Feed on tubers and leaves. |
| 41. | Garlic (*Allium sativum*) | (*a*) *Aeolothrips collaris*, Pries. | Thysanoptera Aeolothripidae | Adult | Suck sap from the tender portion of plants. |
| | | (*b*) *Thrips tabaci*. L. | Thripidae | Adult | Suck sap from the leaves. |
| | | (*c*) *Spodoptera* spp. | Lepidoptera Noctuiidae | Larva | Feed on the leaves. |
| 42. | Onion (*Allium cepa*) | (*a*) *Aeolothrips collaris*, Pries. | Thysanoptera Aeolothripidae | Adult | Suck sap from the tender foliage. |
| | | (*b*) Thrips tabaci, L | | | |

(*Contd.*)

| 1 | 2 | 3 | 4 | 5 | 6 |
|---|---|---|---|---|---|
| | | | Thysanoptera Aeolothripidae | Adult | Suck sap from the tender foliage. |
| | | (*c*) *Spodoptera*, spp. | Lepidoptera Noctuidae | Adult | Feed on the tender leaves. |
| | | (*d*) *Prodenia litura*, Fabr. | Lepidoptera Noctuidae | Adult | Feed on the tender leaves. |
| | | (*e*) *Alphitobius leavigatus*, Fb. | Coleoptera Tenebrionidae | Grub | Feed on the seeds in storeroom. |
| 43. | Chilli (*Capsicum annum*) | (*a*) *Aphis gossypii*, Gen. | Hemiptera Aphididae | Nymph & adult | Suck sap from the leaves, and immature fruits. |
| | | (*b*) *Bemisia tabaci*, Gen. | Aleyrodidae | Nymph & adult | Suck sap from the leaves and later get confined to the flowers. |
| | | (*c*) *Scirtothrips dorsalis*, Hood. | Thripidae | Nymph & adult | Suck sap from the leaves. |
| | | (*d*) *Euzophera perticella* | Lepidoptera Phycitidae | Larva | Bore into stems and feed softer internal portion of stems. |
| | | (*e*) *Spodoptera* spp. | Noctuiidae | Larva | Feed on the immature leaves, stems and fruits. |
| | | (*f*) *Anomala bengalensis* | Coleoptera Noctuidae | Larva | Feed on the roots and stems. |

(*Contd.*)

| 1 | 2 | 3 | 4 | 5 | 6 |
|---|---|---|---|---|---|
| 44. | Arhar (Pigeon pea) *Cajanus cajan*. | (*a*) *Thrips hawaiiensis*, Morg. | Thysanoptera Thripidae | Larva & adult | Infest tender shoots, buds and flowers. |
| | | (*b*) *Tricentrus bicolar*, Dist. | Hemiptera Membricidae | Nymph & adult | Suck sap from the leaves. |
| | | (*c*) *Deptocentrus obliqus*, Fabr. | Hemiptera Membricidae | Nymph & adult | Suck sap from the leaves. |
| | | (*d*) *Riptortus pedestris*, F. | Coreidae | Nymph & adult | Suck sap from the tender stems leaves and immature pods. |
| | | (*e*) *Eurybrachys tomentosa*, Fb. | Cicadellidae | Nymph & adult | Suck sap from the tender stems, leaves and immature pods. |
| | | (*f*) *Catochrysops cnejus*, Fabr. | Lepidoptera Lycaenidae | Larva | Feed on buds and seeds inside the pod and make them unfit for consumption. |
| | | (*g*) *Lampides bocticus*, L. | Lepidoptera Lycaenidae | Larva | Feed on buds and seeds inside the pod and make them unfit for consumption. |

(*Contd.*)

| 1 | 2 | 3 | 4 | 5 | 6 |
|---|---|---|---|---|---|
| | | (*h*) *Eucosma critica*, Meyr. | Eucosmidae | Larva | Bore into pods and feed on seeds. |
| | | (*i*) *E. melanaula*, Meyr. | Eucosmidae | Larva | Bore into pods and feed on seeds. |
| | | (*j*) *Heliothis armigera*, Hb. | Noctuidae | Larva | Feed on the immature leaves and fruiting bodies. |
| | | (*k*) *Melanagromyza obtusa*, Mall. | Diptera Agromyzidae | Maggot | Feed on immature seeds. |
| | | (*l*) *Ceuthorrhynchus asperulus*, Fast. | Coleoptera Curculionidae | Grub | Feed on buds and drop it to the ground. |
| 45. | Mung (Black gram) (*Phaseolus mungo*). | (*a*) *Aphis craccivora*, Koch. | Hemiptera Aphididae | Nymph & adult | Suck sap from the tender shoots, leaves flowers and immature pods. |
| | | (*b*) *Riptortus pedestris*, Fb. | Coreidae | Nymph & adult | Suck sap from the immature seeds. |
| | | (*c*) *Catochrysops cnejus*, Fabr. | Lepidoptera Lycaenidae | Larva | Feed on buds and immature pods. |
| | | (*d*) *Diacrisia obligua*, wlk. | Arctiidae | Larva | Feed on buds and immature pods. |
| | | (*e*) *Melanagromyza phaseoli*, cog. | Diptera Agromyzidae | Maggot | Feed on immature seeds. |

(*Contd.*)

| 1 | 2 | 3 | 4 | 5 | 6 |
|---|---|---|---|---|---|
| 46. | Urd (Green gram) (*Phaseolus radiatus*) | (*a*) *Aphis craccivora*, Koch. | Hemiptera Aphididae | Nymph & adult | Suck sap from the leaves, tender shoots and pods. |
| | | (*b*) *Chauliops fallax*. scoot. | Lygaenidae | Nymph & adult | Suck sap from the immature leaves. |
| | | (*c*) *Tricentrus bicolor*, Dist. | Membracidae | Nymph & adult | Suck sap from the pods and make them unfit for consumption. |
| | | (*d*) *Eucosma melanaula*, Meyr. | Lepidoptera Eucosmidae | Larva | Feed on immature seeds. |
| | | (*e*) *Lamprosema indicata*, Fb. | Pyraustidae | Larva | Feed on leaves. |
| | | (*f*) *Lampides bocticus*, L. | Lycaenidae | Larva | Bore into pods and feed on immature seeds. |
| | | (*g*) *Catochrysops cnejus*, F. | Lycaenidae | Larva | Bore into pods and feed on immature seeds. |
| 47. | Pea (*Pisium sativum*) | (*a*) *Tricentrus bicolor* | Hemiptera Membracidae | Nymph & adult | Suck sap from the pods. |
| | | (*b*) *Caliothrips indicus*, Bon. | Thysanoptera Thripidae | Nymph & adult | Feed on tender foliage and leaves turn pale spotted. |

(*Contd.*)

| 1 | 2 | 3 | 4 | 5 | 6 |
|---|---|---|---|---|---|
| | | (c) *Agrotis ypsilon*, Rott. | Lepidoptera Noctuidae | Larva | Feed on leaves and tender shoots. |
| | | (d) *Euxoa spinifera*, Hubn. | Lepidoptera Noctuidae | Larva | Feed on leaves and tender shoots. |
| | | (e) *Etiella zinckenella*. Treil. | Phycitidae | Larva | Bore into the pods and feeds on immature seeds. |
| | | (f) *Leucinodes arbonalis*, Guen. | Pyraustidae | Larva | Bore into the flowers, buds and fruits. |
| | | (g) *Diacrisia obligua*, wlk. | Arctiidae | Larva | Feed on the lower surface of leaves. |
| | | (h) *Phytomyza articornis*, Meign. | Diptera Agromyzidae | Maggot | Feed on leaves and tender shoots. |
| | | (i) *Melanagromyza* pha-seoli, coq. | Diptera Agromyzidae | Maggot | Feed on leaves and tender shoots. |

Ornithologists round the world have not been able to make up-to-date assessments of damage caused by depredatory birds in different regions of the country. Birds are privileged in the sense that they have wings to fly, high intelligence and high sense of adaptability. This leads itself to the fact that methods of controlling birds are different from those of others. Pest status of birds undergoes changes from one region to another. Hence birds can be controlled differently in different regions.

**Important bird pest of under agroforestry system**

We may see birds, alone or in company, moving hither and thither and, at times, to remote places to find food for themselves. It is in the nature of bird-pest to flock to farmers fields and play havoc with the crop at different stages from sowing up to the thrashing yard.

**Rice (Oryza sativa)**

Rice is the principal food crop of about half of the world population. This crop also falls prey to bird pests at day break and sun set. They avoid visiting fields during midday. They find it easy to pick up the soaked seed from the soil and feed themselves on it. Birds like parakeet, myna, spotted munia feast on germinating seed and expose them to the atmosphere. The damage, caused by birds to the grain in the standing crop, begins with the beginning of its milky stage. Birds like house sparrow, munias etc., suck the milky substance and reduce the grain to a chaffy state. As the milky grain develops into ripe corn, roseringed parakeet, house sparrow, munias etc. feed on these shaking grain from panicle.

**Maize (Zea mays)**

It is one of the most important Kharif crops of India. The mature grains are produced in rows on the cob. A few hostile birds like parakeet, house crow or blue rock pigeon etc., damage crop at different stages from sowing up to the thrashing yard. After sowing is done, they are keen at finding soaked seeds in the field and, for three days, feed on them. These birds attack the crop at the milky stage, make a slit in the middle of a cob with their beak and individually feed on grains. Crows and parakeet are also by nature unsparing in damaging maize crops while the cob is at its milky and dough stage. (See Fig. 91)

**Fig. 91 :** The *Psittacula eupatria* peacking into the pulpy grains of the cob.

**Common Birds pests of Crops, Vegetables and fruits**

| Sl. No. | Crop | Common name | Scientific name | Nature of damage |
|---|---|---|---|---|
| 1 | 2 | 3 | 4 | 5 |
| 1. | Rice (*Oryza sativa*) | 1. Common Myna | *Acridotheres tristis*, Linn. | Growing seeds of paddy and uncovered shown seed etc. |
| | | 2. Kabutar (Pigeon) | *Columba livia*, Gmelin. | -do- |
| | | 3. Gauriya (House sparrow) | *Passar domesticus* Linn. | Uncovered grains, and at the stage of earhead too. |
| | | 4. Baba | *Ploceus philippinus*, Linn. | Alongwith sparrow at the earhead stage. |
| | | 5. Sugga or Tota (Red checked parakeet) | *Psittacula longicauda*, Bod. | In vast flock, at the earhead stage. |
| 2. | Wheat (*Triticum sativum*) | 1. Gauriya (House sparrow) | *Passer domesticus*, Linn. | After thrashing, during the sun-rays period and also in the field. |
| | | 2. Kabutar | *Columba livia*, Gmelin. | At the sowing, sunning and at the time of thrashing. |
| | | 3. Kowwa (Crows) | *Corvus splendens*, Viel. | Seedling. |
| | | 4. Kala Gandam (Black headed bunting) | *Emberiza melanocephala*, scop. | Standing crops at the time of thrashing. |
| | | 5. Chhota Fakhta (Dove) | *Streptopelia senegalensis*, Linn. | Gleaned grain. |
| | | 6. Sat Bhai | *Turdoides striatus*, Dum. | Uncovered seed (Grain). |
| 3. | Maize (*Zea mays*) | 1. Kabutar | *Columba livia*, Gmel. | At the time of maturity and during sunning. |
| | | 2. Chhota fakhta (Dove) | *Streptopelia senegalensis*, Linn. | Milky and matured grain. |

(*Contd.*)

| 1 | 2 | 3 | 4 | 5 |
|---|---|---|---|---|
| | | 3. Hiraman Tota (Large Indian Parakeet) | *Psittacula eupatria*, Linn. | From milky stage to the ripening. |
| | | 4. Tota (Roseringed Parakeet) | *Psittacula krameriscop*. | -do- |
| | | 5. Desi Myna (common Myna) | *Acridotheres tristis*, Linn. | Grain in the field but damage is not severe. |
| | | 6. Sat Bhai (Jungle babbler) | *Turdoides striatus*, Dum. | Seedling and grain at the milky stage. |
| 4. | Bajra (*Sorghum vulgare*) | 1. Kabutar (Pigeon) | *Columba livia*, Gmel. | At the sowing stage. |
| | | 2. Parki/Panduk | *Streptopelia decaocto*, Friva. | Shown, ripe and harvested grain. |
| | | 3. Blossom headed Parakeet. | *Psittacula cyanocephala*, Linn. | Legumes of Bajra of standing crop. |
| | | 4. Begeri/Jangichiri Yellow throated sparrow | *Petronia xanthocollis*, Burt. | Feeds on grain. |
| 5. | Pulses. | 1. Kabutar (Pigeon) | *Columba livia*, Gmel. | Showing, sunning and at the time of maturity. |
| | | 2. Tota | *Psittacula krameri*, Scopoli. | Uncovered grain in the field. |
| | | 3. Gauriyya (House sparrow) | *Passer domesticus*, Linn. | At the time of harvesting, thrashing and drying. |
| 6. | Mustard (*Brassicca* spp.) | 1. Hiraman Tota | *Psittacula eupatria*, Linn. | Destroy oily seeds. |
| | | 2. Parki (Indian ring Dove) | *Streptopelia decaocto*, Friv. | From ripening thrashing. |

(*Contd.*)

| 1 | 2 | 3 | 4 | 5 |
|---|---|---|---|---|
| 7 | Groundnut (*Arachis hypogaea*) | 1. Kowwa (Crows) | *Corvus splendens*, Vie. | Sowing and during ripening stage by digging through beak. |
| | | 2. Kabutar | *Columba livia*, Gmel. | At the time of sowing. |
| | | 3. Hiraman Tota | *Psittacula eupatria*, Linn. | From ripening to thrashing. |
| | | 4. Common Myna | *Acridotheres tristis*, Linn. | -do- |
| 8. | Sun Flower | 1. Tota (Roseringed parakeet) | *Psittacula krameri*, Scop. | Feed and damage seeds of standing crop. |
| 9. | Chillies (*Capsicum annunm*) | 1. Hiraman Tota | *Psittacula eupatria*, Linn. | Cut and feed fresh pods. |
| 10. | Maithi | 1. Gauriya (House sparrow) | *Passer domesticus*, Linn. | At the sowing stage. |
| 11. | Mango (*Mangifera indica*) | 1. Hiraman Tota | *Psittacula eupatria*, Linn. | Feed on the ripened fruits. |
| | | 2. Blossom headed parakeet | *Psittacula cyanocephala*, Linn. | Feed on the pulp of ripened fruits. |
| | | 3. Kowwa | *Corvus splendens*, Viell. | -do- |
| 12. | Litchi (*Litchi chinensis*) | 1. Blossom headed parakeet | *Psittacula cyanocephala*, Linn. | Feed on the pulp of fruit, sometime due to attack, fruits fall prematurely. |
| | | 2. Hirman Tota | *P. eupatria*, Linn. | -do- |
| | | 3. Myna | *Acridotheres tristis*, Linn. | -do- |
| 13. | Guava (*Psidium guajava*). | 1. White checked Bulbul | *Pycnonotus leucogenya*, Gray. | At the ripening and premature-stage |
| | | 2. Red vented Bulbul. | *P. cafer*, Linn. | -do- |
| | | 3. Hirman Tota | *Psittacula eupatri*, Linn. | -do- |
| | | 4. Red breasted Tota | *P. alexandri*, Linn. | -do- |

(*Contd.*)

| 1 | 2 | 3 | 4 | 5 |
|---|---|---|---|---|
| | | 5. Blossom headed Tota | *P. cyanocephala*, Linn. | At the ripening and premature-stage. |
| | | 6. Kokari | *Dendrocitta vagabunda*, Lath | At the ripening stage. |
| 14. | Grape (*Vitis vinifera*) | 1. Gauriya | *Passer domesticus*, Linn. | Makes the fruit rot in immature and ripened stage. |
| | | 2. Myna | *Acridotheres tristis*, Linn. | -do- |
| 15. | Ber (*Zizyphus species.*) | 1. Hiraman Tota | *Psittacula eupatri*, Linn. | Feeds on the pulp of ripened fruit. |
| | | 2. Myna | *Sturnus malaparicus*, Gmel. | -do- |
| 16. | Apple (Mallus sylvestris) | 1. Tota | *Psittacula himalayana*, Less. | -do- |

Apart from these crops, other crops of cereals like pulses and oilseeds are also subject to destruction by birds. A few birds also habitually feed on vegetables, even flower, buds and fruits.

Even when it is made into a pile and left sunning on open ground, a few birds are cunning enough to come flying unseen or half seen and feed on the grain. If calculation of loss caused by birds are made on the basis of intake by the bird pests, the quantity is bound to be grievously huge. It is, thus, obvious that a great deal of food grains, fruits and vegetables that we fail to bring into consumption go into the mouth of bird pests. These bird-pests seek shelter on trees and make roost on trees or buildings. These birds are called socially gregarious bird. In view of fighting the bird pests, it is of great importance to make a study of bird behaviour and breeding biology. The cultivators are, on an average, put to a great loss of crop because of the bird pest.

**Method used to protect the crops from birds pest in agroforestry system**

A few method can well be devised to protect the cereals from the enemy birds. Cultural practice also goes a long way towards freeing the crop from bird pest. Cultivation of bird resistant crop varieties appear equally promising as cultural practice for protection. Aldrich (1962) and Nichaus (1966) have discovered it in Sorghum grains containing higher tanin can withstand the ravage caused by birds. Apart from the above mentioned methods, deep sowing, sowing sometimes before or after the due date, concentration on quick mature variety, as also late maturity varieties help fight the depredation.

The contrivances to scare away birds are one of the age-old methods of bird management and control. The methods of frightening away the birds by the beating drums and slinging shots were invented by our forefathers. With the development of human science, newer methods on scientific lines were invented and applied of which scaring guns, pyrotechniques and bioacoustics are best examples. On the recommendations of Fitzwater and Prakash (1973) and Bindra and Tur (1972) bird scaring guns began to be widely brought into play to intimidate birds out of the fields. It is also effective in keeping off birds migrating from other regions and damaging the crop for a short period. In cases where these methods do not prevail upon the birds, we choose crackers as substitutes for them. As crackers burst out, it makes sudden frightful noise with flashes of fire which is surety to disperse the flock of birds. It has proved to be of great use in frightening large flocks of migrant birds. Sound of warning or cries of distress is purposely raised to cause them to disperse. Bird's call can be of avail in manifold ways, as for collecting them at a desired source by replaying their feeding and assembly calls and inversely, they may be scattered by emulating their distress and warning cries through playing recorded tapes. Major pest species like parakeets, sparrows and crows can be put under control by means of these methods. Frings and Jumber displayed in 1953 that starlings could be frightened out of their nests by using their distress calls. It is regrettable that India still lags behind in updating these methods and is still far from solving the problem.

Netting method is not so useful from practical point of view as the cost of net is increasing day by day and, moreover, it is inconvenient to hang net in the field. Its practical use can be felt only in protecting costly seed producing farms and orchards.

**Chemical method**

Chemicals used to evade the bird pests are inductive of corporal effects which eventually evoke behavioural responses and frighten and affect individual so as to cause dispersal of the entire flock. Chemical methods provide practical get-through to manage and exercise control over some species of birds. Even then, this method labours under certain limitations, and its effect lasts only for a limited period and is narrowed down to the area of applications only. In this way, as far as these chemicals produce responses in birds, these can be classified as stupefying, frightening and repellents. Daude (1942) and Hariere (1943) have reported that stupefying substance effects the cerebrocortical region of brain. Quantity of this chemical brings about behavioural changes like erratic flying and fluttering. As reported by Schafer *et al.* (1967), the effect of N-methyl carbamates on dispersing the crows, pigeons and sparrows has been wonderfully quick. In India, ornithologists have not diverted their attention to the studies of stupefying compound and it is time for them to make a study of biological activity of stupefying substance on Indian species.

The use of frightening agents is responsible for inducing erratic behaviour and eventually they become too affrighted to remain in flock. Goodhue *et al.* (1964) observed, that the effect of 4-

aminopyridine on *Passer domesticus* is to frighten. Mehrotra *et al.* (1979) has expressed in their report that besides aminopyridine compounds as well as other insecticides like Methoxychlor, carabaryl and delphen too, at low concentration, can be fruitful in the induction of fright. Mehrotra *et al.* (1979) reported that in U.S.A. 4-amino-pyridine N-Oxide and 4-aminopyradine under trade names of Avitol-100 and Avitol-200 respectively are registered for use. According to Besser (1976) baits containing 4-aminopyradine have been tested and tried on crops in different nooks and corners of America, and a number of black-birds feeding on pea-nuts was minimised to a great degree by baiting. Use of repellants serve twofold purposes *viz.*, affording repellancy and at the same time, bringing about some behavioural change, force them to disperse. It is of significant value to understand mode of action of repellants. Although, a series of repellents have been brought forward in view of seed treatment, none has proved so effective as bird repellent as Tetramethyl thiuram disulphite (Young and Zavallous, 1960). Royall and Ferguson (1962) studies on repellency activity of Tetramethyl thiuram disulphite against birds affecting wheat crop and found it fruitful. Other repellants brought into light till date are termed as lindane, Authraquinone, Malathion and Mathiocarb.

## RAT MENACE AND ITS CONTROL IN AGROFORESTRY SYSTEM

Apart from insect-pests and birds, rats also harm crops in many ways.

Rats are gnawing animals because they have chisel-like teeth at the front of their mouth. They are omnivorous and can eat all types of grains, fruits, vegetables, milk and its products. An adult rat consumes 56.7 gms of food in a day. 11 species of rats found in India are such pests which go on attacking and damaging grains from the time it is sown till it is harvested and consumed. The rampage caused by rats can easily be estimated with the knowledge of the breeding capacity of female rat. It has been proved through experiment in the laboratory that one female rat along with its young ones breeds approximately 1250 rats in a year. In every case of invisibility of rats, but with apparent signs of them, they are estimably understood to be one to hundred in number. Whereas, if seen only during night, their number is estimated to be one hundred to five hundred. If many rats are seen at night and several of them during the day, the number of offensive creatures may be calculably fixed at one thousand to five thousands in a spacious area.

A loss of about 40 crore quintal of grain is caused in India on account of rats. Hence it leads to the fact that we may save 40 crore quintal of grain, if only we could succeed in controlling the rats.

We may keep off the rats from our fields if we are alert and get acquainted with the nature and inclinations of rats. It has been experienced that rats are mostly active at night. It is when the world is laid to sleep that it finds itself free to toy with things unnoticed. It can also be seen at play in broad-day-light inside fields. It is interesting to know about the hides and haunts of rats. It is a very cunning creature, too cunning to be easily entrapped. It makes its stay in secret places or holes out of the reach of human ken. They can jump and climb three feet or so or climb wherever there is good footing. A study of the character of rats has shown that it is highly susceptible to fear or suspicion. Everyday is festival day for rats, till the crop stands in the fields. They infest the standing crops in thousands, dig holes here and there, collect ears, separate grain from them and inside their earthen tent, masticate them to their full.

**Control over rats in cornfields**

Rats can be kept under control in the cropfields but it requires a great deal of labour. If we try to control them by sprinkling chemicals here and there, we would fail in our attempt. First of all, for satisfactory control, it is very essential to make a survey of the area severely infested with rats. All the holes detected in the area should be plugged with dry soil. After a lapse of twenty four hours, those holes should be observed. The holes with damp or fragile soil at end about its mouth should be understood as living hole and populated one. In these very living holes, rats reside. The holes with no damp or fragile soil on the surface of it are to be known as dead holes or depopulated ones. It leaves nothing for doubt that rats have vacated these holes. The number of living holes will be one fourth of all the holes in the area. Dead holes should be plugged carefully and no rodenticides is needed to be used in these holes. The use of chemicals can be useful only in case of living holes. Chemical control of rat appears most practical and has great potential. Chemicals mentioned below are most commonly used to control rat population.

**Cyanogas**

It is the most common fumigant used for controlling rats in the cropfields. It is a volatile colourless liquid. Cyanogas contains 40 to 50 per cent CaCN and the gas is liberated slowly. On breaking the sealed tin, gas is emitted, which is known as Hydrocyanic acid gas. This very gas kills the rats. This is toxic to warm blooded animals as it combines with iron atoms of Cytochrome oxidase. This cyanogas is to be used with a good deal of caution. The tinned chemicals should be exposed to the air only when it is to be put to use. Soon after the chemical is thrust with foot pump, the tin should be properly closed. After all this, all the holes should be plugged well. The chemically affected holes should be continuously observed for three or four days.

It is very necessary to keep an eye on the weather and the dampness of soil before using the chemical. In time of rains or at hours when clouds are present in the sky, fumigant must not be put in to the holes. At such places where the land is damper, one should not use cyanogas. The application of fumigant on the damp soil does not bear any fruit. If fumigant is properly used in dry weather and on dry soil rats would be satisfactorily controlled. In case of godown infested with rats, the growth of the latter can be curbed by the proper use of Cyanogas.

Only experienced persons should be allowed to handle fumigant. Gas masks should invariably be used.

**Aluminium phosphide** (Calphos or phostoxin)

Aluminium phosphide is the most effective in killing rats. It is available in the market in the name of calphos and phostoxin. The tablet of celphos is of one gram and that of phostoxin is of 3 grams in weight. Every hole should be plugged after dropping into it two tablets of celphos or one tablet of phostoxin. A gas named phosphin is liberated, which kills all the rats inside the hole. It is better to make use of this tablet in the morning hour when all the rats remain inside the holes.

**Control of rats in the godown**

Besides the chemicals mentioned above a few other chemicals mentioned below may be used to control rats in the house or godown.

**Zinc phosphide ($Zn_3P_2$)**

This chemical is dusty in form whose colour is as black as coal. This is deadly poisonous. Rats die as soon as they consume it. The chemical can be used by mixing it with any food-stuff. A dose of chemical mixed with forty times more of food-stuff should be taken for use. As food-stuff, we may make use of bread, rice, flour, grampowder, piece of tomato, ripe banana, which serve as baits for rats. For prebating only foodstuff should be left for two to three nights in the house or godown. In two or three days, rats are tempted towards the foodstuff. On the third or fourth day, the foodstuff, properly mixed with chemical, should be kept at night in the godown. The baits should be placed beside the wall, behind almirah and in the granary. As usual, rats consume the baits and a large number of them die of eating it. The zinc phosphine reacts with the hydrochloric acid present in the stomach and release phosphine gas which is lethal to the rats.

**Rodofarin**

Rodofarin or warfarin is a readymade rat-killer available in the market, chemically known as 3-($\alpha$-Acetonylbenzyl)-4-hydroxycocumarin. Marketed rodofarin is a mixture of 5 per cent warfarin and its nineteen times of foodstuff. At every place infested with rats, a spoonful of rodofarin should be placed at night. Rats will die of eating it. It is the most effective chemical for controlling rats in godowns.

**Cautions**

Zinc phosphide and Rodofarin are highly poisonous chemicals, which may prove fatal not only to rats, but also to domestic cats, dogs, hens, goats and children and adults. Hence, every care and caution should be taken at the time of using these chemicals. After mixing the chemicals with food and placing it at proper place, hands should be washed thoroughly. Chemicals should be kept in counted places. The person that places the foodstuff mixed with chemical at night should be remembered and the ramaining stuff may be lifted in the morning and bury then so that cats, dogs or men may not die of eating it. The use of chemicals should be discontinued for a few days after using the chemicals at night. If it does not succeed to the extent of completely controlling the rats, then, after a lapse of a fortnight, chemicals mixed with changed foodstuff should be put to use. This way of using the chemicals twice or thrice may lead to a satisfactory control of rats.

**Use of Chemosterilant**

Chemosterilisation brings a new device of rat's control. The original concept of sterility method to hold insect population in check was first, suggested by Serebrovsky (1940). It is identical to that of Knipling's (1959) sterile male release method.

On the basis of research, several compounds were tested and found effective as sterilant on rats.

These chemicals have opened up a new avenue of research in rat control technology. The chemosterilants act in dissimilar ways. The physiological effect of individual compound and their mechanism of action are extremely variable.

For the last few years, considerable report has appeared on the chemosterilants effects on the gonad of rats. Several scientists have worked in this field but prominent ones are Cattanach (1959) ; Jackson *et al.*, (1959 and 1964) ; Steinberger (1962) ; Hemsworth *et al.*, (1963 a, b) etc.

On the basis of their findings, it may be concluded that gonad of treated rat showed complete or partial arrest of spermatogenesis and oogenesis. As a result of it, sperms and ova are not produced.

Furadantin and Colchicin chemicals have been tested as chemosterilants in C.S. Azad University, Kanpur. Oral administration of a single dose consisting of a mixture of 50 mg. Furadantin and 2 mg. Colchicine to either male or female lesser bandicot, (*Bandicota bengalensis bengalensis*) induced complete sterility to both sexes.

The compound has been extensively tested in laboratory. Yet the fact remains that further study on evaluation of compound against Indian species and their trial outside the laboratory should be given paramountcy.

**Rat trap as controlling device**

The use of mechanical rat-trap does not endanger anybody's life but its use on a large scale is most expensive. After using chemical trap, mechanical trap may be used to kill the rest in the following manners.

Rat-traps should be kept in a place which is mostly infested with rats. It is better to use a large number of rat-traps at a time at one night than to use it singly for many nights.

The well-set-rat-traps should be used for four or five days consequentively so that rats may enter into the trap and come out of it after eating the food. Thus they become fearless and get accustomed to enter into the traps in numbers. In order to add to the attraction of rats, foodstuff should be changed every night. Rats are easily attracted towards ripe banana, stale bread etc. It is worth mentioning that all the chemicals described above are available in the plant protection office and also in the market.

**Source :** Applied Agril. Entomology, New Central Book Agency, Calcutta.

**Research and Extension needs**

The following research and extension need require immediate attention :

1. The complex ecosystem of agroforestry system should be analysed from the socio-economic aspect and technical point of view.
2. All India Co-ordinated Project on Pest Management in agroforestry should be established.
   (*a*) To understand the pest interaction in agroforestry system and their suppression.
   (*b*) To establish a feasible and compatible crop rotation scheme in agroforestry system to reduce the population of pest in small scale farm.
3. In order to understand and solve the pest problems farmers, extension worker and scientist should work together.

## REFERENCES

Aldrich, D.J, Jr. (1962) People pests and some plants, *In Proc. vertebrate Pest control conf.* National Pest control Ass. ; Elizabeth, N.Y. pp. 213-226.

Beeson, C.F.C. (1941). *The ecology and control of the forest Insects of Indian and neighbouring countries*, Govt. of India Publication, pp. 172-173 and 489. (Reprinted 1961).

Beri, Y.P., Raizada, M. and Bhatnagar, R.K. (1972). Food and adults sparrow, *P. domesticus*, Linn, vol.2, 10 :65 (cross Ref.).

Besser, J.F. (1976). 4-aminopyridine for protecting crops from birds—a current review, *Proc. 7th vertebrate pest control : conf.*, p. 11.

Bhalla, O.P and Pawar, A.D. (1977). *A survey of insect and non insect pests of economic importance of Himachal Pradesh*, pp. 1-80. Deptt. of Entomology-Zoology, College of Agril, Chambaghat Solan, H.P.

Bhandari, R.S. and Sushil Kumar (1992). Insect pest management under agroforestry system of India, Cited from *Agroforestry Indian perspective*, Jha, L.K. and Sen-Sarma, P.K., Eds., Ashish Publishing House, Delhi.

Bhatnagar, R.K. (1976a). Significance of Bird management and control, *Pesticides*, pp. 74-83.

Bhatnagar, R.K. (1976 b). Bird pests of agriculture and their control, *Proc. Nat. Acad. Sci.*, India, vol. (46 B), I, II, pp. 245-256.

Bindra, O.S. and Tur, H.S. (1972). The common harmful birds of Punjab and their control, Punjab Agril. Univ., *Farm Bull.*; pp. 1-7.

Cattanach, B.M. (1959). The effect of triethyline melamine on the fertility of female mice, *J. Rad. Biol*; 3 : 288.

Chaudhuri, M.C. (1925). The defoliation of Gamhar in the Chittagong Hill tract, Bengal, *Indian For* : L I, pp. 57-60.

Daude, J.L. (1942). Capture et destruction des carbeaux, pies autres oiseaux nuisible aux recoltes. *Bull Acad. Med. Paris*, Vol. 126, pp. 452-454.

Deabreu, E.A.D. (1920). Some insect prey of birds in central provinces, pp. 859-874. Cited from Report of the *Proc. III Ento. Meeting* (1919), Pusa, Vol. III, Supdt., Govt. Press., Calcutta.

Fitzwater, W.D. & Prakash, I (1973). *Hand Book on Vertebrate Pest Control*, I.C.A.R., New Delhi, pp. I-III, 1-89.

Garthwaite, P.F. (1939). Biology of *Calopepla leayana* and the possibilities of control, *Indian Rec.*, V, No. 2, pp. 237-275.

Goodhue, L.D., Renert, A.J. and Williams, R.D. (1964). *Amino substituted heterocyclic nitrogen compounds as bird and rodent repellent*, U.S. Pat., 3, 150 041, 6pp. (C.R.).

Hariere, A.A.K. (1943). Contribution a L' etude toxicological chloralose. These docteur en pharmacie university de paris (In thearle, R.J.P.; 1968). (Cross Ref.)

Hedge, N.G. and Relwani. L.L. (1988). Psyllids attack *Albizia lebbeck* (Linn). Benth. in India, *Nitrogen fixing tree Research Report*, Vol. 6, pp. 43.

Hemsworth, B.N. and Jackson, H. (1963 a). Effect of Busulphan on the developing Gonads of the male rat. I., *Reprod Fertil.* 5 : 187.

Hemsworth, B.N. and Jackson, H. (1963 b). Effect of Busulphan on the developing ovary in the rat, *J. Reprod. Fertil.*, 6 : 229.

Jackson, H; Fox, B.W. and Graig, A.W. (1959). Antifertility substances and their assessment in the male rodent, *J. Reprod. Fertile*, 2 : 447.

Jackson, H; Partington, M. and Walpole, A.L. (1964). Production of heritable partial sterility in the mouse by methyl methanesulphonate, *Brit. J. Pharmacol.*, 23 : 521.

Jha, L.K. (1987). Pests of tropical fruits. Cited from *Applied agricultural Entomology*, New Central Book Agencies, Calcutta, p. 104.

Jha, L.K. (1987). Pests of oil seeds, cited from *Applied agricultural Entomology*, New Central Book Agencies, Calcutta, pp. 79-88.

Jha, L.K. & Singh Bina (1990). *Annual report on pest management in agroforestry system*, State council Adhoc Research Scheme B.A.U., Ranchi, pp. 5-6.

Knipling, E.F. (1959). Sterile male method of population control, *Science*, 130 : 901-904.

Mathur, R.N. (1961). Important Problems in forest entomology in India, *XI Inter. Cong. Ent. Vienna* pp. 277-283.

Masson, C.W. and Lefray, M.H. (1912). *Food of birds in India*, Memoirs Department. Agric. in India, Entomology Ser. 3, p. 367.

Mehrotra, K.N. and Bhatnagar, R.K. (1979). *Status of Economic Ornithology*, ICAR Publication, pp. 1-75.

Mehrotra, K.N.; Beri, Y.P.; Mishra, S.S. and Phockela, A. (1967). Physiological effects of malathion on house sparrow *P. domesticus* L. *Indian. J. Exptt. Biology* ; Vol. 5, No. 14, pp. 219-222.

Nayar, K.K., Ananthakrishnan, T.N. and David, B. V (1984). *General and applied entomology*, Tata Mc Grow Hill Publishing Company Limited, p. 273.

Nichaus, M.H. (1966). Bird resistant sorghum. *Ohio Rep. Res. and Dev. Biol. Agric. Home Econ*, Vol. 51, No. 5, pp. 70-71.

Raychaudhuri, S.P. (1983). Role of vector transmission for developments of sandal spike with reference to forest and Agri-Ecosystem. In : *Insect, Interelation in forest and agro-ecosystem*, P.K. Sen Sarma, S.K. Kulshrestha and S.K. Sangal, eds., Pub. by Jugal Kishor and Co ; Dehradun pp. 1-5.

Royall, W.C. (Jr.) and Ferugoson, E.R. (1962). Controlling bird and mammal damage in direct seedling loblotty pine in east Texas, *J. Forestry*, Vol, 60, No. I, pp. 37-39.

Schafer, E.W. (Jr.); Starr. R.I.; Cunningham, D.J. and Decino, T.J. (1967). Substituted Phenyl N-methylcarbamates as temporary immobilising agents for birds, *J. Agr. Food Chem.*, Vol. 15, No. 2, pp. 287-289.

Sen Sarma, P.K. (1975). Spike disease of sandal, a yellow type disease, *Proceeding of the 2nd world consultation of pests and Diseases*. IUFRO, New Delhi.

Sen Sarma, P.K. (1983). Preface In : *Insect interrelations in forest and Agro-ecosystem*, P.K. Sen-Sarma, S.K. Kulshrestha and S.K. Sanyal, Eds., p.-V.

Sen Sarma, P.K. (1992) Forest Entomology in India, *Proc. Zool. Soc.* Calcutta, pp. 1-5.

Serebrovsky, A.S. (1940). On the possibility of a new method of insect control, *Zool. zh.*, (USSR), 19 : 618-630.

Sevastopulo, D.G. (1939). The early stages of India lepidoptera. *Journ. Bomb. Nat. Hist. Soc.*, XL, pp-684.

Singh, and Singh, R. (1978). Two new pests of Khasi pine, *pinus, Kesiya, Indian Forester*, Vol II, p. 185.

Sivaramkrishnan, V.R. (1977). Experimental transmission of spike disease of sandal (*Santalum album*), by the leaf. hopper *Nephotettix virescens* (Distant). (Homeptera ; cicadellidae), *Indian Forestor*. Vol. 104, pp. 202-205.

Sivaramkrishnan, V.R. and Sen Sarma, P.K. (1978). Experimental transmission of spike disease of sandal, *Santalum album*, by the leaf-hopper *Nephotettix virescens* (Distant), *Proceedings All India* Sandal Seminar, Bangalore, 7-8th February 1977. Karnataka Forest Department ; India : 140-142.

Steinberger, E. (1962). A quantitative study of the effect of an alkylating agent. (Triethyienemelamine) on the seminiferos epithelium of Rats, *J. Reprod. Fertill.* ; 3 : 250.

Stebbings, F.P. (1914). *Indian Forest Insects of Economic Importance coleoptera*, Published by order of H.M's Secretary of state for India in Council, p. 401.

Subramaniam, T.V. and Anantanarayanan, K.P., (1938). *Journal Bomb. Nat. Hist. Soci.* XL. pp. 257-263. Cited from *Ecology of forest Insects*, C.F.C. Beeson. (1941), Govt. of India publication, pp. 477-478.

VanTyne, J. and Berger, A.J. (1959). *Fundamentals of ornithology*. John Wiley and Sons, Inc. London.

Waage, J. and Huxley, P.A. (1990). Leucaena Psyllid. Is this deadly pest on its way to Africa. *Agroforestry today*, Vol. 2(3), p. 13.

Young, W.R. and Zavallos, D.C. (1960). Studies with chemical seed treatments as bird repellents for protection of germinating maize in the mexican tropics, *F.A.O. plant. prof. Bull.*, Vol. 8, No. 4, pp. 38-42.

# 15

# *Pathogen Management in Agroforestry System*

There is no gainsaying that plant protection works is an important aspect for growing more and more corn, good quality of timber and other forest produce under agroforestry system. One cannot hope for desired output, if pests and pathogens are not kept under control. At times, the damage is so huge that the sickless of the farmers remain unused and they have to wring their hands in repentance. If farmers, take care control of pests and diseases, things would not come to a sorry pass.

The damage caused by plant diseases have affected man all through the ages. Besides the recurrence of rusts and mildews of crop plants, there have been periodic outbreaks of certain diseases which caused large scale death and famine. The severe epidemic occurrence of the late blight of potato in Europe during 1830-1840 and later in Ireland in 1845 caused the famous Irish Famine resulting in death of about a million people and large scale migrations of the people to other countries, particularly to North America. The severe outbreak of helminthosporise of rice in north Eastern India, causing a heavy loss of food grains, is reported to have caused the Bengal Famine of 1943, leading to several thousands deaths and political unrest in the country (Rangasawami, 1984).

In India, ecosystem in the traditional as well as in scientific agroforestry practices varies widely in different agroclimatic zones. The disease situation varies with the nature of the agroforestry ecosystem and divergent cultural practices. Thus, pathogen management in agroforestry system is different from the management of pathogen in pure agricultural crops or in pure forest. Thus, there is need to study in-depth about the nature of pathogen problems in different agroforestry systems and development of appropriate technology to combat pathogen problems. The study of pathogen problems in agroforestry system would be interesting and quite important in order to understand the ecology of pathogens to tree and agro-ecosystem. Though, only a few reports are available in regard to pathogen management in agroforestry system. This chapter deals the pathogen problems in different agroforestry systems and the suitable methods for their management.

## (A) PATHOGEN INTERACTION IN AGROFORESTRY SYSTEM

It is apparent that multiplication of pathogen depends upon the type of plant communities. The parasitic diseases are characterized by a cycle of events starting from infection of the host by the pathogen, its multiplication, fructification, dissemination and re-infection of the host. The time required for completion of the cycle varies with the pathogen, host, environmental condition and their interactions. First, the pathogen must be virulent and present in sufficient number to provide the minimal inoculum potential. Secondly, the host must be susceptible to infection by the establishment of the pathogen. Thirdly, the environmental conditions must be favourable to the activity of the pathogen as well as to susceptibility of the host and, more than all, conducive to the development of a host-pathogen relationship (Rangaswami, 1984). Agroforestry system has very complex ecosystem. Tree species and crop plants in agroforestry system may influence the pathogenic inter-relationship between crop plants and tree species. Only a few reports are available about the pathogen interaction in agroforestry system.

### *Sclerotium rolfsii* Sacc. State of *Corticium rolfsii* Curzi

*S. rolfsii* is a common soil-borne pathogen which is known to parasitize seedlings of various tree species (Browne, 1968). The sclerotium leaf spot of teak caused by *Sclerotium rolfsii* occurs in all the tropical and subtropical areas with high moisture and temperature.

Sharma *et al.*, (1985) reported that *S. rolfsii* infect *Tectona grandix* (teak), *Eucalyptus grandis* and *Eucalyptus tereticornis*. In teak, the disease is characterised by the initial dull-brown coloured, small leaf spots, 2-3 mm across, which undergo prolonged wet periods, increase rapidly in size irregularly, often covering more than one-fourth of the lamina. Frequently, when the mid rib of petiole is attacked, leaves wilt and dry up; rarely tender parts of the stem are also infected and growing shoots are killed. A profuse mycelial growth is generally observed on the lower surface of the necrotic area of leaf. Abundant white coloured sclerotia are produced on the mycelial stands, which become dark when mature and disperse by rain drops and wind to lower leaves where they cause fresh infection.

In eucalyptus *S. rolfsii* forms a white weft of mycelium at the base of stem, spreading up to the leaves. The general wilting of eucalyptus seedlings was accompanied by the development of round, off-white sclerotia on the affected leaves and stems. The sclerotia becomes light brown with age. Wilted seedlings turn brown and die. *Eucalyptus grandis*, which is succeptable to wilt at a younger stage, become affected with root-rot when seedlings are 4-5 month old. The root rot intensity can be reduced considerably by drenching affected seed beds with Errisan - 6(0.005% a.i) (Sharma, *et al.*, 1985).

Ragi or Finger millet, one of the important food crops, of Karnataka and Andhra Pradesh is severely infected by Mosaic disease complex caused by one or more species of *Helminthosporium* and *Sclerotium rolfsii.* After 4 to 6 weeks entire plant become pale, dwarfed and sterile. It also cause wilt, when crops are attacked by insect and nematode. The fungus enters through the wounded root and becomes active during rainy season and causing severe damage to the plants.

*S. rolfsii* has also been reported to cause tuber rot in sweet potato (*Ipomea batatus*), rhizome rot in Kachu (*Colocasia esculents*), sclerotium rot in potato (*Solanum tuberosum*) and several other crops.

Presently farmers are taking much interest in planting teak, eucalyptus on the bund and in farm land. If above crops are grown along with teak and eucalyptus, chance of infection, multiplication, re-infection of the pathogen is higher.

***Fusarium solani***

It has been reported that *Fusarium solani* cause wilt disease in sissoo (*Dalbergia sissoo*), seedling blight in Gamhar (*Gmelina arborea*) and wilt and root rots of brinjal (*Solanum melongena*) and other crops.

Singh (1986) has reported that sissoo suffers from wilt disease caused by *fusarium solani*, if the species is raised on stiff and clayey soils with poor drainage as on such sites finer roots are killed due to asphaxiation and are subsequently colonized by the wilt pathogen which then progresses to conducting tissues and causes the disease. Sissoo also suffers from wilt disease in taungya plantations raised on sites with high water table, as on such sites also finer roots are killed by repeated inundation in subsoil water.

Same pathogen also causes wilt and root rot of brinjal. The younger leaves die in first phase and after few days entire plants wilt and die. If vegetables like brinjal are grown in the interspace of *Dalbergia sissoo* under agroforestry system, potential of multiplication of pathogen will be higher, because fungus remains present saprophytically in the soil. When host plants are grown, it attacks root region and spreads upwards. Under such condition proper management of pathogen is essentially required.

***Fusarium semitectum***

Under alley cropping system, *Leucaena leucocephala* is widely planted and crops are grown in the interspace. *Fusarium semitectum* infect subabool and cause gummosis and canker disease. The pathogen is seed borne and also infect many other intercrops like wheat. Seed infection can be successfully eradicated by giving a dip treatment to seeds in 0.1% solution of emisan for 15 min. before sowing (Singh *et al.*, 1983). Thus in order to get desired yield, control of above seed borne disease is essential under agroforestry system.

*Rhizoctonia solani* **Kuhn.**

*R. solani* is a common facultative parasite known to cause varied kinds of diseases in a number of host.

The main tree host plants are *Eucalyptus, Ailanthus triphysa and Bombax ceiba.* This pathogen is also considered important pathogen of few agricultural crops and vegetables like potato (*Solanum tuberosum*), wheat (*Triticum* species), Spinach (*Spinacea oleracea*).

The Black scurf of potato is caused by *Rhizoctonia solani,* is characterised by black root. Black sclerotial bodies are formed on the diseased stem and tubers. Same pathogen cause seedling blight of wheat and root rot in Spinach (*Spinacea oleracea*).

In *Bombax ceiba*, initial symptom of the disease manifests as watersoaked lessions on the hypocotyle. Later on these lessions turn brown in colour and the infected area gets decayed. Affected seedlings usually collapse from the decayed portion fall over the ground and die. This pathogen causes collar rot in *Ailanthus triphysa* (Sharma *et al.*, 1985).

This pathogen is considered to be occasional pathogen of agroforestry nurseries. Under agroforestry system this pathogen may be a threat to the crops. Thus, judicious management to control the pathogen is essential.

*Corticium salmonicolar* **Berk and Br.**

Sharma, *et al.*, (1985) reported that pink disease was widespread in plantation of *Eucalyptus tereticornis* throughout Kerala. The site of the plantation appeared to have a considerable practice, raising of tapioca as a taungya crop for the first two to three years of establishment also had considerable impact on the incidence of pink disease. Since the tapioca grows faster and forms a close canopy even if the above described conditions were absent, it create conducive microclimatic conditions, which facilitated pink disease infection. Under identical conditions a crop of either, sesame or paddy as taungya had low incidence of pink disease.

***Pseudomonas solanacearum***

The pathogen infects the vegetables and fruits like *Solannum tuberosum, Lycopersicon esculentum* (tomato), brinjal, *Musa* species (banana) and trees like *Casuarina* and *Tectona grandis*.

The bacterium *Pseudomonas solanacearun* causes heavy loss in these crops and tree species. In tree species initially the small patches of brown tissues appear between veins. This is followed by wilting of leaves and seedlings. The dark discolouration of vascular tissue confirms the disease (Kaushik & Bangarwa, 1992).

In vegetables like potato and tomato the disease spread during cultural operations and through irrigation and rainwater. It remains saprophytically on plant residues left over in the field. In potato *Pseudomonas solanacearum* cause brown rot. When it attacks potato crops, dwarfing of the shoots by brown discolouration of the leaves are observed and finally plant wilt. The tubers also rot in the soil. In tomato this virulent pathogen causes vascular wilt disease. This is considered a serious pathogen of potato and tomato. Moko disease of banana is prevalent in Kerala and Tamil Nadu. In the initial stages, the bacterial wilt is characterised by the development of yellowish discolouration of the inner leaf lamina close to the petiole. This is followed by the collapse of the lamina near the junction of lamina with the petiole. Within a week, most of the leaves shows wilt symptoms, and consequently, wilting of the inner leaves with the exception of heart leaf, if common although in some cases the heart leaf may also collapse (Rangaswami, 1984). (See Fig. 92)

In multicropping system trees are planted on bund, and banana and vegetables are grown in the field. Under this condition, pathogen will multiply and may cause great loss in the presence of seasonal and perennial food crop and trees. Thus, under agroforestry system, all suspected plants should be destructed to minimise the spread of pathogens in new areas.

***Macrophomina phaseoli***

Kaushik and Bangarwa (1992), pointed out that tree species like mulbery, popular, acacia, albizia, fig, etc., suffers from *Macrophomina phaseoli* seriously under agroforestry system. This pathogen also cause root and stem rot in tomato (*Lycopersicon esculentum*), brinjal (*Solanum melongena*), Gram (*Cicer arietinum*), chilli (*Capsicum annuum*), cluster beans (*Cyamopsis tetragonoloba*) and many other legumes. The infested leaves and roots turn brown and finally plant wither. In potato it causes charcoal rot. This disease is characterised by darkened patches on the tubers. In agroforestry system when above crops and trees are grown on same unit of land, multiplication of the pathogen should be stopped by judicious management practice.

***Ceratocystis fimbriata***

Kaushik and Toky (1990), have reported that *Ceratocystis fimbriata* causes stem cankers on poplar and neem. Both tree species are widely planted in agroforestry system. This also causes black rot in sweet potato (*Ipomea batatas*) and other vegetables. It is essential to conduct further research to understand pathogen host interaction in agroforestry system.

***Phytophthora palmivora* and *P. parasitica***

This pathogen causes rotting of rootlets, girdling of the trunk and defoliation. The first symptom is dark staining of the bark and the infected bark is at first light olive brown in colour, firm, with a water-soaked appearance. The fungus affects the bark and can penetrate in a limited way into the wood. In advance stages, the bark

**Fig. 92 :** Potato Brown rot (Ring like vascular discolouration) caused by *Pseudonomas solanacearum*.

cracks, spreads in length-wise strips as it dries and gum starts exuding, particularly in the late spring. (Ghosh, 1985). Sometime fruits are also attacked, thus causing weakening of the stalk and resulting severe fruits drop. (See Fig. 93)

*Phytopthora palmivora* is also associated with fruit and stem rot of tomato. In most of the warm and humid citrus growing belt Pine apple (*Ananas comosus*) are grown in the interspace. *Phytophthora parasitica* and *P. palmivora*, causes root rot disease in pineapple. The leaves turns olive brown coloured and root appears dark. Fruit development ceases. In the late stage entire plant wilts. Thus, in citrus growing belt good quality of pine apple fruits is only possible by suppression of this soil-borne disease. Further research is required to find out effective method of control.

**Cladosporium fulvum**

In agrihorticulture practice when ber trees are planted, intercrops can be grown for 2 to 3 years. In the north Bihar, farmers grow wheat and tomato, etc. *Cladosporium fulvum* causes light brown spot on the upper surface of ber leaves and it also causes seedling blight in wheat and leaf mould in tomato. Multiplication of pathogen can be controlled by spraying dithane Z-78 (0.2%) at an interval of three weeks. The field should be kept clean.

**Sandle spike disease**

The disease causes reduction in size of leaves and shortening of internodes resulting in a 'spike' like appearance of branches. Shivaramkrishnan and Sen Sarma (1978), were the first to discover that *Nephotettix virescens* also transmits sandle spike and 12 out of 15 test plant got infected by using 60 infective leaf hopper per test plant. Disease transmission by *N. Virescens* was confirmed that *Nephotettix* induced diseased plants passed on the disease symptoms *via* bad grafting and *cuscuta*. Roy Chaudhuri reported that rice yellow draft is caused by mycoplasma, which is also transmitted by *Nephotettix virescens*. Rice is grown in close proximity of sandle tree plantation in Karnataka, Tamil Nadu and Kerala. It will be worth-while testing (*i*) if *N.virescens* is the natural vector of sandle spike; (*ii*) whether or not the disease, can be transmitted by the vector from the spike sandle wood to healthy rice developing into yellow dwarf disease and *vice versa*. Further research in this direction is essential to understand the role of this insect in transmitting disease in sandle tree and rice field. If the fact is established, we have to find

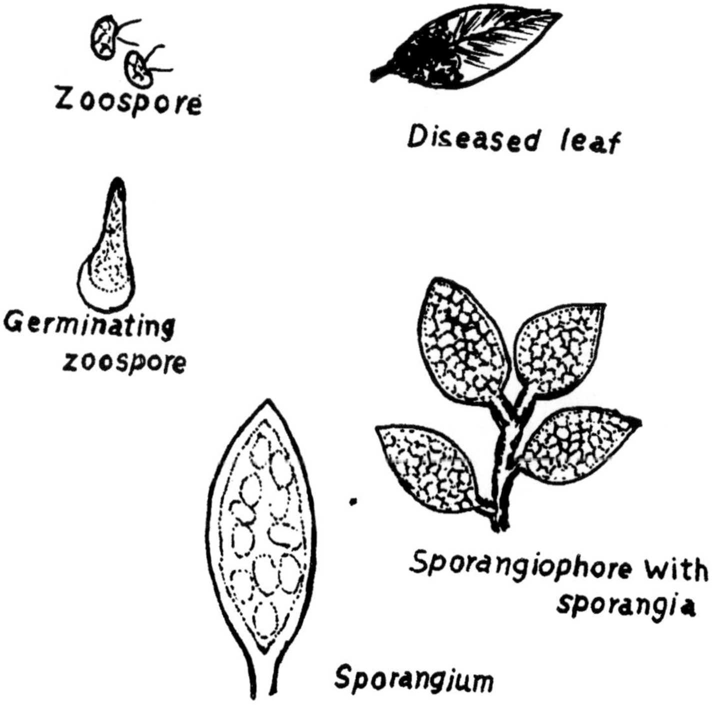


**Fig. 93 :** *Phytophthora palmivora.*

out suitable insect suppression technique, in order to protect costly sandle wood stand and to maintain the productivity of rice.

## (B) PLANTATION DISEASES AND THEIR MANAGEMENT

The incidence and nature of damage caused by pathogens in agroforestry plantation are lacking. Bakshi *et al.*, (1972), Sharma, *et al.*, (1985) and others reported about the pathogen problems of forest trees in forest area. The pathogen problems to agroforestry trees are discussed here under subheadings root pathogens, stem pathogens and foliage pathogens.

### (*a*) *ROOT PATHOGENS*

#### *Ganoderma lucidum*

Ganoderm root-rot causes serious losses to *Dalbergia sissoo, Acacia catechu, Eucalyptus* species, *Albizia lebbeck, Populus* species, citrus, etc. The pathogen is soil-dweller, attacking the roots and spreading upward direction. Bakshi (1976), reported contact between roots of diseased and healthy trees, was the mode of disease spread and affected trees die.

#### *Cylindrocarpon lucidum*

*Cylindrocarpon* root-rot causes loss to eucalyptus and acacia. In a young *Eucalyptus tereticornis* plantation, the disease appeared to be a potentially serious one. In severe cases most of the roots of the affected trees is dead and it spreads up to the collar area and forms canker.

#### *Cylindrocladium floridanum*

Sharma, *et al.*, (1985) reported *Cylindrocladium* root rot in plantation of *Eucalyptus tereticornis*. The leaves of the affected plants become flaccid and apical shoot shows droping. The leaves turn brown and dry up. The root system of such plants is found to be completely rotted with pronounced brown discolouration. The affected plants fail to survive.

### Management of Root Rot Pathogens

It is apparent that root rots are caused in different tree species by various pathogens. All pathogens may not be controlled by one fungicides. In general, management practices mentioned below should be practiced to suppress the multiplication of pathogen.

1. Removal of dead roots before the growing season of intercrops.
2. If only few trees are infected, trees must be uprooted.
3. Always avoid injuring to root system during cultural operation.
4. Use of effective fungicides at the proper time under the supervision of extension worker.

(*b*) *STEM PATHOGENS*

*Corticium salmonicolor*

The pink disease caused by *Corticium salmonicolor* is recorded on many agroforestry tree species like, *Ailanthus triphysa, Azadirachta indica, Bombax ceiba, Cassia siamea, Eucalyptus* species *Gmelina arborea, populus deltoides, Tamarindus indica, Tectona grandis,* etc. It is widespread in India. The disease is characterised by a distinct pinkish powdery coating on the stem. The pathogen invades the bark and forms abundant spore masses. Gradually tissues of the inner bark, including cambium are killed and canker is formed. The shoot above the canker is killed. During dry period, longitudinal splitting of bark occurs.

*Botryodiploidia theobromae*

Singh, *et al.*, (1983) has reported that *Botryodiploidia theobromae* has been found to colonize in basal wounded portion and develops cankers. In case of severe infection pathogen kill the affected plants. Sharma, *et al.*, (1984) has observed stem (root) canker disease in *Ailanthus triphysa* and *Eucalyptus* species respectively. (See Fig. 94)

In Eucalyptus wilting was recorded. In all the diseased plants the root collar region was typically constricted and compressed with irregular crevices. Often the stem just above the canker was abnormally swollen. This swelling was the result of callus growth beneath the bark associated with the infection. Occasionally the splitting and rupturing of the outer bark at the canker resulted in girdling of the stem, and plants died. The tissue in the canker region was brown and dead or dying; sometimes the browning also extended to roots, causing root decay. (Sharma, *et al.*, 1985).

**Management of canker pathogen**

1. Avoid causing wounds during cultural operation by employing trained labourers in agroforestry field.

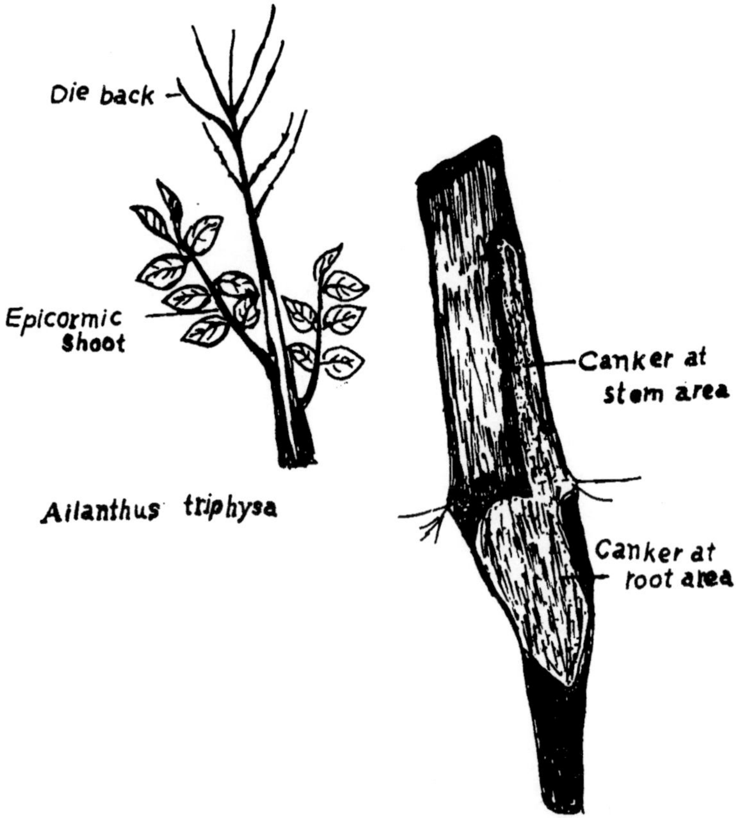


Fig. 94 : *Botryodiplodia theobromae*

2. Remove all infected portion by pruning and apply fungicidal paint over whole wound.
3. Use of recommended doses of fertilizer and regular watering provides congenial atmosphere for proper growth of trees. It also increases resistance to disease.

### (c) *FOLIAGE PATHOGENS*

The most common foliage diseases are leaf spot, leaf blotch, anthracnose, powdery mildew, etc. Some of the foliage diseases have been discussed under heading pathogen interaction in agroforestry system.

Kaushik and Bangarwa (1992) have stated that leaf disease might not threaten the health of agroforestry tree, it could cause the tree to lose its true purpose indirectly. The leaf disease can be controlled by sanitation. It has often been recommended that infected leaves be gathered and burned in the field in order to remove any potential sources of inoculum for the tree. The spores that infect the leaves may also travel great distance by the wind, so that removal of local sources of inoculum may have no effect on reducing disease severity. It that case a farmer has to go for protection against infection using fungicides. Fungicides can be used effectively to protect the tree against infection from most leaf pathogen by spraying at or before bud break after every 2 weeks following until the wet weather prevails.

## (C) NURSERY DISEASES

### (*a*) *Tree seed pathogen*

Social/agroforestry plantation requires large quantity of viable seeds. Seeds can be damaged by pathogen during their formative stage on the mother trees, or those falls on the ground or storage.

Sen Sarma, *et al.*, (1988) have stated that like insects, microrganisms, including fungi play an important role in the infestation and deterioration of seeds in storage and in seed transmission of disease. Most records available on seed borne fungi are invariably based on mere presence of mycoflora on, in, or on accompanying the seeds and are quite often accorded the pathogenic status with preconceived assumption of seeds transmission. While some of the seed borne fungi infect seeds in the field "field fungi", there are some which invade seeds generally after harvest "storage fungi". There is yet another group of fungi which resemble the above groups in their general ecology, but this group invades seeds after harvest "post-harvesting fungi" (Jamaludin, 1980).

Tree seeds, are liable to attack by various microbes, especially the fungi, due to the presence of cellulose, lignin, proteins, carbohydrates, fats, etc. Of the long list of seed born fungi nearly forty-two species of genera *Alternaria, Aspergillus, Cephalosporium, Chaetomium, Cladosporium, Curvularia, Epicocum, Fusarium, Gliocladium, Poecilomyces, Penicillium, Pithomyces, Rhizopus, Trichoderma, Trichothecium* and *Ulocladium* have been found commonly associated with seeds of timber trees (Noble and Richardson, 1968).

Sen Sarma, *et al.*, (1988) have pointed out that of these, the species of genera *Aspergillus, Chaetomium, Fusarium, Penicillium* and *Rhizopus* are more common fungi of tree seeds under storage.

In some cases, seed borne fungi infect seeds, while they are still on the plant which are then carried to storage where they remain dormant but reappear in the filed, *e.g. Collectotrichum* spp. and *Macrophoma* spp. on *Pongamia pinnata* (Jamaludin, 1980).

**Management**

Management practices to protect the pathogen infestation to stored seeds of forest tree species are in progress in various forestry research institutes and forest colleges. The methods devised to protect the agricultural seeds can be applied for protecting tree seeds with considerable degree of success till such time when more data are generated on forest seed pathogens. Sharma, *et al.*, (1988) have stated that various methods adopted and in vogue are cleansing method, sterilisation method, seed treatment with chemicals and storage in chemically treated or air-tight seed containers. In Hot and Cold treatment methods seeds are exposed to a blast of hot air followed by a blast of cold air. A more radical method, usually recommended for sterilisation of seeds, meant for storage, is subjecting the seed stock to lethal agents (other than chemicals) such as ultraviolet rays, X-rays or gama-rays.

*(b) Seedling disease*

For social/agroforestry plantations, seedlings are required in millions. The seedlings are raised in temporary nurseries located near the sides of crop fields. There is considerable loss of nursery stock due to diseases and insect pests which often disrupts the scheduled plantation programmes (Sen Sarma, 1986) The common seedling diseases are as follows :

**Damping-off**

Damping off is a most common disease of tree species nurseries. It may be defined as "rapid rotting of seedlings before germination (Pre-emergence damping off) or rotting of bases of seedling so that they fall over (post-emergence damping-off) (Rao, 1972). Some of the common pathogens associated with damping off disease of tree species nurseries are, *Aspergillus, Botrytis, Botryodiplodia, Cladosporium, Cylindrocladium, Fusarium, Rhizoctonia,* etc. (Boyce, 1961; Browne, 1968; and Sharma, *et al.*, 1985).

*Ailanthus triphysa* and *Eucalyptus* species are planted in the farm yard and home-stead. Sharma, *et al.*, (1985) have reported that *Pythium* spp. and *Rhizoctonia solani* cause damping off disease in *Ailanthus triphysa* and *Eucalyptus* species respectively. Pythium is a common damping-off pathogen capable of causing large scale mortality of young seedlings under favourable condition. Disease occurred within two weeks of germination of seeds, when the first pairs of leaves are just emerging, and causes upto 50-60 per cent mortality of seedlings. The disease appears in the form of irregular patches in seed beds. These patches enlarged rapidly from the periphery affecting the neighbouring healthy, seedlings under high soil moisture regimes. The disease may be controlled by two soil drenches of Dithane M-45 (0.05% and 0.02% a.i.) applied at seven days interval.

**Root rot**

There are a few fungi associated with root rot of seedlings. The chief ones are *Macrophomina phaseoli, Phytophora* spp., *Cylindocladium* spp., *Helicobasidium*, etc. Gupta and Kaushik (1988), observed severe root rot in eucalyptus due to *Cylindrocladium quinqueseptatum*. Sharma, *et al.*, (1988) reported occurrence of root rot disease in *Eucalyptus tereticornis* and *Eucalyptus grandis* caused by *Cylindrocladium eurvatum, Sclerotium rolfsii* and *Rhizoctonia solani.* Root rot causes slow wilting of seedlings in scattered patches. The first symptom is the change of pigmentation in apical leaves from normal green to light purple. The apical portion of seedling showed wilting resulting in death of the plants. The root system of such plants completely damaged due to rotting. Singh and Singh (1985) have pointed out that *Rhizoctonia solani* causes heavy losses in *Populus ciliata* nursery.

Sudden dying of stray seedlings is the first symptom of the disease. The mortality then spreads around the dead seedlings and ultimately patch mortality occurs in the nursery beds. The roots were completely decayed by the pathogen. The pathogen produced globose, hard, light brown sclerotia in the soil as well as on root surface. The Pathogen survives in the soil in the sclerotial stage. Control of *Rhizoctonia* root rot can be achieved either by sterilization of the soil or by mixing the soil with fungicides before inserting the cuttings or by drenching the nursery bed with water suspension of fungicides after sprouting of the cuttings.

**Alternaria tip blight of poplar**

Singh and Singh (1985) have reported that *Alternaria* stage of *Pleospora infectoria* causes death of growing tips of young poplar plants in the nurseries. Main symptom of disease include the development of brown to dark brown spots on young unfolding leaves, blackening of growing tips followed ultimately by death of juvenile leaves of young twigs.

**Bacterial wilt**

This disease was recorded in *Casuarina* and *Tectona grandis*. The bacterium *Psedomonas solanacearum* causes heavy loss. Initially the small patches of brown tissues appear between veins. This is followed by wilting of leaves and seedlings. The dark discolouration of vascular tissues confirm the disease.

In addition to the above disease, many other pathogens causes damage to seedlings of different tree species in nursery.

**Suggestions**

* Great variation in climatic, topographic and edaphic conditions requires study of diseases of crop, and tree species under agroforestry system in each agroclimatic zone.
* Agroforestry system have complex ecosystem. Pathologists should carry out research regarding intensity, nature of damage their recurrence year after year and control of air borne, seed-borne, soil-borne pathogens of crop and tree species under agroforestry system.
* Study on role of meteorological and microclimatic factors in multiplication dissemination and re-infection of the hosts by the pathogens is essential.
* Need to develop new varieties to resist the prevalent diseases.

**MAJOR PATHOGEN(S) ON SELECTED TREES AND CROPS IN AGROFORESTRY SYSTEM**

| Sl No. | Host | Pathogen(s) | Disease |
|---|---|---|---|
| 1 | 2 | 3 | 4 |
| **(A) FODDER TREES** | | | |
| (*i*) | Subabool (*Leucaena leucocephala*) | (*a*) *Clonosterchys compectiuscula* | Infects dead bark and pods. |
| | | (*b*) *Didymium minus* | Develops slime mold on pod. |
| | | (*c*) *Didymium squamulosum* | –do– |
| | | (*d*) *Fusarium semitectum* | Causes gummosis and canker disease. Infected plant remain thin and stunded. |
| | | (*e*) *Fusarium acuminatum* | Causes gummosis. Affected plants showed stunted growth. Heavy gum produced from gum slits from the main trunk. |
| (*ii*) | Siris (*Albizia lebbeck*) | (*a*) *Ravenelia sessilis* | Causes severe incidence of rust. |
| | | (*b*) *Ganoderma lucidum* | Causes root - rot. |
| | | (*c*) *Cercospora albiziae* | Causes leaf spot and premature defoliation. |
| (*iii*) | Sesbania (*Sesbania grandiflora*) | (*a*) *Pseudocerospora sesbaniae* | Causes grey leaf spot on leaves. |
| | | (*b*) *Uromyces poonesis* | Causes rust on leaves. |
| (*iv*) | Sissoo (*Dalbergia* sissoo) | (*a*) *Ureco sissoo* | Causes rust on leaves. |
| | | (*b*) *Fusarium solani* | Roots are killed due to asphyxiation. |
| | | (*c*) *Ganoderma lucidum* | Causes root - rot. |
| | | (*d*) *Phyllactinia delbegias* | Causes powdery mildew. |
| | | (*e*) *Cercospora sissoo* | Causes leaf spot. |
| | | (*f*) *Colletogloeum sissoo* | Causes leaf spot and premature defoliation. |

(*Contd.*)

| 1 | 2 | 3 | 4 |
|---|---|---|---|
| (*v*) | Khair (*Acacia catechu*) | (*a*) *Ganoderma lucidum* | Causes root - rot. |
| | | (*b*) *Ravenelia tandonii* | Causes foliage and twig rust. |
| | | (*c*) *Erysiphae acaciae* | Causes powdery mildew. |
| (*vi*) | Neem (*Azadirachta indica*) | (*a*) *Corticium salmonicolor* | Causes pink disease. |
| | | (*b*) *Cercospora meliae* | Causes leaf spot and premature defoliation. |
| | | (*c*) *Cercospora subsellis* | Causes leaf spot. |
| **(B)** | **TIMBER TREE SPECIES** | | |
| (*i*) | Teak (*Tectona grandis*) | (*a*) *Pseudomonas* species | Causes bacterial wilt. Systematic infection of vessels. Plants turned yellow. |
| | | (*b*) *Uncinula tectonae* | Causes powdery mildew. Irregular white patch on leaves. |
| | | (*c*) *Olivea tectonae* | Causes leaf rust. Severe infestation caused defoliation. |
| | | (*d*) *Corticium salmonicolor* | Causes pink disease. Canker formed on the stem with pink encrustation. |
| | | (*e*) *Phialophora richardsiae* | Causes die-back. |
| | | (*f*) *Marasmiellus ignobilis* | Causes stem rot. Infected portion becomes soft and spongy. |
| (*ii*) | Gamhar (*Gmelina arborea*) | (*a*) *Thyronectria pseudotricha* | Causes stem canker. |
| | | (*b*) *Corticium salmonicolor* | Causes pink disease near the upper portion of stem. |
| | | (*c*) Sclerotium species | Causes stem rot. Infection appeares at the base of the stem. |
| (*iii*) | Poplar (*Populus* species) | (*a*) *Rhizoctonia solani* | Causes root-rot. |
| | | (*b*) *Botryodiplodia palmarum* | Causes bark decay both in sprouted and unsprouted cuttings. |

(*Contd.*)

| 1 | 2 | 3 | 4 |
|---|---|---|---|
| | | (*c*) *Ganoderma lucidum* | Causes root-rot. |
| | | (*d*) *Corticium salmonicolor* | Causes stem and branch cankers. |
| | | (*e*) *Melempsora ciliatea* | Causes leaf rust and premature defoliation. |
| | | (*f*) *Cladosporium humile* | Causes serious leaf spot disease. |
| (*iv*) | Eucalyptus (*Eucalyptus tereticornis* | (*a*) *Botryodiplodia theobromae* | Colonize in basal wounds and causes canker. |
| | | (*b*) *Cryphonectria cubensis* | Causes stem canker and gummosis. |
| | | (*c*) *Corticium salmonicolor* | Causes pink disease. Infected trees appears bushy. |
| | | (*d*) *Cryphonectria gyrosa* | Causes stem canker. |
| | | (*e*) *Cytospora eucalypticola* | Causes branch canker. |
| | | (*f*) *Nattrassa toruloidea* | Causes stem canker. Complete girdling of stem causes death of plants. |
| | | (*g*) *Coniella fragariae* | Causes leaf spot and premature defoliation of leaves. |
| **(C)** | **FRUIT TREE SPECIES** | | |
| (*i*) | Citrus | (*a*) *Phytophthora parasitica* | Causes phytophthora root rot disease. |
| | | (*b*) *Phytophthora citrophthora* | -do- |
| | | (*c*) *Phytophthora palmivora* | -do- |
| | | (*d*) *Oidium tingtaninum* | Causes powdery mildew. Severe infection cause premature leaf falling and premature fruit dropping. |
| | | (*e*) *Corticium salmonicolor* | Causes pink disease. |
| (*ii*) | Guava (*Psidium guajava*) | (*a*) *Fusarium solani* | Causes yellowing of leaves and complete wilting of trees in advance stage. |
| | | (*b*) *Macrophomina phaseoli* | -do- |

(*Contd.*)

| 1 | 2 | 3 | 4 |
|---|---|---|---|
| | | (*c*) *Gloeosporium psidi* | Causes die - back. |
| | | (*d*) *Physalospora psidi* | Causes stem canker. |
| (*iii*) | Coconut palm (*Cocos nucifera*) | (*a*) *Phytophthora palmivora* | Causes bud-rot. |
| | | (*b*) *Bipolaris halodes* | Causes leaf rot. |
| | | (*c*) *Gliocladium roseum* | Causes leaf rot. |
| | | (*d*) *Ganoderma lucidum* | Causes trunk and root rot. |
| (*iv*) | Sapota (*Achras sapota*) | (*a*) *Phaeophleospora indica* | Causes pinkish to brown spots on leaf. |
| | | (*b*) *Botryodiplodia theobromae* | Causes flattening of branches. |
| (*v*) | Avocado (*Persea americana*) | (*a*) *Phytophthora cinnamomi* | Causes root rot. |
| | | (*b*) *Fusarium solani* | Causes softening of fruits. |
| | | (*c*) *Colletotrichum gloeosporioides* | Causes shedding of young fruits. |
| (*vi*) | Ber (*Zizyphus mauritiana*) | (*a*) *Isariopsis* species | Causes black leaf spot. |
| | | (*b*) *Oidium* species | Causes powdery mildew. |
| | | (*c*) *Phoma* species | Causes fruit rot. |
| (*vii*) | Phalsa (*Grewia asiatica*) | (*a*) *Cercospora grewiae* | Causes premature leaf fall. |
| (*viii*) | Aonala (*Emblica officinalis*) | (*a*) *Penicillium islandicum* | Causes blue mold. |
| | | (*b*) *Revenellia emblicae* | Causes rust on leaves and fruits. |
| **(D) FRUIT PLANTS** | | | |
| (*i*) | Pineapple (*Ananas comosus*) | (*a*) *Phytophthora parasitica* | Causes root rot disease. |
| | | (*b*) *Ceratocystis paradoxa* | Causes fruit rot disease. |
| (*ii*) | Banana (*Musa* spp.) | (*a*) *Fusarium oxysporum* | Causes banana wilt disease. |
| | | (*b*) *Pseudomonas solanacearum* | Causes moko or wilt disease. |
| | | (*c*) *Gloeosporium musarum* | Causes anthracnose. |
| | | (*d*) *Xanthomonas musarum* | Causes leaf blight disease. |
| **(E) CASH CROPS** | | | |
| (*i*) | Sugarcane (*Saccharum officinarum*) | (*a*) *Colletotrichum falcatum* | Causes red rot disease. |
| | | (*b*) *Ustilago scitaminea* | Causes smut disease. Plants are stunted |

| 1 | 2 | 3 | 4 |
|---|---|---|---|
| | | (*c*) *Ceratocystis paradoxa* | Infested shoots die. |
| | | (*d*) *Marmor sacchari* | Causes sugarcane mosaic. |
| (*ii*) | Tobacco (*Nicotiona tabacum*) | (*a*) *Pythium aphanidermatum* | Causes damping off disease. |
| | | (*b*) *Phytophthora parasitica* | Causes black shank disease. |
| | | (*c*) *Erysiphae cichoracearum* | Causes powdery mildew disease. |
| **(F)** | **CEREALS** | | |
| (*i*) | Rice (*Oryza sativa*) | (*a*) *Pyricularia oryzae* | Causes blast disease. |
| | | (*b*) *Helminthosporium oryzae* | Causes brown spot on leaves. |
| | | (*c*) *Xanthomonas oryzae* | Causes bacterial blight on leaves. |
| | | (*d*) *Fusarium moniliforme* | Leaves of the infected plant dry up. |
| | | (*e*) *Trichoconis padwickii* | Causes leaf and grain spot disease. |
| (*ii*) | Wheat (*Triticum spp.*) | (*a*) *Puccinia recondita* | Causes leaf brown rust. |
| | | (*b*) *Puccinia graminis tritici* | Causes black stem rust. |
| | | (*c*) *Ustilago nuda* | Causes loose smut. Spikelets transformed into a mass of black powdery spore. |
| | | (*d*) *Pythium graminicolum* | Causes root rot disease. |
| | | (*e*) *Pythium arrhenomanes* | Causes root rot disease. |
| (*iii*) | Sorghum (*Sorghum vulgare*) | (*a*) *Sphacelotheca cruenta* | Causes loose smut. Most grains smutted. |
| | | (*b*) *Sphacelotheca sorghi* | Causes grain smut. Most grain smutted. |
| | | (*c*) *Tolyposporium ehrenbergii* | Causes long smut. |
| | | (*d*) *Sphacelotheca reiliana* | Causes head smut. Entire head smutted. |
| | | (*e*) *Puccinia purpurea* | Causes rust and older leaves dry prematurely. |
| | | (*f*) *Cercospora sorghi* | Causes red or dark brown irregular leaf spot. |

(*Contd.*)

| 1 | 2 | 3 | 4 |
|---|---|---|---|
| (*iv*) | Maize (*Zea mays*) | (*a*) *Sphacelotheca reiliana* | Causes head smut. |
| | | (*b*) *Ustilago maydis* | Produces of gall on the ears, buds and stalk. |
| | | (*c*) *Puccinia sorghi* | Causes rust disease. |
| | | (*d*) *Cephalosporium acremonium* | Causes ear rot disease. |
| **(G) PULSES** | | | |
| (*i*) | Arhar (*Cajanus cajan*) | (*a*) *Fusarium oxysporum* | Causes wilt disease. |
| | | (*b*) *Xanthomonas cajani* | Causes leaf spot disease. |
| | | (*c*) *Cercospora indica* | Causes leaf spot disease. |
| | | (*d*) *Diplodia cajani* | Causes cankerous lesions on the stem. |
| (*ii*) | Peas (*Pisum sativum*) | (*a*) *Uromyces fabae* | Causes rust on leaves. |
| | | (*b*) *Peronospora viciae* | Causes downy mildew disease on the under surface of leaves. |
| | | (*c*) *Aschochyta spp.* | Causes foot rot and blight disease. |
| **(H) VEGETABLES** | | | |
| (*i*) | Cauliflower (*Brassica oleracea*) | (*a*) *Alternaria brassicae* | Causes leaf spot and blight disease. |
| | | (*b*) *Albugo candida* | Causes white rust disease. |
| | | (*c*) *Peronospora parasitica* | Causes downy mildew disease. |
| (*ii*) | Carrot (*Daucas carota*) | (*a*) *Erwinia carotovora* | Causes soft rot disease of the root. |
| | | (*b*) *Erysiphe polygoni* | Causes powdery mildew disease. Leaves turn brown and wilt. |
| | | (*c*) *Cercospora apii* | Causes leaf spot disease. |
| (*iii*) | Onion (*Allium cepa*) | (*a*) *Puccinia porri* | Causes rust disease. |
| | | (*b*) *Alternaria palandui* | Causes leaf blight disease. If infection spreads to the bulb, gradually it decay. |
| (*iv*) | Tomato (*Lycopersicon esculentum*) | (*a*) *Pythium aphanidermatum* | Causes damping off. |
| | | (*b*) *Fuserium oxysporum* | Causes wilt and plant dies. |

| 1 | 2 | 3 | 4 |
|---|---|---|---|
| | | (*c*) *Phytophthora infestans* | Causes fruit rot disease. |
| | | (*d*) *Phytophthora parasitica* | Causes fruit rot disease. |
| | | (*e*) *Corynebacterium michiganense* | Causes canker like infection. |
| (*v*) | Brinjal (*Solanum melongena*) | (*a*) *Cercospora solani* | Causes leaf spot disease. |
| | | (*b*) *Alterneria melongenae* | Causes leaf spot disease. |
| | | (*c*) *Puccinia penniseti* | Causes rust disease. |
| | | (*d*) *Fusarium solani* | Causes wilt and root rot disease. |
| (*vi*) | Chilli (*Capsicum annuum*) | (*a*) *Cercospora capsici* | Causes leaf spot disease. |
| | | (*b*) *Colletotrichum capsici* | Causes fruit rot disease. |
| (*vii*) | Lady's finger (*Hibiscus esculentus*) | (*a*) *Vein-clearing virus* | Causes vein clearing disease. |
| | | (*b*) *Erysiphe cichoracearum* | Causes powdery mildew disease. |
| | | (*c*) *Uromyces heterogeneus* | Causes rust disease. |
| (*viii*) | Potato (*Solanum tuberosum*) | (*a*) *Phytophthora infestans* | Causes late blight disease. |
| | | (*b*) *Alternaria solani* | Causes early blight disease. |
| | | (*c*) *Pseudomonas solanacearum* | Causes brown rot disease. |

## REFERENCES


Bakshi, B.K., Reddy, M.A.R., Puri, Y.N. and Singh, S. (1972). Forest disease survey, (*Final technical Report*, FRI, Dehradun, pp. 1-175.

Bakshi, B.K. (1976). *Forest Pathology—Principles and practices in forestry*, Delhi, pp. 1-400.

Boyce, J.S. (1961). *Forest Pathology*. IIIrd ed; *Mc. Graw Book* Co. Inc. New York, pp. 1-572.

Browne, F.G. (1968). *Pests and Diseases of Forest plantation trees*, Clarendon Press, Oxford, pp. 1-330.

Ghosh, S.P. (1985) Citrus, In : *Fruits of India tropical and subtropical*, Bose, T.K. Eds., *Naya Prokash*, Calcutta.

Gupta, V.K. and Kaushik, J.C. (1988). *Cylindrocladium* spp. associated with eucalyptus in Himachal Pradesh, *Jour. Tree Science*, 7(1), pp. 17-19.

Jamaludin, (1980). Problems and prospects of seed pathology of forest trees, *Second For. conf.*, Dehra Dun (16-19 January, 1980), p. 6. (cyclostyled).

Kaushik, J.C. and Toky, O.P. (1990). *Ceratocystis fimbriata*. A new record on poplar from India, (communicated) cited from pathogen management under Agroforestry In. *Agroforestry - Indian perspective*, Jhā, L.K. & Sen Sarma, P.K. eds., Ashish Publication, Delhi.

Kaushik, J.C. and Bangarwa, K.S. (1992). Pathogen management under Agroforestry In *Agroforestry - Indian perspective* L.K. Jha and Sen Sarma, P.K. Ashish Publication, Delhi.

Rangaswami, G. (1984). *Diseases of crop plants in India,* Printice - hall of India Private Limited, New Delhi, pp. 1-520.

Rao, D.S. (1972). *A hand-book of plant protection,* Rangaswami and Co. Private Limited, pp. 1-842.

Sen Sarma, P.K. (1975). Spike disease of sandal, a yellow type disease, *Proceeding of the 2nd world consultation of pests and diseases, IUFRO*, New Delhi.

Sen Sarma, P.K. (1986). Forest entomological problems in India and their management in the year 2000 *A.D., Proc. Indian Nath. Sci. Adac*, 852, pp. 134-147.

Sen Sarma, P.K., Thakur, M.L. and Sehgal, H.S. (1988). Protection of forest seeds against insect pests and fungi during storage, *Journal of Tropical Forestry*, Vol. 4 (iv), pp. 350-357. •

Sharma, J.K., Mohanan, C. & Maria Florence, E.U. (1985). Disease survey in nurseries and plantation of forest tree species grown in Kerala, *K.F.R.I* Research Report 36.

Singh, Khan, S.N. and Mishra, B.M. (1983). Some new note worthy diseases of poplars in India, *Indian Forestor*, 108: pp.653-659.

Singh, P. and Singh, S. (1985). Insect pests and diseases of poplars, *F.R.I. and Colleges*, Dehra Dun, pp. 1-74.

Singh, Sujan (1986). Mortality of sissoo (*Dalbergia sissoo* Roxb) in plantations, *Second forestry Conference,* Dehra Dun, 1980. (cited from *Pests and pathogen management in Agroforestry system*) Singh, P & Singh, S., Workshop proceedings, *Agroforestry for rural needs*, Vol-I, Khosla and Khurana, Eds., I.S. TS, pp. 153-176.

Sivaramkrishnan, V.R. (1977). Experimental transmission of spike disease of sandal (*Santalum album*), by the leaf-hopper, *Nephotettix virescens* (distant) (Homoptera, Cicadellidae), *Indian Forestor*, 104; pp. 202-205.

Sivaramkrishnan, V.R. & Sen Sarma, P.K. (1978). Experimental transmission of spike disease of *sandal* (*Santalum album*) by the leaf hopper *Nephotattix virescens* (Distant), *Proceedings, All India Sandal Seminar*, Bangalore, 7-8th February, 1977, Karnataka Forest Department, India, pp. 140-142.

# 16

# *Silviculture of MPTs Suitable for Agroforestry System*

## 1. AUSTRALIAN PHYLLODE ACACIA

*Acacia auriculiformis* A. Cunn. ex. Benth.

*FAMILY* : LEGUMINOSAE (Mimosoideae)

*Common name* : Akash moni

*Distribution* : It is an introduced species from North Australia and Papua New Guinea and commonly planted in degraded farmlands and wastelands in Bihar, Uttar Pradesh, Karnataka, West Bengal, Orissa, Rajasthan, Gujarat and Maharasthra. It is one of the very popular species in Karnataka for high altitude grasslands afforestation programme (Kushalapa, 1988).

*Climate* : It has wide climatic adaptability. In areas in India where it has become naturalised, the mean annual temperatures varies from 25° to 30° C. and the normal rainfall varies from 500 to 1500mm. In India this species is well adapted to drought.

*Soil* : *Acacia auriculiformis* grows on most kinds of soil, Clayey grey soils, lateritic soils, laterite soils, gravely brown soils, sand dunes, etc.

*Description* : A very handsome evergreen tree with pendulous branches. The place of leaves is taken by phyllodes which consits of the leaf stalk modified into a flattened blade,

**Fig. 95 :** *Acacia auriculiformis.*

phyllode is 8-18 cm. long, 1.2-3.7 cm broad, narrowly oblong, slightly curved, tapered at each end, rather leathery, dully green, with 4-6 principal parallel veins. Flowers are nearly 4 mm long, tiny, white, sessile, rich yellow, fragrant, crowded in slender axillary spikes 4-9 cm long, Pods are nearly 1.2 cm wide, flat, almost woody, much twisted in irregular coils when ripe (Randhawa, 1983 and Joshi *et al.* 1983).

**Silvicultural characters**

*Light* : *Acacia auriculiformis* is a light demander, but tolerates shade during early years.

*Frost* : It can tolerate frost.

*Fire* : It is a fire resistant species.

*Drought* : It is drought resistant and can withstand drought conditions like *Eucalyptus tereticornis*.

*Root suckers* : Roots of felled trees often produces root suckers, especially in dry localities. The tree has a superficial spreading root system (Banerjee, 1973).

*Establishment* : Mature pods are collected in February to March and sundried for 5-10 days. A kg contains about 30,800 to 42,800 seeds. Soaking the seeds in cold or hot water improves the germinative capacity. The germinative capacity is 40 per cent. Seeds are sown directly in the field or nursery. Overhead shade is essential during germination period. The distance between sowing lines depends upon the objective of plantation. Polythene bags plants are planted after 3-4 months. It attains more than 1-5m height by the end of first growing season. Jha *et al.* (1989) have reported that for energy plantation plant should be planted at the spacement of 1 m x 1 m and in agroforestry farm 2 m x 1 m or more. After third year alternate rows are harvested in order to facilitate the penetration of sunlight and to reduce competition between trees and intercrops.

In agroforestry farm, good quality of timber are produced by regular pruning of side branches of trees. Pruned side branches are used as fuelwood.

To get maximum yield proper management is essential. Application of fertilizers accelerates the yield per unit area. Kushalapa reported comparative biomass of *Acacia auriculiformis* and *Leucaena leucocephala* trees in moist region of Karnataka. *Acacia auriculiformis* has grown taller (13.16m) than leucaena (11.80m). The yield of wood per ha. (Oven dry), taking the survival percentage as 80, at 2 m x 2 m spacing would be respectively 100.64 t/ha and 260.26 t/ha for ***Acacia auriculiformis*** and ***Leucaena leucocephala.*** In papua New Guinea on abundoned infertile sites the *Acacia auriculiformis* has a height of 17 m in 8 years. In Indonesia and Malaysia, the annual wood production recorded was 17-20 $m^3$ per ha. In poorer soils but moist conditions, it is expected to yield 10$m^3$ per ha per year. In Zanzibar they reached 11m height and 30 cm in diameter in 6 years in deep sand. In shallow soils of semi arid West Bengal, the recorded yield was only 5$m^3$ per ha per year at the 15th year (Ayensu, 1980 and Kushalapa, 1988).

**Utility**

*Fire wood* : *Acacia auriculiformis*, a very good source of energy as fuelwood or charcoal. The calorific value of wood is 4800-4900 Kcal per kg.

*Tannin* : Tannin from bark is used in leather industry. The bark contains 13 per cent water soluble tannin.

Fig. 96 : Block plantation of *Acacia auriculiformis* (on-farm trial, Kumharia Village, Ranchi).

Fig. 97 : Energy plantation *Acacia auriculiformis*.

*Furniture* : Its wood is not suitable for good quality of furniture.

*Pulp* : We can get high quality of pulps by the neutral sulfite semichemical process. It is a source of wood pulp.

*Ornamental* : Flowers are light yellow. Flowers several times during the summer and rainy season. This species is suitable for parks and open and also for small home gardens (Randhawa, 1983).

## 2. BABUL

*Acacia nilotica* (Linn.) Wild. ex Delile spp.
(*Acacia indica* (Benth.) Brenan)
(*Acacia arabica* Wild.)

*FAMILY* : LEGUMINOSAE (Mimosoidaea)

*Common names* : Babul, kikar, jali, gobalu, babli, karuvai, tuma, gobil.

*Distribution* : Babul is widely distributed almost throughout India. It is grown occasionally in crop fields and village grazing lands. It is commonly planted throughout dry, hot regions of India. It is one of the most common trees in tropical Africa, West and East Sudan and Saudi Arabia.

*Climate* : In its natural habitat the absolute maximum shade temperature ranges from 40° to 47.5°C, the absolute minimum from 1.1° to 15°C, and the average rainfall from 200 to 1270 mm.

*Soil* : It grows on most types of soil, it does well on alluvial soil in riverain areas, black cotton soil, lomey soil etc.

*Description* : *Acacia nilotica* is a moderate size tree, growing to 17m in height and 3m girth. Bark dark brown, almost black much fissured, with 0.8 to 5.1 cm long straight, white, sharply pointed spines. The leaves are bipinnate, bearing 10-20 pair alternate leaflets. The leaflets are 2 to 5 mm long,

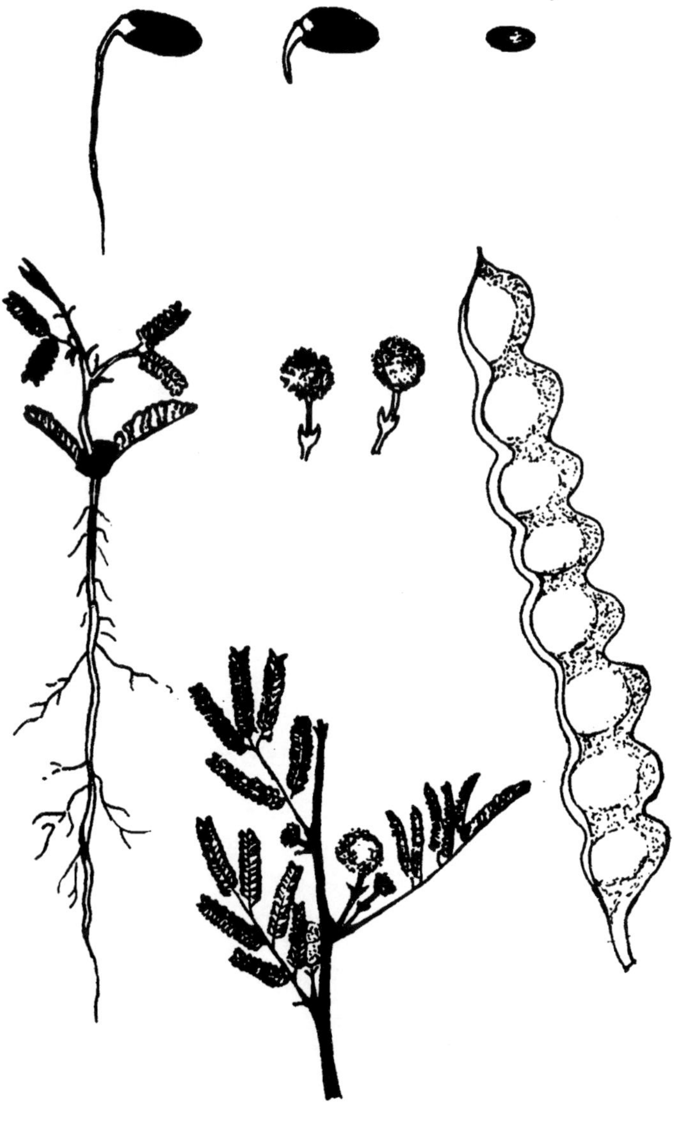

**Fig. 98 :** *Acacia nilotica.*

linear subsessile. The flowers are yellow, fragrant in axillary globose heads which are 13 mm diameter. Each flower has cup-shaped calyx. Corolla is twice as long as the calyx. The pods are solitary, 7.6-15.2 cm, by 1.2 to 1.5 cm, 18-12 seeded, stalked, dark brown, shinning (Kanjilal, 1927).

There are to types of babul. *A. nilotica* var. *indica* has laterally spreading branches whereas var. *cupressiformis* has upright branches (Annon, 1992).

**Silvicultural characters**

*Light* : The tree is strong light demander.

*Frost* : It is frost tender and has power of recovery when injured by frost.

*Drought* : It is drought resistant but does not stand excessive drought.

*Coppicing power* : The coppicing power of babul varies greatly. Throughout the greater part of its habitat it coppices poorly. Certain important exceptions however, have been recorded where babul coppices well, among which may be mentioned babul occurring in Anantpur, Bellary, Chingleput and Gunter districts in Southern India and under favourable conditions in ravine areas of Doab districts in Uttar Pradesh. The precise reason why babul coppices in some localities and not in others is not satisfactorily explained (Annon, 1981).

*Establishment* : Mature pods are harvested in April or May and the seeds are separated from pod walls. If the seeds are to be used for immediate planting, no seed treatment is necessary. Stored seeds require hot water treatment or soaking in cold water for 48 hours. Feeding babul pods to goats and collecting the seeds from dung is also method of seed treatment. Seeds germinate in 10-14 days.

Direct sowing of seed is common, but seedlings are also raised in nursery. Direct sowing in contour trenches is done at the beginning of the rainy season. About 2.0 kg of seeds is required per ha for direct seeding at close spacing and subsequent thinning. Nursery raised seedlings are planted at 2.0 x 2.0 m spacing when they are 3-5 months old. Seedlings must be protected from browsing animals and weed competition (Annon, 1992).

The sowing of babul in conjunction with field crops has been carried out in Maharashtra for more than 60 years and has proved more successful and economical. In the first and second year cultivation with field crops alone is carried out and in the third year a mixture of babul and neem seeds is sown in lines along with field crops. The field crops raised are chiefly cotton, sesamun and arhar (*Cajanus cajan*). The cultivation of juar (*Sorghum vulgare*) is allowed in the first two years of tree cultivation only. The young plants are spaced out annually during the first few years (Annon, 1983).

Plantation of *Acacia nilotica* in agroforestry system is suitable in the drier belt.

**Utility**

*Fuelwood*

*Acacia nilotica* is considered to be one of the best sources of energy as fuelwood or charcoal. It is harvested at 15-20 year rotations. The calorific value of moisture free sapwood is 4,793 Kcl per kg and that of heart wood being 4,946 Kcl per kg (Krishna and Ramaswamy, 1932). Varma and Dey in 1942 reported that ash content of the wood showed a mean of 0.91 per cent ash on dry basis.

*Wood* — The sapwood and heartwood of Babul is yellowish white and pinkish white respectively. The weight of wood is 830 kg per cubic metre. The wood is considered best where hardness and toughness are required. The wood is extensively used for making agricultural implements.

*Bark* — It is one of the important source of tannin materials. The tannin content of the bark varies considerably; sometimes it is as high as 20 per cent. The tannin and non-tannin ratio is 1.5 : 1 (Edward *et al.*, 1952).

*Pods* — The entire pods contains around 12-19 per cent tannin and after removal of seeds the pods contain 19 to 27 per cent tannin. On dry basis, the pods yield crude protein 15.8, other extract 0.8, N-free extract 65.5, fibre 12.4 and total ash 15.5 per cent (Annon, 1948).

*Leaves* — The leaves and twigs are good fodder and branches are used to fence the agricultural fields.

*Lac Host Plants* — *Acacia nilotica* is one of the host plant of *Laccifer lacca*. It is desirable to carry our research in this host plant, introduction and evaluation with reference to particular environment.

## 3. GUM ARABIC TREE

*Acacia senegal* Willd.
(*Acacia rupestris* Stocks)

*FAMILY*: LEGUMINOSAE (Mimosoideae)

*Common names* : Khore, bavel, kumta, kagar

*Distribution* : *Acacia senegal* is one of the popular species of the arid zones of India. It is now common in Rajasthan, Gujarat, Maharashtra etc. It is also common in Sind and tropical Africa.

*Climate* : In its natural habitat, the absolute maximum shade temperature may rise to 48.3°C and the absolute minimum 2.5°C to 5°C. The annual precipitation ranging from 250 to 600 mm.

*Soil* : It grows on poorest soils, like rock hills, sandy tracts, sand dunes. These soil contains high percentage of soluble salts.

*Description* : *Acacia senegal* is a small thorny deciduous tree, usually gnarled and of straggling form when young but developing into a fairly straight tree after 15-20 years. It attains height of 3-5 (rarely 9 m) and girth of 30-60 (rarely 90 cm) (Annon, 1983). The bark is pale, greenish-grey becoming rough and then peeling. The leaves are compound pinnate, bearing 8-15 pairs leaflets. The leaflets are 2.5m long, greyish green. The flowers are creamy white, fragnant. The pods are 5-8 cm by 1.8-2.5 cm, thin flat and straight. There are 5-6 seeds in each pod. Maximum yield of gum is expected within a period of 20 to 25 years of planting and therefore the trees are considered as over mature and are felled and replanted with fresh seedling (Muthana, 1988). The best period for gum tapping is when the trees start shedding their leaves after turning yellowish colour, i.e., usually when the winter season sets in, which coincides after mid October or early November. At that times it is also observed that tips of the branches remain leafless (Muthana, 1988) Gupta and Tindal (1985) has reported that leaf sample of *Acacia senegal* contains protein 14.88 to 21.87 per cent, NDF 22.8 to 31.0 per cent, lignin 2.8 to 6.0 per cent, silica 0.2 to 0.4 per cent and hemi cellulose 11.8 to 16.8 per cent. In *Acacia senegal*, tannins were absent.

**Fig. 99 :** *Acacia senegal.*

**Silvicultural characters**

*Light* — It is a strong light demander.

*Drought* — It is drought resistant and can with stand extremes of temperature and very poor soil. It survive even on sand-dunes in West Rajasthan where very few tree species can survive (Annon., 1983).

Anderson (1987) has reported that in Sahelian zone the devastating droughts of 1984-85 led to extensive losses of gum-bearing trees and a drastic decrease in gum production.

*Frost* — It is somewhat tender at seedling stage. The bole and pole crops are frost hardly.

*Coppice* — The tree coppices well. Observations on the growth of coppice crops in Ajmer—Merqara Forest Division showed that at 30 years rotation percentage of successful stools were 81.94, whereas in 40 years rotation 92.11 percentage recorded (Annon., 1935-1936).

*Establishment* — Ripe pods of *Acacia senegal* are collected in October to November. The seeds are separated by gently thrashing. Seeds are 5-6 per pod and 8,200-10,900 per kg. The germinative capacity varies from place to place. In general the germination capacity is more than 50 per cent. The seeds are soaked in cold water to get good result. The seedlings can be raised in nursery for transplantation. Direct sowing of seeds in the arid tract at the beginning of the rains gives good result.

Solanki *et al.* (1985) studied the performance of 22 years old *Acacia senegal*, *Prosopis cineraria* tree species at the CAZRI, Jodhpur. In *Acacia senegal* tree height ranged from 3.57-6.30 m, basal diameter from 4.05-6.95 cm and estimated volume from 3,27.94-26,156.25 $cm^3$ (7 times). In *Prosopis cineraria* tree height ranged from 3.2-5.8 cm, basal diameter from 6.6-17.5 cm, dbh from 3.4-14.3 cm, canopy diameter from 1.75-3.65 m and estimated volume from 161.60 - 14, 802.08 $cm^3$ (13 times). In *A senegal* individual trees grew as much as 6.3 m with 16 cm dbh where as *P. cineraria* trees grew as much as 5.8 m with 14.3 cm dbh at 22 years of age. Canopy diameter of a tree in *A. senegal* was almost double at this age. Higher values of all the parameters in *A. senegal* may probably be due to the presence of more fibrous roots in the 40 cm zone as compared to *P. cineraria* (Muthana *et al.*, 1984). The coefficient of variability was more in *P. cineraria* for all the character except volume.

**Utility**

*Wood* : The heart wood is black and sapwood is pale creamy. It is strong and durable. Its weight is about 865-913 kg per cum (Gamble, 1922). The wood takes a beautiful polish and is suitable for small articles, agricultural implement, cart wheel, posts etc. The wood is highly resistant to termites.

*Fuelwood* : Branches are also used as fuelwood.

*Living fences* : A row of this tree species makes a very effective living fence. This will last for many years without maintenance.

*Gum arabic* : The only Acacia exudate (Gum arabic) to have been subjected, to modern standards of toxicological safety evaluation has been that from *Acacia senegal.*

The modern world wide regulatory difinition of gum arabic requires that it must be the exudate from *Acacia senegal* (L) Wild. or related species of Acacia and that it will give lacvorotatory (*i.e.* negative) optical rotations in solution. Clearly the word "related" is intended to denote some restriction on the number of permissible *Acacia* spp. (Anderson, 1987). Average gum yield of mature tree is 100 kg per hectare of 500 trees (*i.e.* 80 per cent of stocking). Average gum yield per tree is about 0.225 gm -1/2 pound approximately per season (Muthana, 1986). The gum arabic is used in manufacture of chewing gums, pastilles, in the finishing of silk and crape, as adhesive in match industry and etc. (Mac Donald, 1942). Muthana (1988) has reported that in India, gum arabic is used mainly for confectionaries glues, pharmaceuticals and also for preparing some sweets for feeding ladies after their confinement as it is said to bind their stomach muscles.

*Vegetables* : The seeds are palatable and commonly used by the local people as a vegetable.

## 4. ISRAELI BABOOL

*Acacia tortilis* (Forsk.) Hayne
(*Acacia raddiana* Savi.)

*FAMILY*: LEGUMINOSAE (Mimosoideae)

*Common names* : Israeli babool.

*Distribution* : The Israeli babool is believed to be a native of Israeli, Saudi Arabia, Kenya, Ethopia and Tanzania. It is now common and naturalized in arid zones of Rajasthan, Gujarat, Haryana, Tamil Nadu and Andhra Pradesh. *Acacia tortilis* has four distinct sub-species. There are considerable ecological variation and tolerance. In India, sub-species raddina has been introduced to Jodhpur during 1950.

*Climate* : In the area where it has been introduced (Jodhpur), the mean annual temperature is 26.2°C, and the mean annual rainfall is 360 mm or more.

*Soil* : Israeli babool is common on shallow soils. This species fabours alkaline soil. The lithological formation of Jodhpur is sandstone and effusive rocks. The soils are light brown in colour, sandy in texture, with little or no calcium carbonate up to 195 cm soil depth and pH ranges 7.95 to 8.30. The water holding capacity of the soil is about 26% (Annon., 1983). Man and Muthana (1984) reported that study was carried out to determine the performance and suitability of species on saline soils. The depth of soil varies from 7.5 cm to 9.0 cm at 0.1 per cent slope with pH varying from 8 to 9 and salinity ranging from 12,000 to 28,000 mm at 25°C. Out of the six trees species tried in such saline path, it has been observed that *Eucalyptus camaldulensis* has given the highest height in growth (299.0 cm) within a period of 2-years followed by *Prosopis juliflora* (Israel variety) (244.0 cm), *Acacia tortilis* (186.0 cm), *Tamarix articulata* (167.0 cm) and poorest by *Albizia lebbeck* (94.0 cm) and *Azadirachta indica* (90.0 cm).

*Description* : It is a small to medium-sized tree, reaching 8-10 m in height. In unfavourable localities, it grows like shrubs. Its spines are distinguishing feature. It has two kinds of spines (a) the long, straight, slender, white, about 6 cm in length, (b) the hooked, small and brownish in colour. The leaves are 1.25 to 3.5 cm long, petioles pubescent pinnate 4-10 pairs. The leaflets are very small and oblong in shape. The flowers are white fragrant, and born singly or in clusters. Flowers buds appearing in the last weak of April. The pods are yellowish brown and become spiral as they mature. The pods start ripening from middle of November and continue to ripen till the end of February. A 6-8 year old tree yields 5-6 kg of clean seed on an average (Muthana and Arora, 1973).

**Silvicultural characters**

*Light* : *Acacia tortilis* is a strong light demander and is intolerant of suppression.

*Moisture* : It has long tap root and number of side roots. This tree effectively utilizes limited soil moisture of arid soils.

*Frost* : Young plants are frost hard but seedlings are tender to frost.

*Rainfall interception* : The rainfall interception studies revealed that the interception under moderate to heavy rainfall (20 to 35 cm) conditions was high under *Acacia tortilis* tree cover and varies between 23 to 33 per cent and 14 to 19 per cent under 1964 and 1970 plantation of *A. tortilis* as against 3 to 8 per cent rainfall interception recorded under *Holoptelia integrifolia* (1969) tree covers. This difference was attributable to the plant shape and dense crown canopy of *Acacia tortilis* trees (Mann & Muthana, 1984).

*Coppice* : It coppices very well and the coppice shoots growing vigorously. Twelve year old *Acacia tortilis* trees were cut back from 3 different spacing experiments to study their coppicing behaviour. It was observed that 12.6 coppice shoots per stump from 6 x 6m, 12.6 from 4.5 x 4.5m and 9.7 from 3 x 3m spacing and mean height of coppice shoots recorded were 130 cm, 76 cm and 107 cm, respectively, under these three spacings in the first year. To attain the mean height of two years old coppice shoots, the seedlings originally transplanted took four years whereas two years old coppice shoots attained mean height of 325, 302 and 286 cm from 5 x 6m, 4.5 x 4.5m and 3 x 3m spacing, respectively (Mann and Muthana, 1984).

*Establishment* : Seeds should be collected in April to May and sown in the month of July or August. The seed weight varies from 11, 765 to 25,000 kg. Kaul (1965) has recommended 24 hours soaking in water before sowing. Muthana and Arora (1973) has reported placement of seeds for 40 minutes in 50 per cent concentrate sulphuric acid, followed by washing in cold running water and then drying in shade overnight accelerates germination.

Germination in usually 42-43 per cent and is complete within 10 days after sowing (Kaul and Ganguli, 1965 ; Gupta and Balara, 1972). Sowing of seeds in cylindrical metallic containers have proved economic in watering over other containers to the extent of 20-30 per cent and it last many years. The seedling will be over 1m high in next year. Planting can be done in pits of 60 cm x 60 cm x 60 cm filled with weathered soil. The soil around the stump should be throughly rammed and crescent shaped ridge of 15-20 cm high to be formed across the slope to harvest rain water.

Mann and Muthana (1984) reported that application of nitrogenous (20 kg, 40 kg and 80 kg as $NH_4 SO_4$/ha) and phosphate (30 kg, 60 kg and 120 kg as $P_2O_5$/ha) fertilizer in different does when administered for 3 consecutive years after planting seedlings at sites did not exhibit any significant response in height growth and seedling establishment in both the species. Feasibility study on agroforestry was initiated to study the competition between forage crops on forest trees and vice-versa. The forage crops, *viz.*, cowpeas, sorghum guar and *Cenchrus ciliaris* were raised in the interspace of 12 year old *Acacia tortilis* plantation. Significantly higher green and dry forage was produced by cowpeas, 28.2, 43.3 and 32.5 q/ha, respectively, in 1976 and 1977 and 1978 followed by sorghum (Mann and Muthana, 1984).

**Utility**

*Fuelwood* : In Jodhpur the plants attained the maximum height of 6.40 m and 7.85m in 10 and 14 years, respectively. Collar diameter was recorded 24.1 cm and 28.7 cm in 10 and 14 years, respectively (Mann and Muthana, 1973).

A 12 year-old plantation yeilded 54-t of fuel per ha. The dense, red heartwood of this species has high calorific value (4,400 Kcal per kg) and makes superior firewood and charcoal (Annon., 1980).

*Wood* : The wood is much used for packing cases, post, interior fitting of buildings.

*Fodder* : The foliage is palatable. Leaves are lopped for fodder.

*Fence materials* : Farmer use small thorny branches as fence material to protect their farmland from cattle, sheep and goats.

*Shifting Sand-Dunes* : *stabilization* This species is one of the best species for sand dune stabilization.

## 5. SIRIS

*Albizia lebbeck* (L) Benth.

(*Mimosa lebbeck* L.)

*FAMILY* : LEGUMINOSAE (Mimosoideae)

*Common names* : Sarshio, Kalshish, siris, koroi, kalsis, sharin, shrirish, vakai, frywood tree, etc.

*Distribution* : The species is one of the common trees of India. It is extensively planted in agroforestry system. It is native also to Bangladesh, Burma and Pakistan and has been cultivated in tropical and subtropical region in North Africa, West Indies, South America, and Southeast Asia (Annon., 1980).

*Climate* : The species prefers annual rainfall between 500 to 2500 mm but is capable to withstand short periods of drought. It is frost hardly. In areas where it has been planted the mean monthly temperature varies from 5°C in the winter to 45°C in the summer.

*Soil* : The tree grows on a variety of soil, *e.g.* drained loam, clayey soil, laterite or black cotton soil. It can grown on poor soils (mild alkaline and saline soils).

*Description* : *Albizia lebbeck* is a tropical hard wood species. The genus name is sometimes spelled *Ablizzia*. It is a moderate to large deciduous tree. It has dark grey irregularly cracked bark, Leaf-rachis 8-20 cm long; leaflets 4-9 pairs, shortly petioled, oblong, oblique, obtuse. Flowers greenish-white, shortly pedicellated; calyx funnel shaped, short-toothed ; corolla twice the length of the calyx. Pod flat, thin, straw coloured, 5-10 seeds, with brown spot (Singh and Srivastra, 1989) The tree bears plenty of foliage readily eaten by cattle sheep and goats.

The analysis of variability of *Albizia lebbeck* for 'pod length' and for the 'number of seeds per fruit' reveaded that the latter character contains higher genetic variability than the former. It was also found out that the plants which are located near busy roads are more variable and also the insect infestation is more. The higher variability thus induced, generates under developed and abortive seeds and the infested seeds are consumed by the insects, resulting in very less availability of goods seeds. A congenial atmosphere produce more good seeds. Thus it was concluded that establishment of seed sources may better be made away from the road side, keeping in mind that the production cost of the improved seeds may be very high and also the availability may be less (Bagchi and Emmanuel, 1984).

**Silvicultural character**

*Light* : The tree is a strong light demander. It can, however, withstand shade when young. Young seedlings tolerate shade for about two season *i.e.*, first or second years.

*Frost* : After the first year trees can tolerate frost. The development of the seedling varies greatly among individual plants even under identical condition, but there are certain factors which stimulate development, and which induce remarkable rapid growth from the commencement ; the most important fac-

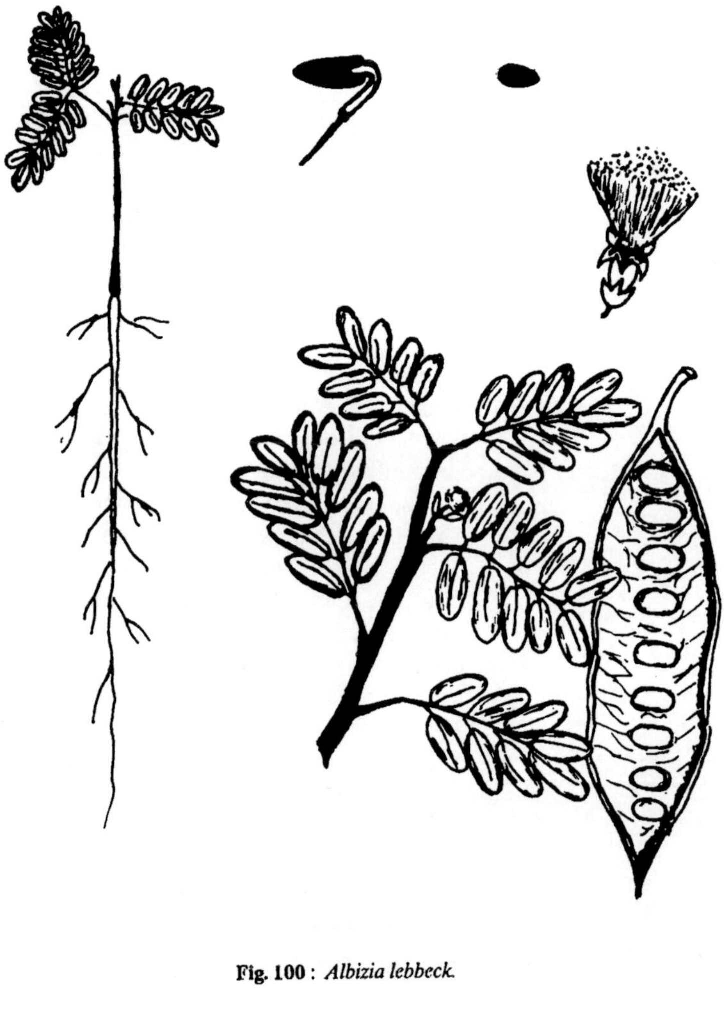

Fig. 100 : *Albizia lebbeck.*

tors are absence of weeds, loose soil, sufficient soil moisture. Numerous experimental plots at Dehra Dun have demonstrated the marked effect of regular weeding and loosening of soil on the growth of seedling, and although watering has a beneficial effect it is of comparatively little value unless weeding is properly carried out (Joshi, 1983).

*Establishment* : *Albizia lebbeck* produces seed prolifically. In Bihar, seeds are sown in the nursery beds from April to July and watered. Generally seeds are immersed in boiling water and allowed to cool and soak for 36 hours. Kaul (1965) has experimented on the chemical and mechanical scarification of seeds for breaking their dormacy. The chemical scarification consisted of immersing the seeds for 5 minutes in concentrated sulphuric acid, washing the seeds with water ; and the mechanical scarification involves soaking the seed in water brought to boiling point for 24 hours. In the control no pre-treatment was given to seed. He has observed that with chemical scarification, the germination percentage is 75 per cent and germination is completed in 10 days after sowing. The corresponding figures for mechanical scarification are 85 per cent and 3 days and those for control 80 per cent and 8 days respectively. Thus soaking in water brought to boiling for 24 hours is an effective method for pretreatment of seeds. Germination commences in about 2 weeks. Bhimaya and Kaul (1966) studied the growth of root system of seedling grown in a cyclindrical container 78 cm high and 50 cm diameter, till the end of first growing season. It has a monopodial root system with a central vertical taproot and lateral secondary ramifications. The lateral roots originate at a depth of about 2.5 cm from the collar and are uniformly distributed on the taproot. The system affords a good means of utilisation of soil moisture. In a period of 153 days the length of tap root grew to 43.2 cm and the number of lateral roots was 40. The length of the shoot portion during this period was 15 cm.

The tree can be raised by the direct sowing. Direct showing of seeds in lines can be done in June or before monsoon showers. Stem cuttings or root-shoot cuttings are also satisfactory. Planting has definite advantages over direct sowing. Mann and Muthana (1984) adopted three periods as time of transplanting *viz.*, first half of July, second half of July and first half of August and there was no marked effect on the seedling survival and growth.

Regarding suitable age of transplants, *viz.*, 6 months, 9 months and 24 months old ; 24 months old transplants exhibited higher seedling survival (66 per cent) and mean growth in height (2.50 cm) when compared with *Prosopis cineraria* and *Acacia senegal.* The trees coppice fairly well. In Bihar, *Albizia lebbeck* was grown successfully along with agricultural crop (Jha, 1989). In Nalambur *Albizia lebbeck* was successfully raised by direct sowing in the open in conjunction with field crops (Brown, 1929).

**Utility**

*Fuelwood* : The wood of *Albizia lebbeck* burns well. Its calorific value is 5,200 Kcal per kg of dry fuel. Coppice rotations provide an effective method of producing high yield of fuelwood. A rotation of 10-15 years or more is considered to be very suitable.

*Wood* : The heartwood is a rich walnut brown, often handsomely streaked with lighter and darker markings, and with a pleasing golden lusture which shows up well under polish.

Albizia is a strong wood, being about the same weight and hardness as teak. The wood is excellent for high class furniture, interior decoration, and panelling (Trotter, 1982).

*Fodder* : Its fodder is a valuable supplement to pasture in North Bihar. Albizia forage has about 20 per cent protein.

*Green manure* : The foliage may be used as green manure or mulches in agroforestry system. The mulch reduces raindrop impact and prevent deterioration of the land. The chopped leaves when used as green manure improves fertility status of soil.

*Erosion control* : *Albizia lebbeck* is good soil binder.

## 6. SAFED SIRIS

*Albizia procera* (Roxb.) Benth.
(*Mimosa elata* Roxb.).

*FAMILY* : LEGUMINOSAE (Mimosoideae)

*Common name* : Modeloa, medeloa, safed siris, bagai, konda, vagai sirsi, chikul, tellachinta, etc.

*Distribution* : *Albizia procera* is one of the important multipurpose trees of India. It is found all over Assam, Bihar, Northern Madhya Pradesh, Andhra Pradesh and Central and Eastern Uttar Pradesh.

*Climate* : In its habitat, the maximum shade temperature varies from 37°C to 46°C and the absolute minimum from 10°C to 18°C. Annual rainfall varies from 1000 to 4000 mm. The trees cannot tolerate frost.

*Soil :* It is found growing on a variety of soils, *e.g.*, alluvial soil, clayeyi and moderately alkaline and saline soils, red sandy and loamy soils. If soil moisture level is low, attains smaller size.

*Description :* *Albizia procera* is large graceful tree with a oval crown bark yellowish or greenish-white, smooth, peeling off in thin flakes, often with horizontal lines. Leaf rachises 25-46 cm long, glabrous or pubescent, with a large gland near the base. Leaf lets 12-14 on each pinnae, obliquely oblong-ovate, glabrous. Flowers yellowish white, sessile, about 5-6 cm long. Stemens a little more than twice the length of the corolla. Pods 10-20 cm by 1.8-2.5 cm. glabrous, brown, 8-12 seeded with darker blotches over the seeds (Kanjilal, 1927). The flowering commences from the first week of August and lasts for about two weeks. Maximum flowers are seen in the last week of August. After the third week of September hardly any flowers are left on the trees. Flowering always commences when the renewal of leaves is complete. The indications are that excess of rainfall during June and July coupled probably with low temperatures delays the flowering while low rainfall and high temperature have apposite effect.

Ripe fruits begin to appear on the trees from the beginning of April. By the end of this month of the pods are ripe. By the third week of May all the pods, ripen and begin to disperse the seed (Krishnaswamy and Muthauda, 1954).

**Silvicultural character**

*Light :* The tree is light demander and becomes suppressed under heavy shade. Young plant can tolerate moderate amount of shade.

*Water :* It is moisture loving. It is idealy suited for planting in low lying moist area of Bihar, U.P., Assam and Bengal.

*Frost* : Young plant are apt to be injured by frost.

*Coppice* : Its capacity to coppice is very sell. 5-10 coppice shoots appear in a stump depending on its size and age. The foliage of this species are quite palatable to cattle and goats.

*Establishment* : Treatment of seed is required before sowing in the nursery bed. In Bihar seeds are soaked in water for 48 hours. Chaudhri (1957) reported soaking of seeds in water for 12 hours before sowing. The seeds are sown in the beds at 7.6 x 7.6 cm in the month of May. Watering is essential and care is taken to see that beds remain wet all through the day for about a week. Germination continued for 3 weeks or more. The seedlings are ready for planting out in July. Root and shoot cutting do better.

The species has been successfully used for plantation in farm/agroforestry system. *Albizia procera* has been successfully introduced in saline area. While sowing with ditch ridges generally fail in such areas, planting of container plants about 1 year old in large pits (60 $cm^3$, 60 $cm^2$ x 120 $cm^2$ and 120 $cm^3$), particularly after replacement of whole or part soil by non-alkaline or non-saline one, addition of vermiculik, gypsum and farm-yard manure and irrigation succeeds, at least in mildly alkaline areas (Mishara, 1973).

Husain, *et al.* (1986) reported that the effect of N.P.K. treatment on this species. Five N.P.K. treatments were compared to the unfertilized control. Differences among fertilizer treatments were statistically significant. All treatments increased plant height as compared to the original height of the saplings, and the maximum increase (66.08 per cent above control) was due to N.P.K. All treatments increased the shoot girth as compared to control, and maximum shoot girth (10.1 per cent above control) was with NPK application.

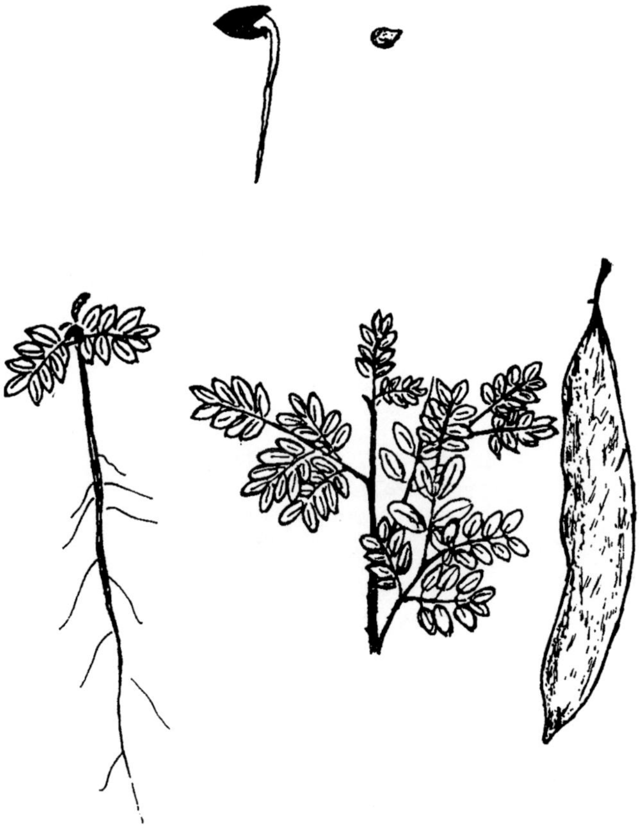

Fig. 101 : *Albizia procera.*

Oven dry weight of the saplings were increased by the fertilizer treatments except where P.K. was applied. Maximum over-dry weight of the saplings was where only N was applied and maximum where PK was applied. All treatments that received N showed adverse effects on nodulation and a significant increase in nodulation was observed in treatments that did not received nitrogen. Maximum nodulation was where PK was applied and minimum where N-alone was applied. Application of P or PK with N reversed the adverse effect of N on nodulation. All fertilizer treatments enhance the uptake of N as compared to control, and maximum uptake (89 per cent above control) was where NPK was applied. Infact improved K nutrition increased the number of rhizosphere bacteria requiring a more complex nutrient medium (Trolldenier, 1970). Thus to increase the nodule formation only P.K. should be applied in the agroforestry farm.

**Utility**

*Fuelwood* : Branches of trees are utilised as fuelwood. Its charcoal is considered very good.

*Wood* : The sap wood is white and often fairly wide. The hardwood is a light or dark brown, straight and even-grained, hard, takes a fairly good polish, seasons well and is very durable (Khanjilal, 1929). The *Albizia procera* wood are excellent for high class furniture. It weights about 640 kg/m$^3$.

*Bark* : The bark is used for tanning.

*Fodder* : Like *Albizia lebbeck* its fodder is valuable supplement to pasture, sheep and goat.

*Mulch and green manure* : The foliage are good green manure. This tree is one of the important nitrogen fixing tree species.

# 7. CALLIANDRA

*Calliandra calothyrsus Meissn.*
(*Calliandra confusa* sprague Riley)

*FAMILY*: LEGUMINOSAE (Mimosoideae)

*Common names*: Calliandra

*Distribution*: It is a native of Central America. It has been planted extensively in Indonesia. This had been introduced in Karnataka in 1981. This species is ideal for agroforestry system and may be one of the most promising exotics for fodder and fuel in humid and moist areas of our country, specially for afforestation of grassy blanks. This species now requires extensive trial in all agroclimatic zones of our country to find out its suitability compared to other indegenous and exotic species (Kushalappa, 1989).

*Climate*: It requires an elevation of 150 to 1500 m. It grows satisfactorily on sites that received rainfall over 1000 mm per year.

*Soil*: It can grow under a wide range of soils. It can grow in the poorest compact clay type soils and stony soils. This species was introduced in Karnataka on red sandy loam soil. This species is most suitable for degraded soils. It grows quickly in eroded soils and with its deep root :.tem it binds the soil effectively (Kaushalappa, 1989).

*Description*: *Calliandra calothyrsus* is a small multibranched tree reaching a height of 10 m and 15-20 cm in diameter.

**Silvicultural character**

*Light*: It can stand moderate shade in its early stage up to the stage of saplings.

*Water*: It is considered one of the most promising species in humid and moist areas.

*Coppice* : Its capacity to coppice is very well. Kaushalappa (1989) has studied the rate of height growth of coppice shoots. The felling was done at 15 cm above the ground. It takes about 7 days for the stump to produce coppice shoots.

In three months the coppice shoots have attained an average height of 1.16 m and in 15 months they have attained an average height of 3.0 m. It is seen that coppicing in November showed highest percentage of leaf (16.3 pe cent) indicating that for fodder production, cutting should be done in November.

*Establishment* : It can easily propagated by seeds. Diangana (1985) studied germination rates of *Calliandra calothyrsus* subjected to three treatments. The treatments included (*i*) hot water (3 min. at 80°C), (*ii*) hydrogen peroxide (30 min.) and (*iii*) sulphuric acid (one hour). Final germination per cents were generally best with hot water. Peroxide treatments appear ineffective and sulphuric acid treatments were too severe. Kaushalappa (1989) has stated that seeds (22,000/ kg) if soaked in cold water for 24 hours before sowing gives good germination in polybags containing farmyard manure, red earth and sand in proportion of 1 :1 :1. Boiled water pre-treatment was not found necessary as germination per cent was almost same in both the types of pre-treatment. The germination was noticed on the 5th days and completed in 20 days Natural regeneration has also been observed during rainy season.

6 months old plants are best for transplantation. The best time of plantation is first fortnight of July. In alley cropping system this species may be planted at closer spacing .5 x .5 m or even less. If spacing is kept more, than interspace may be filled up by any perennial grass species. This system will decrease soils erosion hazards on slopes.

**Utility**

*Fire-wood* : The expected yield after 3 years of formation is about 10 tonnes/ha (air dry), thus giving a mean annual increment of 3.3 t/ha (Kaushallappa, 1989). The calorific values of wood being 4,500-4,750 calories per kg. It produces ash content only 1.8 per cent. It is good fuelwood species.

*Soil Erosion* : Introduction of *Calliandra calothyrsus* in alley cropping system will help in minimising soil erosion, conservation of moisture, increase in infiltration rate, decrease in evaporation from soils and balance nutrient status and finally help in land remaining productive on sustained basis.

*Soil improvement* : The foliage are good green manure. The nitrogen fixation ability helps the land remaining productive.

*Fodder* : Its fodder is a valuable supplement to pasture. In Indonesia annual yeilds of 7-10 t of dry fodder (22 per cent crude protein) per ha have been recorded (Annon., 1980). In degraded land this species may be grown along with grass.

*Ornamental* : The tree is admired for its beautiful red "Powderpuff" flowers. When planted in large groves, the flowers created beautiful effect.

## 8. SUBABUL

*Leucaena leucocephala* (Lam.) de Wit.
[*Leucaena glauca* (L.) Benth.]

*FAMILY* : LEGUMINOSAE (Mimosoideae)

*Common names* : Ku-babul, white babool, kaniti, subabul, ipl-ipil, Vilayatibaral.

*Distribution* : The plant is native of Southern Mexico and Central America and spread throughout the tropics and subtropics. After sixteenth century plant spread to Asian countries including India and is naturalized in India. It grows vigorously at lower altitudes.

*Climate* : This species prefers annual rainfall between 500 mm and 2000 mm but able to withstand long periods of drought. It can with stand mild frosts. It will survive frosts, provided they are not too heavy and frequent. The woody above-ground portion can be killed by frost, and regeneration will occur from the woody base when temperature rises. The optimum temperatures for leucaena growth appear to be between 25° and 30°C with very little growth occurring below 10°C (Brewbaka & Hutton, 1987).

*Soil* : It is found growing on rock to heavy clay to coral. It grows well in neutral to alkaline soil (pH 6.0 to 8.0). The plant has considerable salt resistance and grows in exposed coastal areas often right down to the high water mark.

*Description* : *Leucaena Leucocephala* is small to medium sized tree up to 10 m in height. It has brownish bark, leathery leaves, small, whitish "powderpuff" flowers and bunches of straight, flat, brown pods. The pods are typically 10-15 cm long, 1-2cm wide. Seeds are dark brown with hard test. Number of seed varies from 10 to 25. It forms symboitic relationship with *Rhizobium loti*, the nitrogen fixing bacteria (Halliday and Somasegaran, 1983). The *Rhizobium* penetrate young rootlets and multiply to form nodules. They absorb atmospheric nitrogen and transform it to ammonia which inturn is converted to other nitrogen containing organic and inorganic compounds. For average growth, leucaena requires no fertilisers/and it can thrive in nitrogen poor soils that are inadequate to sustain many other crops. The nodules usually occur on rootlets in the aerated surface soil layers. In low land tropics, the plant grows luxuriantly in well-drained fertile soils (Pathak *et al.*, 1982; Pathak and Patil, 1985; and Balsundaram and Mohamed Ali 1987).

**Silvicultural characters**

Abundant whitish, small flowers and pods are produced in the first or second year. The trees readily coppice and growth of coppice shoot is faster than the seedlings. One year old coppice shoot attains height of 6m and a breast height diameter of 5-7 cm. The length of coppice period depends upon the specific tree products that are needed. This species can be coppice on a yearly rotation for more than 30 years in humid zones (Meelink, 1989). Pathak *et al.*, (1981) reported that the coppice shoot growth in 1.5 years has been found to be equal to the original 3 years growth of trees thus indicating high coppice growth potentiality and forage yield in this plant. In Philippines, *Leucaena leucocephala* has recorded highest annual quantities of wood than any species measured. The average annual increments expected to be ranging from 30 to 40 $m^3$ per ha. (*Ayensu*, 1980). The effect of various pruning regimes on biomass (Leaves + small green branches) shows that biomass yield increased with increasing pruning height and lower pruning frequency. Monthly pruning irrespective of pruning height resulted in significantly lower biomass production, while lower pruning frequency particularly with six-monthly prunings produced significantly higher biomass. The lowest yield of 1.61/t/ha was observed with monthly pruning at 25 cm height, while the highest biomass yield of 13.3/t/ha was obtained with six-monthly pruning at 150 cm height. With monthly prunings the increases in biomass yield with increasing height of pruning were small. Similarly with bimonthly prunings, biomass yield differences between 25 cm and 50 cm pruning heights were significant.

The pruning-biomass yield during the dry season was lower than that during the main cropping season. With six-monthly pruning at 150 cm height, the biomass yield during the dry season was less than the half that measured during the rainy season. While during the rainy season the highest pruning biomass yield was obtained with a 150 cm pruning height irrespective of pruning frequencies, during the dry season higher biomass yield was observed at lower pruning height. With monthly and bi-monthly prunings, highest biomass yeilds were obtained with 100 cm and 75 cm pruning heights respectively. In the tri and six-monthly prunings, highest biomass yield was observed with 100 cm pruning height. This phenomena may be caused by limitations in moisture transport to plant tops during the dry season (Duguma *et al.*, 1988).

Duguma, *et al.* (1988) also reported that the prunings showed higher nitrogen percentage with higher pruning frequency as seen from the significantly higher nitrogen values with monthly than with tri-monthly prunings. The pruning height had no effect on nitrogen percentage. However, the nitrogen yield of the loppings showed the same trend as that of biomass yield. The nitrogen yield was significantly higher with trimonthly than monthly prunins. Increasing pruning height with tri-monthly pruning also significantly increased nitrogen yield from the loppings. The highest nitrogen yield was obtained with six monthly pollarding at 150 cm which yielded 757 kg N/ha/year. The lowest nitrogen yield of 114 kg N/ha/year was obtained with monthly pruning at the 25 cm height.

**Nursery technique**

To increase the germination more than 80 per cent seeds are treated with hot water (80°C) for 2-3 minutes. Seeds are sown 4 cm apart by

**Figs. 102 & 103 :** Block plantation of *Leucaena leucocephala* (Two years old).

drills in parallel rows, spaced 24 cm apart. 10 gms/sq m of complete fertilizer should be adequate, if throughly mixed with the soil. Preferably, the seed beds should be trenched and filled with sand, then bedded to provide adequate drainage and to minimize root damage when lifted for planting out.

The seedlings may be transferred or permanent locations at an optimum age of 2-4 months. No pruning is necessary if planted out from polythene bags. Long tap roots should be pruned to a length of 12-15 cm (Joshi *et al.*, 1983).

**Utility**

*Fuelwood* : In agroforestry system, short coppice rotations provide an effective method of producing high yield of fuelwood. Its calorific value is 4,200-4,600 Kcal per kg. In one acre land subabul could provide fuel for at least five families.

*Leucaena wood* : Leucaena wood has the potential to become a source of timber. Timber is suitable for agricultural implements. It has great commercial potential especially in the field of paper manufacture. It produces short-fibre pulp. It may be used in paper manufacture in combination with long fibre pulp. Leucaena leaves are used as much/cover to check run-off of water in hill area. It reduces rain drop impact and reduce runoff due to sponge effect caused by mulch. Thus, prevent deterioration of the land, reduce the soil loss and maintain the soil productivity on sustained basis.

**Soil improvement**

Leucaena is a tree ideally suited for agroforestry system. In agroforestry system tree foliage are cut and utilised as green manure. The nutrient are released from prunings. Pathak and Gupta (1987) studied organic matter addition through leaf litter in 2-year old

Leucaena plantation to be 5-6 t/ha annually, which improved tilth, cation exchange capacity, water holding capacity, bulk density, brings down soil pH from alkaline to normal and improved the yield of successive crops.

**Fodder production**

Leucaena foliage is a valuable supplement to grass pasture. The foliage and pods of the plant are nutritious. Leucaena forage has high contents of crude protein and N-free extract. Leucaena leaf meal is a rich source of potassium, calcium, carotene and vitamins. Its foliage also provides vitamin K and riboflavin. But the mymosine content of leucaena leaves has injurious effect on the animals. Mymosine contents can be reduced by soaking leaf meal overnight in water or adding other feeds to dilute the mymosine content. Low mymosine content Leucaena cultivar has not been evolved.

## 9. VILAYTI BABUL

*Prosopis juliflora* (Swartz) D.C.
(*Mimosa juliflora* Swartz)

*FAMILY*: LEGUMINOSAE (Mimosoideae)

*Common names* : Vilayati kikkar, Vilaytibabul, Vilayati babool, Vilayati jand, ganda-babool.

*Distribution* : The vilayati babul is believed to be a native to Central America and Northern South America. It is widely planted in arid zones of India, Chiefly Rajasthan, Gujarat, Uttar Pradesh, Delhi, Karnataka, Madhya Pradesh, Haryana, Tamil Nadu and Andhra Pradesh. This species was introduced in India during 1899-1900 *i.e.* nearly 100 years ago.

*Climate* : It does well under arid and semi-arid condition. In its habitat, the absolute maximum shade temperature ranges from 40.6° to 45.6°C. The annual precipitation ranging from 100 to 762 mm.

*Soil* : It can grow on a wide range of soils. It is at home on rocky hills, on sandy and saline flats, on shifting sand dunes and on coastal sand, although it attains its largest size in localities protected from wind and having the water-table not far below the surface. It does not tolerate water logged situations. Soils of shallow depth with kankar underneath are relatively unfavourable, as are also soils highly impregnated with salt; although it has been fairly successfully used in the reboisement of such localities. It has been grown on saline and alkaline soils with pH upto 11.5 in Central and Western U.P. In tenacity of life and struggle against adverse condition it has hardly any equal (Annon., 1963).

*Description* : *Prosopis juliflora* is a small sized to medium semi-evergreen tree. It usually attains a height of 10-18 m and a girth of 80-100 cm. The bark is reddish-brown with shallow fissures. The branches are covered with subulate thorn, upto 5 cm long. The leaflets are 10-46 pairs, 5 to 50 mm in length and linear or oblong. The flowers are creamy white. Flowering occurs in February-March and September-October. The pods are yellow or straw-coloured when ripe 12 to 25 cm long and 0.7 to 2 cm wide. Seeds are about 12-34 pre pod with light brown lustrous testa.

In India five forms of *Prosopis juliflora*, namely the Argentine form, the Arid form, the Mexican form, the Peruvian form and the Australian form are recognised. The different forms of *Prosopis juliflora* appear to hybridize freely. It is therefore, not easy to distinguish all the forms. The variety *glandulosa* Torr. is almost deciduous in habit (Annon., 1983). The morphology of the five forms are studied by Raizada and *Chatterji 1954*. They have developed a key to facilitate their identification.

**Silvicultural character**

*Light* : Young plant is strong light-demander. In the later stages it can tolerate moderate shade.

*Frost* : In general this species is frost hardy but the peruvian and Australian forms are frost tender.

*Water* : The tree is drought hardy. It grows in the area where the water table is lower than 25m. It is liable to be killed by excessive moisture.

*Fire* : This species is fire-tender.

*Coppice* : Due to its good growth and enormous capacity to coppice it has great future as biomass for fuel (Tiwari, 1983). Sugur (1988) reported that cutting *Prosopis* at 90 days interval yielded on average 1.97 tons/ha. fuel for a corrected stand density of about 4500 plants/ha. This in terms of M.A.I. worked out to 7.861/ha/year in terms of material suitable for fuel. These results are to be interpreted in light of the fertility of area and observe of any cultural or fertilization whatsoever. From this point the coppicing potential is quite remarkable. Cutting at six months interval yielded 4.94t/ha and a MAI of 9.88 t/ha. The increased total production could be due to bigger size and high density branch materials. The MAI slightly declined for cutting at 270 days interval (7.73 t/ha/yr) and also at yearly interval (8.73 t/ha/yr). He has also suggested that cutting of Prosopis at an interval of 6 months during the period October to April and cutting at three months interval from April to October seems an optimum combination.

*Establishment* : Ripe pod of *Prosopis juliflora* (Australian form) are collected in March or July. The pods are sun dried. To facilitate extraction of seeds from the pods, soaking the pods in water, extracting the pericarp and seeds by slight pounding, and then treating the seeds with dilute sulphuric acid facilitated final extraction of seeds with out injury to their germinative capacity (Kandaswamy and Venakata Row, 1966). Germinative capacity varies 80-90 per cent. The seed is sown in the nursery in May or June. 28 gm seeds are required for sowing a line 27 m to in length at an espacement of 2.5 cm. A second thinning is done to space them 15 to 20 cm apart when plants are 30 cm tall. The seedlings will be ready for plantation at the age of 4-5 months. After 12-15 months seedlings becomes suitable for preparing stump. It can be propagated by direct sowing, planting out entire seedlings or stumps or, vegetatively; by planting root or branch cutting (Annon., 1983).

**Utility**

*Wood* : The wood of *Prosopis juliflora* is strong and durable. It is used for fence, posts, door and window frames etc. Sekhar and Rawat (1960) have recommended it for tool handles of all sorts, crushers and such other purposes where strength and hardness are of importance.

*Fuelwood* : The wood is used as fuel and produce good quality charcoal. Because of its high heat value, the wood has been termed "Wooden anthracite". It burns slowly and evenly and holds heat well (Annon., 1980). Tiwari (1983) stated that large scale energy plantations for power generation using fluidized bed combustion technique could be devised for providing motive power. The fuelwood crisis in rural areas particularly where large chunks of alkaline and semi-arid lands are available, can be removed by planting of this species.

*Honey* : Its flowers produces nectar for high quality honey. In agroforestry system farmers can also start apiculture in their farmland.

*Pods* : Its pods has high nutritive value. The pods are edible.

## 10. CHAKUNDI

*Cassia siamea* Lam.
(*Cassia florida* Vah.)
(*Senna sumatrana* Roxb.)

*FAMILY*: LEGUMINOSAE (Caesalpinioideae)

*Common names*: Iron wood, vakai, ponavari, kassod, minjri, ponavari.

*Distribution*: It is native to south-east Asia from Indonesia to Sri Lanka. Now *Cassia siamea* is widely distributed almost throughout India.

*Climate*: Chakundi can grow in a wide range of rainfall conditions (750-2,500 mm) with 4 or 5 months dry season. Absolute maximum shade temperature in its native habitat ranges from 38°C to 49°C.

*Soil*: The plant grows well in deep, moist, well-drained, hight fertility soils. Experiments carried out in Tamil Nadu show that sowing seeds in mounds was successful while sowing in pits was a failure (Annon., 1983).

*Description*: It is a medium to moderate-sized evergreen tree with a fairly straight bole and dense crown. The bark is brown, with straight longitudinal fissured. Leaves are 15-30 cm long, and with 6-14 pairs elliptic oblong leaflets. The flowers are yellow. Each flower has 5 petals. The pods are brown, 15 to 25 cm long and ribbon like when young. The pods many seeded and splits along both margins.

**Silvicultural characters**

*Light*: It is light demander. It is found growing in open area with complete oberhead light.

*Frost*: It is somewhat frost tender.

*Drought*: It can withstand the effect of drought. In case, continuous two year or more drought, seedlings growth stunted. The early deep rooting

**Fig. 104 :** *Cassia siamea.*

pattern of the *Cassia siamea* would be advantageous under moisture stress conditions (Fayemilihin & Kang, 1988).

*Coppice* : The tree coppices exceptionally well. It produces on an average, 2-4 vigorous shoots per stool.

*Establishment* : Natural regeneration establishes itself freely under favourable condition.

The seeds are collected from March to April. Seeds are separated from the pod by hand.

In a set of tests germinative capacity was found to be 50 per cent for untreated seeds, for seed treated with cold, hot and boiling water respectively it was found to be 37 per cent, 47 per cent and 5 per cent. Seeds soaked in sulphuric acid for 10 minutes gave slightly better results than seed soaked in cold water for 2 days. For artificial regeneration seeds without treatment are recommended (Annon., 1983). The seed weight varies in different regions from 37,000 to 41,000 per kg. This species is best sown on prepared mounds of loose soil at 3m x 3m spacing at the beginning of rains, using 5-6 seeds per mound. (Jha, *et al.*, 1988). One year old nursery grown seedlings are suitable for stump planting. Chakundi is being successfully raised in agroforestry plantation in Ranchi, Bihar under A.I.C.R.P. agroforestry/programme at 3m x 3m spacing. The survived per cent of stump was found to be 40 per cent and average height and girth of the 3 years old plants at Rarha Research Station, Ranchi was 6.8 m and 23.6 cm, respectively (Jha, *et al.* 1989).

**Utility**

(*a*) *Wood* : The heartwood is dark-brown, moderately hard and durable. Its pole and branches are used for temporary buildings. In south Bihar it is one of

Fig. 105 : Block plantation of *Cassia siamea*.

the most favourite timbers for agricultural implements.

(*b*) *Fuelwood* : As a fuelwood, it is excellent and is also made into charcoal. The tree is managed under coppice system to produce small sized wood suitable for fuel. It continue yielding well upto 4th rotations.

(*c*) *Fodder* : The leaves are good fodder for animals, goats and sheep. The pods and leaves contain an alkaloid, which is fatal to pigs.

(*d*) *Vegetables* : The flowers are eaten as Vegetables.

(*e*) *Tannin* : The tannin content of pods, leaves and barks are 6%, 7% and 9% respectively (Annon., 1983).

(*f*) *As host* : *Cassia siamea* is recommended as best host for *Santalum album*. It prevents spreading of sandal spike disease (Annon., 1983).

(*g*) *Ornamental* : It is also planted as an ornamental tree in the garden.

(*h*) *Wind-break* : It is often planted as a wind-break especially in places much exposed to strong winds.

## 11. SUNKESULA

*Delonix elata* (Linn.) Gamble
(*Poinciana elata* Linn.)

*FAMILY* : LEGUMINOSAE (Caesalpinioideae)

*Common* names : Sankesar, sidhasaru, sunkesula, padenarayan, sankasura, sidhsaru.

*Distribution* : *Delonix elata* is native of Ethopia. It is widely distributed throughout India. In Bihar, Orissa, and states of southern India planted on the bound of agricultural field and backyards of houses (Annon., 1983).

*Climate* : In the area where it is growing, the absolute maximum temperature remains 47.6°C and the absolute minimum temperature 0°C. It grows satisfactorily on sites that receive normal rainfall varies from 600 to 1900 mm or more.

*Soil* : It grows on most kinds of soils. This species is suitable for degraded soils. It can grow in the poorest dry soil.

*Description* : Sunkesula is small to medium sized evergreen tree, up to 14 m in height and 20-30 cm diameter. It has ash coloured bark. Leaves are bipinnate 10 to 20 cm long; pinnate opposite, 4 to 8 pairs, leaflets 10 to 20 pairs. The flowers are yellowish, pedicels stout, finally pubescent. Petals 4 white, about 2.5 cm long, on short clows with a rounded lamina and much curled on margin. The pods are 15-20 cm by 2 to 3 cm. Seeds are transverse and oblong. Number of seeds varies from 4 to 8.

**Silvicultural characters**

It is light demander and drought hardy species.

*Establishment* : Mature, riped pods are collected off the trees in the cold season. Seeds are separated from pod wall. The clean seeds give over 65 per cent germination. There is no need of specific treatment to increase germinative capacity of the seeds. The optimum depth of sowing is about 1.5 cm. The beds are irrigated frequently at short interval after sowing. Regular irrigation is essential for good germination and for proper growth of seedlings. Spacing out of seedling in nursery is necessary to avoid over croding and competition.

Besides above this species can be propagated successfully by planting large sized stem or branch cuttings during monsoon. In Karnataka three to six year old plants are uprooted and made into stumps over 1.5 m high after removal of leaves and branches and trimming of tap and lateral roots. The stumps are immediately shifted and planted out. Profuse watering is done at least for ten days (Annon., 1983).

**Utillity**

*Wood* : It has normally yellowish white to irregular red heartwood. The weight of wood is 689 to 721 kg/cum. The wood is extensively used for cot legs, combs, furnitures, etc.,

*Firewood* : Trees planted on bunds are also utilized as firewood in South India.

## 12. ANJAN

*Hardwickia binata* Roxb.

*FAMILY:* LEGUMINOSAE ( Caesalpinioideae)

*Common names* : Anjan, kamte, acha, anjau, veppa, karachu, ura, korachu.

*Distribution*: It has no regular distribution in India and is found in patches in the drier parts of the Indian peninsula. It is common in Uttar Pradesh (Mirzapur and Banda Districts), Bihar (Palamau and Santhalpargana Districts), Maharashtra (Dhulia Jalgaon Nasik, Akola, Amroati, Chandrapur Districts), Andhra Pradesh (East and West Godavari, Krishna Kurnool districts), and Tamil Nadu ( Salem Coimbatore districts) (Pathak, 1985 and Annon., 1983).

*Climate* : In its habitat, the absolute maximum shade temperature ranges from 43°C to 48°C and the absolute minimum from 1.1°C to 10°C. It grows satisfactorily on sites that receive normal rainfall from 200 to 1500 mm.

*Soil*: It is found growing on wide range of Geological formations, schist, conglomerates, limestone, shale and slate. The rock and soil on which the tree grows have a marked influence on its growth. The best growth is attained on sandstone, conglomerate, quartzite and schist with an overlying soils of sandy loam, or on what is a very characteristic soil for *Hardwickia*—a quartzose, reddish, gravely sand.

On deep sandy soil overlying these formations, a height of 24-30 m and a girth of 1.8 to 3. m, with a clean cycylindrical bole 12-15 metres in length, are attained (Annon., 1983). Pathak *et al.*, (1985) have stated that due to its capacity to thrive on marginal soils makes it more desirable in the present context of planting of ideal and useful tree species on wasted lands.

*Description:* Anjan is a moderate sized tree. It grows upto 15-20 m height. The crown are large conical shaped. In late stage the branches extends laterally and sometimes droop slightly low. It has dark grey coloured bark. The bark is characterised by irregular vertical cracks. Leaves are greyish green, paripinnate and leaflets sessile, entire, obtuse, obliquely ovate or semi-cordate. The flower are small and greenish yellow in colour.The flowers appear in the month of July and August. The pods are flat, oblong lanceolate, coriaceous and flat.

The tree seeds sporadically to some extent every year, but gregarious seeding takes place on an average every three to five years according to locality. The local extent and the periodicity of general seeding varies and a good seed-year in one locality need not necessarily to be a good one in another. When a general seeding takes place, the crop of seed is often profuse, the trees being laden with pods. The precise cause of gregarious seeding of this kinds is not entirely clear, but there is a presumption, borne out to some extent by meteorological statistics that it is induced by a season of drough the year before the seeding (Annon., 1983).

**Silivicultural characters**

*Light:* The tree is a moderate light-demander, though it can stand a slight amount of shade in early stage of growth. In full light its growth is better than under partial shed.

*Frost:* The tree is non-sensitive to frost.

*Fire:* It is fire-hardy species.

*Damp:* The seedlings are sensitive to damp.

*Drought:* This leguminous tree tolerate long period of drought. Thus it is common only in drier parts.

*Root system:* Anjan is an efficient water user. It has deep tap root system penetrates hard ground. It has been noticed that length of the tap roots depends upon the edaphic condition of the particular agroclimatic zone.

**Coppicing and pollarding power**

It pollards vigorously. When trees are pollarded in suitable season abundant shoots arise from edge of cut surface.

This tree coppices indifferently. Old trees which send out vigorous pollard-shoots if cut a few metres above the ground produce no coppice shoots if cut flush with the ground-level never coppice. The several experiments and observations indicates that coppicing should in any case not be done at heights lower than about 30 cm (Annon., 1983)

**Establishment**

Seeds are collected by the end of April and sundried for 3-4 days and are soaked in water for 24 hours before sowing. Seeds are sown in the nursery bed at the commencement of the monsoon. Regular irrigation is essential for good growth of seedlings.

Pathak and Gupta (1985) conducted an experiment to study seed longevity under storage. Open seeds and pods were stored at the ambient temperature of room and low temperature 10°C in a refrigerator in polythene packet. Seeds germinated at various intervals have indicated no loss of viability up to 14 months. At 24 months there was a loss of 9% in germinability. Germination was completed in 3 days in all the cases.

Under moisture stress conditions up to 12 atm 60 per cent of the seeds germinated indicating its initial capacity to cope with the erratic moisture supply of red gravelly soils. At 0.33 atm 97.8 per cent germination with healthiest seedlings was obtained. At higher moisture stress seeds germinated after the 3rd day, as compared to the control treatment, where seeds germinated within 24 hours. Further, seeds could tolerate up to 5000 ppm of sodium carbonate solution giving 60 per cent germination. Beyond this, germination was drastically reduced. However, germination was delayed in concentrations past 2000 ppm.

They have also studied effect of ascorbic acid on seedling growth. Seeds gave higher emergency at 100 ppm with least mortality in large seeds. Smaller seeds gave very high mortality but at 500 ppm it was less. Seedlings from larger seeds showed greater growth rate compared to those from smaller ones, but ascorbic acid treatment improved the seedling growth in both the groups. At 100 ppm the seedlings from large seeds nodulated after 45 days. Nodulation occurred in both the control treatment and in small seeds. Internodes were also longer compared to others. Thus, larger seeds coupled with ascorbic acids pretreatment can be used for speedy nursery growth of seedlings.

**Utility**

*Wood:* Anjan is an extremely hard, heavy and fairly strong wood. It is nearly twice as hard as teak, but in actual strength as a beam or in stiffness, it does not come up to the strength of teak. It is difficult to saw and work. It should be converted green. The average weight of wood at 12 per cent moisture content is 849 kg. It is used for naves of cart wheels, plough for posts and beams and mine props (Trotter, 1982). The heartwood was found to be very resistent to decay and classified as a class I timber (Bakshi, 1976).

*Fuelwood:* It is good fuelwood species. Its calorific values is 4,952 Cal/gm (Krishna and Ramaswamy, 1932). Its wood yield good quality of charcoal.

*Green Manure:* The chopped leaves when used as green manure improves fertility status of soil.

*Fodder:* The leaves are used as fodder during lean period.

*Fibre:* Its bark yield good quality of fibre. The ropes made of Anjan fibre are preferred for use in well as they last better if kept moist.

## 13. SHISHAM

*Dalbergia sissoo* Roxb.

*FAMILY*: LEGUMINOSAE (Papilionioideae)

*Common names:* Sissu, tanach, shisham, simsapa, shisha, aguru, sissu-karra, erresissu.

*Distribution :* The tree is widely distributed. It is found in sub-Himalayan tract and also in the Himalayan valleys, usually upto 1500 m. It has been planted in many parts of Andhra Pradesh, Bihar, Uttar Pradesh, Gujarat, Maharashtra, Tamil Nadu and Haryana.

*Climate* This species prefers an annual rainfall between 760 to 2500 mm but able to withstand short period drought. In areas where it has been planted, the absolute maximum shade temperature varies from 39°C to 49°C and absolute minimum from - 4°C to 6°C. The species is frost hardy even in the sapling stage.

*Soil:* The wide distribution of the tree also shows it is capable of growing on a wide variety of soils. It grows on sandy alluvial, porous soil of sand, pebbles and boulders. It is also found on freshly exposed soil along roads and streams. It can grow on very poor soils.

*Description:* It is a fairly large and fast growing multipurpose tree species. It attains a height of upto 30 m. Bark grey, somewhat reticulately longitudinally furrowed, exfoliating in narrow strips. Leaflets 3-5, distinctly alternate, 2 to 6 cm in diameter, brod-ovate or rhomboid, tough acuminate, glabrescent. Flowers 5-8 mm long, pale white in racemes 2 to 4 cm long, which are arranged in short axillary panicles. Calyx about half the length of flower. Stamens 9, united in a sheath. Ovary pubescent, style much shorter than the ovary, Stigma large. Pod 5 to 7 cm long, glabrous pale-brown. Seeds kidney shaped, flat (Kanjilal, 1927).

The tree sheds its pods during December-March and seed germinates at the commencement of rains. Ripe pods are collected in December-January, dried in the sun and broken up by beating with sticks. About 18 pods weigh one gram. The ratio of dry pod to clean seed is approximately 125:1 and about 50 seeds in one gram. Seed yield per tree average about 1 to 3 kg. (Annon, 1984).

**Silvicultural character**

*Light* The tree is strong light demander, though it can stand a slight amount of shade in early stage of growth. In full light growth of saplings are better than under partial shed.

*Fire:* The Sissoo is not fire resistant. Its capacity to throw out coppice shoots and root suckers off sets the risk to some extent, provided the calamity takes place before other tree species have captured the ground. When a fire takes place, pole crops may be entirely destroyed (Joshi, *et al.*, 1984).

*Frost:* Tree is not very sensitive to frost. Sometimes leaves of seedlings are affected by servere frost.

*Water:* It is moisture loving tree. It grows gregariously along strem banks.

*Coppicing power:* It coppices vigorously. After felling, masses of shoots arise from the edge of cut surface and side of the stool. It is not certain upto what age or size the trees will coppice successfully; this probably depends the vigour of individual stools and on other conditions. In Chhanga Manga trees were known to coppice well upto an age of at least twenty years. It was noticed that stumps of some trees cut out in thinning in a plantation thirty years old in the Gorakhpur district (Uttar Pradesh), produced numerous coppice shoots, though many failed to produce them; this is, however, hardly a fair test, since many of the trees removed would naturally have been the less vigorous members of the crop, while the stumps in a thinned crops are not exposed to the favourable light conditions obtaining in a coppice coupe. Observations extending over some years in a sample plots in Chhanga Manga plantation showed that the growth of coppice shoots from untrimmed stools is in no way inferior to those obtained from trimmed stools (Joshi *et al.*, 1984).

*Establishment:* Sowing is done in the nursery beds which are normally 40 cm wide at the base and 30 cm wide at the top and 25 cm high. The seeds are soaked in water for 48 hours before sowing. The optimum depth of sowing is about 1.5 cm. The beds are irrigated soon after sowing. The germination percentage is about 60-80 per cent. Regular irrigation is necessary for germination and for good growth of seedlings. Under North Indian condition two irrigation are needed per week till the completion of germination. Generally one irrigation are

**Fig. 106 :** *Dalbergia sissoo.*

needed per week till the completion of germination. One irrigation thereafter till the seedlings attain a height of about 10 cm and from that stage onwards fortnightly irrigation till the commencement of monsoon rains. During the monsoon season the nursery beds are not irrigated. As and when necessary weeding is done manually.

Spacing out of seedlings in nursery is necessary to avoid over crowding and competition. The seedlings are spaced 5 cm apart in lines, when they attain a height of 5 cm spaced out seedlings are put in polythene bags.

Polythene bag plants obtained from February sowing become fit for planting out in July but planting out of one year old polythene plants or two years ball plants give good results (Annon., 1984).

Besides above this species can be propagated through almost all the common methods. Multiplication by vegetative means offers the advantage of greater genetic uniformity and imature availability of superior clones for plantation in agroforestry programmes. Stem cuttings of *Delbergia sissoo* was made from one year old branches of plants (10-12 years old).

These were divided into 12 groups with 10 cuttings in each group. Group-I has treated with distilled water to serve as control. Group 2-9 were treated with 2,000-5,000, and 10,000 mg/1 of NAA. The chemical treatment was given as quick dip (2-3 sec.) method. The cuttings were then transferred in polythene bags (5x10 cm) filled with river sand and kept in the mist chamber unit. Misting is done at regular intervals. The air temperature was maintained at 20-25°C with slightly higher temperature of rooting bed. Relative humidity was maintained at 85-95%. The experiment was kept under observations till 45 days.

All Chemical treatments promoted adventitious root formations as compared with control. However, rooting percentage was much higher in cutting treated with 2,000 mg/1 NAA, 5000 mg/1 IBA and 10,000 mg/1 of IAA. 2000 mg/1 NAA resulted in 70 per cent rooting in cuttings (Shamet and Kumar, 1988)

Sinha (1984) reported that sissoo should be planted in East-West orientation to facilitate maximum sunlight penetration to the inter-crops.

It is found that sissoo raised in the plots with intercropping have given better growth than in the control. The height growth of sissoo in case of oil seeds particularly the groundnut and soyabean is more than grown with the food crop. Thus raising of cash crops accelerates the rate of growth of tree species. Muhammad (1984) has pointed out that sissoo is most suitable tree species to be used in agroforestry on uplands of Chhotanagpur and Santhal Parganas. For planting in fields with agricultural or other crops, a spacing of 10' x 10' could be adopted in which cereals could be harvested upto 3-5 years depending on the site and growth of tree crop. On fertile soils growth of sissoo will be fast and cereals can be harvested for 3 years. By judicious canopy management crop can be harvested upto 6 years or more.

**Utility**

*Fuel wood* — The branches are pollarded at an interval of 2-years and used as fuelwood. Sissoo is a good valuable fire wood. Its sap and heart wood has a calorific value of about 4908 and 5181 calories.

*Wood* — Sissoo has normally a golden-brown to dark-brown wood. The weight of wood at 12 per cent moisture content ranges from 625 to 930 kg/cum. The wood is extensively used in furniture industry, musical instruments, farm implements, cart carriage and wheel works industry of rural areas and also for carving by small artisans (Tiwari, 1985).

*Fodder:* During scarcity period the leaves may also be used as fodder. The chopped leaves when used as green manure improves fertility status of soil.

*Drug Properties:* The bark is used to check Cholera.

## 14. KARANJ

*Derris intica* Bennet
*(Pongamia glabra* Vent.*)*
*(Pongamia pinnata (L.) Pierre).*

*FAMILY : LEGUMINOSAE (Papilionioideae)*

*Common names :* Karanj, punku, paper, kanji, punku, pungu, unju, konja, sukhchain, honge, kerun, kagu, punga.

*Distribution:* *Derris intica* is native to Indian subcontinent. This tree species is grown in farm land in Bihar, Bengal, Uttar Pradesh and Madhya Pradesh. It is indigenous throughout India from the foot-hills of the Himalayas down to the South of the peninsula especially not far from the sea-coast. It is particularly common along banks of rivers, nalas and streams (Santapau, 1966).

*Climate:* In its habitat, the absolute maximum shade temperature varies from 37 to 49°C and the absolute minimum from 10 to 16°C. It grows satisfactorily on sites that receive rainfall from 500 mm to 2500 mm or more.

*Soil:* It is found growing under a wide range of soils. It shows a preference for deep sandy loams with abundant supply of moisture. It is also planted in drier parts and poor soils of the Chhotanagpur plateau of Bihar. The trees grows wild on murramy and slightly saline or alkaline black-cotton soils.

Fig. 107 : *Dalbergia sissoo* new sapling and old tree.

*Description:* *Derris indica* is a moderate-sized nearly evergreen tree with a short bole and spreading shady crown and a pale bark. Trifoliate leaves are bright green and glossy. Leaves are oval in shape and upto 10 cm long. Flowers 1.5 cm long, in racemes in the axils of leaves, the racemes up to 20 cm long, calyx about 3-4 mm long, purplish, the edge of the calyx-cup smooth or shallowly toothed. Corolla up to 12 mm long, the general colour scheme of the flowers is purplish or a lilac. Flowers come out in April-June when the tree is leafless or nearly so. Fruits develop rather slowly, generally reaching maturity at the beginning of the monsoon of the following year. Pods are 4-6 cm long, about 2-2.5 cm broad, rather thick though compressed, narrowed at the base and apex (Randhawa, 1983 and Santapau, 1966).

**Silvicultural characters**

*Light:* It can tolerate moderate amount of shade.

*Water:* It prefers abundant supply of water. It can tolerate drier condition for shorter period. The result of studies carried out on six tree species clearly showed that the average consumption of water/g/of biomass was 0.51 for the *Eucalyptus tereticornis* compared to 1.30 for *Derris indica* (Chaturvedi 1984).

*Frost:* This trees is moderately frost hardy. Under normal winter conditions in the plains of Northern India the leaflets may be partly burnt leaving the young stems intact. It is, however, susceptible to severe frosts (Joshi, *et al.*, 1983).

*Coppice:* Its capacity to coppice is very well. Under agroforestry system 8-10 years coppice rotation are considered good. It produces root suckers. The trees are pollarded to collect foliage and used as green manure.

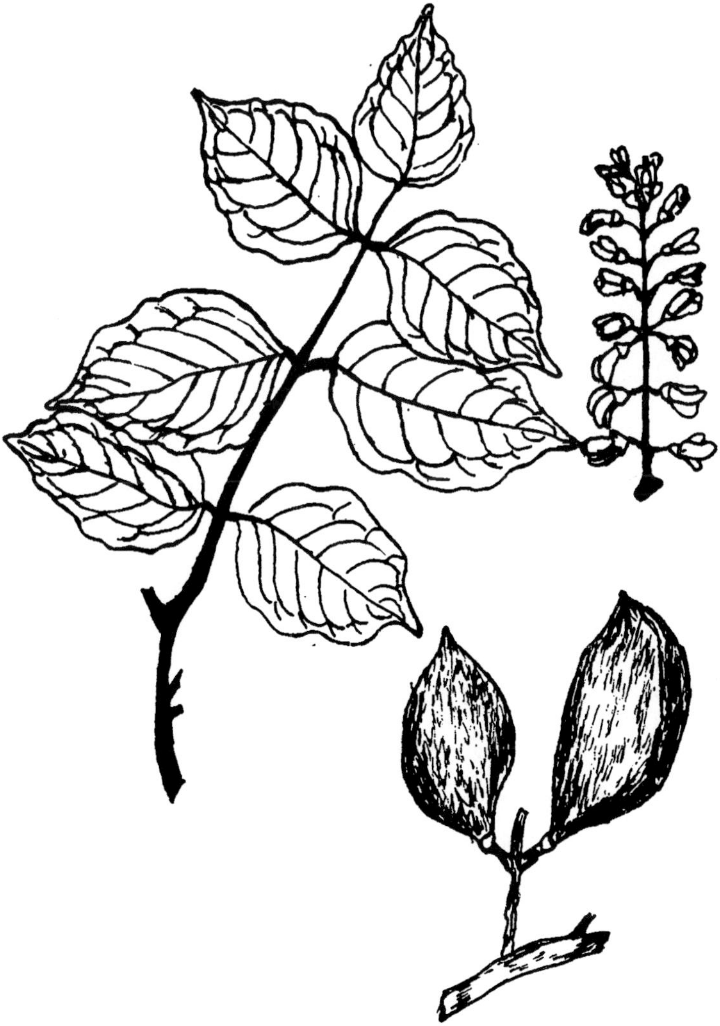

**Fig. 108 :** *Derris indica.*

*Establishment:* It can be propagated easily by seeds or stem cuttings. Seeds should be collected in April to May off the ground. The seeds are allowed to dry in the sun. Seeds are extracted by pressing along the sutures. Seeds remain viable a long period. Seeds are sown in the beginning of hot weather. Germination of seed commences about 10 days after sowing. The seedlings attain a height of 20-30 cm in the first growing season. Plants up to 1 m height have been planted out with success.

Planting of root and shoot cuttings is also convenient for large scale plantation. Stumps 1 to 2 cm in diameter at the collar have been found to give best result at Dehra Dun. Young plants need 2-3 weedings per year for the first 2-3 years of their life (Joshi, *et al.,* 1983). Dayama (1985) reported that foliar sprays of sucrose showed good response on the growth, nodulation and nitrogen content of *Derris indica*. Foliar spray of sucrose enhance the growth (5% best) and nodulation (2.0% best) as the content of nitrogen in the plant.

**Utility**

*Fuelwood:* In agroforestry system plantation of *Derris indica* managed under short coppice rotation and provides effective methods of producing high yield of fuelwood. It is important fuelwood trees of Chhotanagpur plateau of Bihar. The calorific values of wood being 4839 calories (Krishna and Ramaswami, 1932).

*Wood:* Wood is yellowish white and coarse textured. It is moderately strong light wood with sp. gr. of 0.59-0.79 and widely used for farm implement, furnitures, carts, etc.

*Oil:* This plant profusely sets seeds which contains 35 per cent fatly oil used in leather dressing, soap making and lubrication industry. The seed cakes can be processed to develop biofertilizer industry (Tiwari, 1983).

*Repelents:* Dried leaves of *Derris indica* are used as insect repelents to protect the stored grains.

*Green manure:* The foliage are good green manure. The foliage are chopped and incorporated in the soil to improve the fertility status of soil. It also reduce nematode infestations.

*Medicine:* The roots, barks and leaves have medicinal properties. The flesh of bark is used internally in bleeding piles. The juice of roots with coconut milk and lime is used for the cure of gonorrhoea (Singh and Srivastra, 1989)

## 15. BLACK LOCUST

*Robinia pseudacacia* Linn.

*FAMILY : LEGUMINOCEAE (Papilionioideae)*

*Common names :* Black locust, robinia, kikar, false acacia.

*Distribution:* It is believed that this tree species is native to North America. This species was introduced in India during 1890-1900. During the last 20 years, being planted in different places in North-West Himalayas between 1500 and 2000 m elevation. (Puri and Shamet, 1988). This species is now one of the promising exotics for fodder, fuelwood and timber in North-West Himalayas. It has been planted and naturalised in France, Italy, Hungary, Belgium.

*Climate:* In its habitat, the absolute maximum shade temperature is 29.4°C and the absolute minimum shade temperature 3.8°C. The annual precipitation and humidity ranging from 700-2000 m and 33 to 89 per cent respectively.

*Soils:* It can grow on wide range of soils. It is at home on deep, rich, gravelly, well drained loamy soil. In U.P. it grows on alluvial or lime-derived soils. It grows particularly well in porus soils, especially on new embankments, spoil banks, loose hill sides, lanslips, old alluvial flat lands, provided the situation is not too hot (Annon., 1983).

*Description:* *Robinia pseudacacia* is medium sized, deciduous tree. It usually attains a height of 20-25 m and a girth of 1.8-3.6 cm. The bark is dark grey, rough and with irregular longitudinal cracks. Leaves are 10-12 cm long with swelling at the base of petiole. Stipules modified in to spines. The leaflets are 9-19 pairs, 2.5 to 5.0 cm in length. The flowers are white and fragrant. Corolla nearly 1.5 cm long with a specific greenish-white spot. The pods are dark brown outside and silvery white inside, 2.5-10 cm x 0.9-1. 8 cm in size. Seeds are olive green. Each pod contains 10-15 seeds.

**Silvicultural characters**

*Light:* Black locust may be classified as a strong light demander. It can tolerate light shade on southern slopes.

*Drought:* The tree is drought-resistant but seedlings suffer considerably in their early stage.

*Frost:* The tree is frost hardy but seedlings up to one year suffer from frost.

*Root system:* It has deep as well as surface spreading root system. The deep root system transport nutrient from the lower surface by pumping action and makes minerals etc. available to the inter-crops.

*Nitrogen fixation:* It can fix atmospheric nitrogen with the aid of bacteria that live in nodules on its root-hairs and for this reason, it can enrich and grow on infertile soil (Annon, 1983).

*Coppice:* It coppices well. From stools many coppice shoots come up. Root suckers are produced from the superficial roots.

*Establishment:* The *Robinia pseudacacia* bears pods when the trees are 5 to 6 years old. The ripe pods are usually collected from the branches of standing

**Fig. 109 :** *Robinia pseudacacia*

trees in September-October and dried in the sun. Seeds are extracted from the pods by beating them. The seeds have good viability. They weigh of 60,000 to 80,000 seeds are one kg. Pretreatment of seed is necessary. When seeds are soaked in water at room temperature for 24 hours gives good result. The seed completes its germination within two weeks. The seeds are sown in the carefully levelled beds from March to June. The transplanting of one year old seedlings with balls of earth from January to February has proved satisfactory. Only seedlings of 1 m or above height are selected for plantation.

Branch root and root-and-shoot cuttings (stumps) have been tried successfully as planting material in Uttar Pardesh. Branch cutting generally dried out after sprouting (Annon., 1983).

Puri and Shamet (1988) have reported about propagation of *Robinia pseudocacia* by stem cutting; Stem cuttings (22 cm in length) were divided in to 16 groups of 20 cuttings each. Half of the cuttings in each group received at 1 cm vertical cut that "Split" their lower end. Group 1 was treated with water to serve as control. Groups 2-16 were treated with different concentrations of IAA, IBA, NAA.

The lower portion of the cuttings was dipped in the test solution for 24 hours, transferred into polytubes ( 6 x 12.5 cm.), and kept in a mist chamber unit with a bed temperature maintained at 25°C. Observations on rooting of cuttings were made after an interval of 7 days till 28 days.

The hormonal treatment to the stem cuttings promoted adventitious root formation IAA 10 $\mu$g./1 triggerred rooting in the regular cuttings. For the cuttings split at the lower end, IAA inhibited root formation, but lower concentrations (10, 100 and 250 $\mu$g/1 of IBA and NAA promoted rooting. NAA (250 μg/1) caused 90 per cent of the split cuttings to root. The average length of root was observed to be maximum in both the types of cuttings treated with 10 μg/1 of IBA.

The rate of growth of *Robinia* varies in different soils and climates in the different regions. Where conditions of soil and climate are very favourable, the rate of growth average 0.6-1.2 m in height and 0.6-1.2 cm in diameter annually. Growth, however, slackens when 15-20 year old (Annon., 1983).

**Utility**

*Wood:* The sapwood is white and the heartwood is golden yellow to yellowish brown in colour. The wood is used for agricultural implements, fence-posts, spoke of cart wheel and furniture.

*Firewood:* As fuelwood, it is excellent. It is successfully grown in hilly tracts and managed under coppice system. Thus, good yield of fuelwoods is produced.

*Fodder:* The leaves are palatable and are considered as good fodder. It contains crude protein 20.45 to 25.5 per cent and total ash 6.09 per cent. Its seed also contains crude protein 38.8 to 39.5 per cent and ash 4.7 percent. Seeds are good cattle feed.

In a silvipastoral study, Sharma *et al.*, (1988) reported mean grass production of 7.6 t/ha under *Robinia pseudocasia* compared to 5.2 t/ha under *Bauhinia veriegata* and 3 to 5 t/ha in natural grass land and concluded that tree component in natural grassland improved the grass productivity in addition to availibility of tree leaf fodder and fuelwood. Silvipastoral technology also helps in improving the palatability of poor quality forages and the palatability of local grasses and popular leaf fodder can be increased by mixing these fodder with *Robinia* psedoacacia leaf fodder and/or treating with salt, molasses or salt molasses additives (Arja, 1988).

*Honey:* During spring *Robinia* flowers provide nectar. Thus in hilly tract if under agroforestry system tree component is *Robinia pseudacacia* farmers can start bee keeping.

*Soil Conservation:* In hill-area this tree is considered best species because its superficial and deep root system makes this species suitable for soil conservation. This tree is suitable for agroforestry in temperate zone of middle and western Himalayas.

## 16. AGASTYA

*Sesbania grandiflora* (L.) Poir.
(*Agati grandiflora* (L.) Desv.)

*FAMILY*: LEGUMINOSAE (Papilionioideae)

*Common names :* Agasti, agase, akatthi, bak, bagphal, agasta, anari, avasinana, swamp pea.

*Distribution :* *Sesbania grandiflora* is native to Malaysia. It is found in many states such as Bihar, West Bengal, Assam, Punjab, Tamil Nadu, Karnataka and the Andamans. It is commonly grown in the rice field, garden and on farm bund.

*Climate :* In India it grows in the areas having rainfall more than 1000 mm. It grows well up to 1000 m.

*Soil :* The tree grows on a variety of soils. It prefers well drained soil, but can grow on black cotton soil. It is self productive where the surface soil is loose and uneven. (Joshi, *et al.* 1983).

*Description :* *Sesbania grandiflora* is a soft-wooded tree of quick growth reaching up to 8 metre or more in height. The bole is straight with light crown. The leaves are pinnate, petiolated, and green. Flowers large, pea shaped, borne-singly or in short recemes in the axil of the leaves. There are two varieties with salman pink flowers and cream coloured flowers. The flowering period is September-January. Flowers of pink variety open at dusk when they are pale greenish white but by the morning turn pink. Fruit (pods) are long and narrow (Randhawa, 1983).

**Silvicultural characters**

*Light* : It is light demander. It can tolerate moderate amount of shade. For proper flowering and pod formation, tree requires full overhead light.

*Water* : It is an efficient user of water. It can tolerate drought for shorter span.

*Frost* : This species is moderately frost hardly.

**Establishment**

It can be propagated by any methods; namely direct sowing of stump planting. Owing to low cost, direct sowing seems preferable. The seeds should be collected from trees in the month of Feb.-March. The clean seeds give over 90 per cent germination. Tests carried out with a seed sample one year old have shown that the seeds commence germinating on the 3rd day after sowing and nearly 80 per cent of them germinate within a week. The germination per cent was 98 (Joshi *et al.*, 1983). Mishra and Singh (1985) studied the level of salt tolerance of this species at germination stage at F.R.L., Kanpur. Healthy and uniform seeds were used for the study. Length (Avg.0.61 cm), breadth (0.39 cm) and weight (0.05 g/seed) were measured. The viability was assessed with the help of T.T.C. Salt solution (100%). The seeds were pre-soaked in moist filter paper in sterilized petri dishes. Salt solution of 5 different concentrations (2, 4, 8, 12, and 15 E.C) were prepared along with a control. The final germination count was made after a week. With increasing concentration of salts even up to 8 E.C. minor decline in germination occurred. When concentration increased up to 15 E.C. a sharp decline in germination was

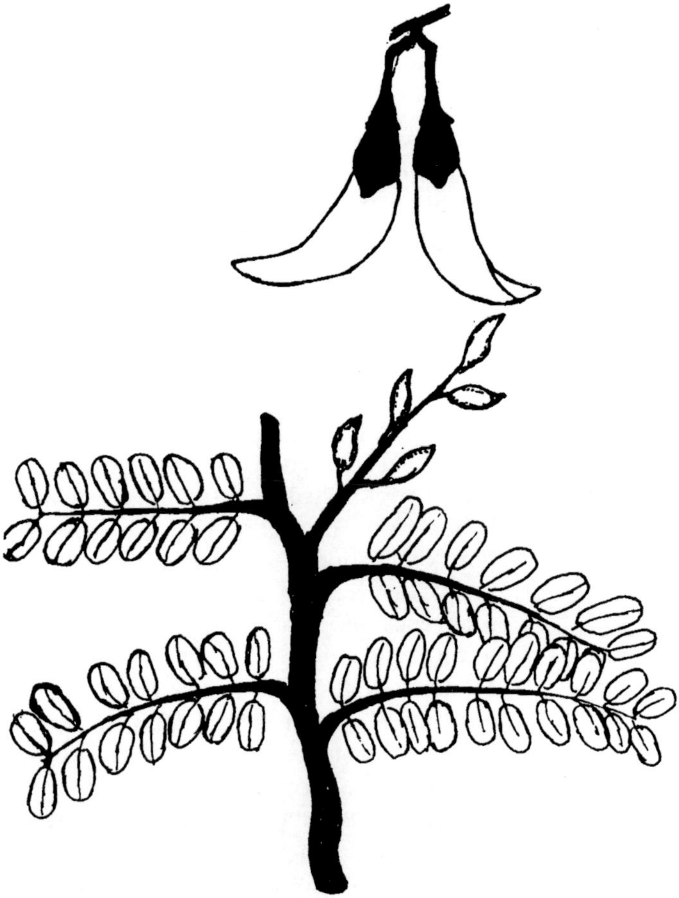

**Fig. 110 :** *Sesbania grandiflora.*

observed. They concluded that this plant, could tolerate low salanity levels and could be planted in mild 'usar' areas.

Growth rate of this species depends upon the site and method of soil preparation. At good site, it attains average height of 2 m with 3 cm girth. In agroforestry system, this plant can be managed under 3 yr. rotation. This species yields a fine hedge in a year. This tree species can be planted in the alley cropping system.

**Utility**

*Green manure* : Tree forms fine hedge within two years. The foliage is consiuered as excellent green manure. The foliage is chopped and incorporated within soil, nearly one month prior to the sowing period. This period facilitates proper decomposion of foliage. The species can be introduced in hilly area to control soil erosion etc.

*Food* : The tender leaves and pods are good vegetables The flowers and tender leaves are fried. The leaves contain 36 per cent crude protein (dry weight) and other minerals.

*Pulp* : The wood's fibre length and composition (high cellulosic raw material production) makes this species suitable for pulp. This species can be raised on large scale under agroforestry system for production of pulpwood.

*Support and shade* : It is good support plant for pepper and betel vines and is also planted in coconut nursery to provide shade to seedlings.

*Gum and Tannin* : The bark contain tannin. The gum produced by agastya may be a good substitute of gum arabic.

**Fuelwood and fodder production**

Gill (1986) studied fuel and fodder production under agroforestry system on cultivated land by using *Sesbania grandiflora* as tree component at IGFRI, Jhansi. Maximum fuel production (375 q/ha dry wt.) was registered with the pure *Sesbania* plantation treatment, closely followed by the *Sesbania* and Nandi grass treatment (373 q/ha dry wt). Maximum forage production was registered with hybrid napier (pure planting) during 1984 and 1985. Interestingly, per plant yield among cereal forage crops was the highest for the hybrid napier raised in combination with *Sesbania grandiflora* during 1985 (after the removal of tree for fuel production). This exibited the positive role of *Sesbania grandiflora*. Taking into consideration the monetary benefits, the highest gross return during 1984 was registered with the tree+Hybrid napier intercropping (fuel-fodder production system). During 1985 (after removal of the tree component), the gross return was the highest in the hybrid napier pure-plot treatment as it had the highest plant population. Such a study was brought forward the idea of combining *Sesbania* or any leguminous tree raised for fuelwood with fodder crops for a fuel-fodder production system. This system can raise the per plant productivity of forage crops even after the removal of the tree components.

## 17. PHALSA

*Grewia asiatica* Linn.
(*Grewia elastica* Royle)
(*Grewia vestica* Wall.)

*FAMILY*: TILIACEAE

*Common names*: Dhamani, phalsa, pharsia, bimla, dhaman, phuhura, pheru, dhamni, man-bijal.

*Distribution*: It is widely distributed in India. It is common in Uttar Pradesh, Bihar, Madhya Pradesh, Punjab, Haryana, Gujarat, Andhra Pradesh and West Bengal. In sub-Himalayan tract tree is found up to 1,500 m.

*Climate*: Phalsa grows the best in regions where there is distinct winter and summer. The yield and fruit quality are better in such region. Full grown plants can tolerate freezing temperature for a short period. The plants can tolerate temperature as high as 40°C. It requires clear weather at the time of flowering. In its natural habitat, the rainfall ranges from 2500-4000 mm or even more. Rains at the flowering time affect fruit setting (Annon., 1983 and Shankar, 1985).

*Soil* : It is found growing on all kind of soils but grows well in well drained loamy soils. The plant has resistance to alkaline soils.

Description : Phalsa is a small to medium sized deciduous tree. It grows up to 6-12 m height and 60-90 cm girth. Young branches and petioles are covered with grey tomentum. It has blackish bark with large white blotches. Leaves are obliquely ovate, rounded or slightly cordate at the base acuminate, minutely serrate, sometimes obscurely 3-lobbed. The flowers are yellow, densely crowded, hairy. Petals are yellow much shorter than the sepals. Drupe globose, 0.5-0.8 cm diameter, sometimes indistinctly 2-4 lobbed, dark brown or black when ripe (Kanjilal, 1927).

Two distinct types *i.e.*Tall and Dwarf were recognised at Hissar and the Dwarf type was found to be more productive (Shankar, 1985).

**Silvicultural characters**

*Light* : It is a light demander species. In agroforestry system this species can be planted at 2.5x2.5 m spacing.

*Drought* : It is a drought hardy species.

*Frost* : The species is moderately frost resistant.

*Coppice* : Its capacity to coppice is very well. Nearly 5-10 coppice shoots appear in a stump. It does not produce root-suckers.

*Establishment* : Mature seeds with pulp are collected in the cold season. Sengupta (1937) reported that seeds collected in the month of July weighed

**Fig. 111 :** *Grewia asiatica.*

19,400/kg while seeds, with pulp, collected in the month of Nov. weighed 14,800/kg. There is no need of specific treatment to increase germinative capacity of the seeds. The beds are irrigated at short interval after sowing. It completes germination within 15-20 days. 12 months old seedlings are planted out as entire in prepared field along with balls of earth round the root system.

It can be also propagated by cutting, layering and grafting. A report from Rajasthan Agricultural University, Udaipur claims 76.2 per cent success with hardwood cuttings, treated with 250 ppm Ethrel for five minutes followed by 3,000 ppm IBA for two minutes (Annon, 1981).

Air layering yeilded 50 per cent success when the air-layers were treated with mixture of IBA, NAA, 2,4-D and boron 10,000, 1,000 and 100 ppm respectively (Saxena, 1963). Soft wood grafting for phalsa in which root stock was grown *in situ* for a year or more and pre defoliated (10 days before operation) Scions were grafted and got cent per cent success (Amin, 1978).

In order to get high yield of good quality fruits, annual pruning at the height of 1 metre in the month of December, is suggested by many workers. Shankar (1985) has stated that it is desirable to grow green manure crops such as moongs, urid during the rainy season in early life of orchard. The green manure crop should be turned in towards the end of rainy season. One-two ploughing after pruning the plant is desirable to control weeds and to incorporate farm-yard-manure or compost. It necessary, one ploughing may be given after the first irrigation to mix fertilisers and to control weeds.

**Utility**

*Fruit* : The ripe phalsa fruits are edible and have cooling effect. It is good source of Vitamin A and C and also contains 10-11 per cent sugar.

*Wood* : Wood is normally greyish-white to light yellowish-grey. It is used for picks, handles of hammers, agricultural implements, tennis racket frames, etc.

*Firewood* : The tree is pruned yearly and lopped branches are used as fuelwood.

*Fodder* : Its foliage is a valuable supplement of fodder.

*Bark* : The local people use its bark for rope-making. It yields strong fibre.

## 18. BAHUL

*Grewia optiva* Drumm.

(*Grewia oppositifolia* Roxb. ex Mast.)

*FAMILY* : TILIACEAE

*Common names* : Bahul, biul, behel, thidsal, dhaman, pharwa, bhengal, biung.

*Distribution* : It is widely planted on the terrace and around homesteads of Himachal Pradesh, Punjab, West Bengal and outer hills of Uttar Pradesh. It grows successfully in the mid hills of Western Himalayas to an altitude of about 1800 m. About 50-100 trees are planted over one hectare in villages.

*Soil* : It can grow on a wide range of edaphic and climatic condition.

*Description* : *Grewia optiva* is a small sized to medium, deciduous tree. It usually attains a height of 10-12m and a girth of about 1m. The branch is pale silvery brown and bark thick, dark brown with small woody scales. The leaves are broadly ovate-lanceolate, acuminate, obtusely b... closely serrate scabrous above and pubescent beneath. Leaves are 5-13 cmx3-6 cm in size. The new leaves appear in April or May. One to eight flowers remains together. Sepals 1.0-1.5 cm long, linear oblong and green in outsides. Petals are while or yellow, linear shorter than the sepals. In fact, it has two variety one with while petals and other has yellow petals. Fruits are 1 to 4-lobed, fleshy, dark green and black when ripe (Annon., 1983 and Kanjilal, 1927).

**Silvicultural characters**

*Light* : *Grewia optiva* is strong light demander and is intolerant of suppression.

*Frost* : It is a frost hardy species.

*Pollarding* : It is one of the species which can thrive excessive pollarding. Whippy branches are pollarded in order to get fodder and fuel wood. This species is best suited for agroforestry systems, because it survives many years without loosing its pollarding power. Pollarding of *Grewia optiva* spaced at 2mx2m, beginning at the age of four years yielded 3.3 tonnes fodder and 0.09 tonne fuel per hectare every year (Puri, 1990).

*Establishment* : Ripe fruits are collected in October-December. The seeds collected in the months of October-December weighed 12,000/kg, while the seeds collected in the month of December from Uttar Pradesh weighed 9,500/kg (Sen Gupta, 1937). It can be propagated by direct sowing, by planting out entire seedlings or by stump planting.

The seed is sown in the nursery just after collection before monsoon. The seedling will be ready for plantation at the age of 6 months. After 10-12 months seedlings becomes suitable for preparing stump.

**Utility**

*Wood* : The wood of *Grewia optiva* is hard, tough durable and fine textured. It weights about 801 km/m$^3$. Being tough and elastic, it is used for

Oar-shafts, axe handles, cot frames, banghy-poles, bows etc. (Kanjilal, 1927).

*Fuelwood* : It is a good fuelwood species. Pollarding of *Grewia optiva* spaced at 2 mx2 m, beginning at the age of four years, yielded 8.09 tonnes of fuel per hectare every years (Puri, 1990)

*Fodder* : Its fodder has high nutrative value. *Grewia optiva* is a very important fodder tree of mid-Himalayas and provides highly nutritious green fodder during Nov-December when no other green fodder is available in mid-hills of H.P. (Arya, 1988). Experiments conducted in Dr. Y.S. Parmar University of Horticulture and Forestry, Solan, revealed that the complete replacement of concentrate by *Grewia optiva* leaf fodder in the ration of lactating cows increased fat, protein and S.N.F. contents from 4.46, 3.63, and 8.40 per cent to 5.51, 3.74 and 8.44 per cent, respectively, without any significant reduction in the milk yield. This also helped in tanning up the reproductive system of the cows (Anonymous, 1989 and 1990).

It was also observed that there is no adverse effect on milk yield, its composition and health of the animals even if 100 per cent digestible crude protein (DCP) is substituted by *Grewia optiva* fodder in the ration of lactating cross-bred cows (Arya, 1989).

Lal *et al.* (1985) observed that 50% of the concentrate in rabbit rations may be replaced with protein rich biul (*Crewia optiva*) leaf meal. A fully mature tree is capable of yielding 20-30 kg of fodder. An acre of plantation may well contain about 125 trees, yielding 3-4 tonnes of fodder annually (Annon., 1983).

*Fruits* : Ripe fruits are edible.

*Hedge formation* : In hilly tracts this tree may be best utilised to form hedge under alley cropping system Hedge on contour line will help in formation of natural terrace and soil erosion and water loss will be within permissible limit.

## 19. NEEM

*Azadirachta indica* A. Juss.
(*Melia azadirachta* Linn.)
(*Melia indica* Brandis)

*FAMILY* : MELIACEAE

*Common names* : Neem, nimgachh, nimb, bevina, kiri-bevu, veppu, vepa, yepa, veppam, vempu, khumba, balnimb, bevu.

*Distribution* : The tree is native of India, Indonesia, Sri Lanka, Burma, Pakistan and Thailand. It is common in Bihar, West Bengal, Orissa, Gujarat, Maharashtra, Karnataka, Tamil Nadu and Assam.

*Climate* : In its natural habitat, the absolute maximum shade temperature varies from 30°C to 48°C and the absolute minimum from 1°C to 16°C. It grows satisfactorily on sites that receive rainfall from 450-1200 mm or more.

*Soil* : It is found growing under a wide range of soils. It is planted in black cotton soils, alkaline soils, dry stony shallow soil, well-drained loams. On silty sands and in clayey depressions it remains stunted. On the silt and clayflats, little of the rainfall is absorbed into the soil, which is often quite dry lower down through saturated and marshy superficially. Evaporation loss between rainfalls is very high and the soil rapidly bakes dry after the rainy season so that very little moisture is available to deep-rooted plants (Mackay, 1952).

*Description* : A very handsome evergreen tree with wide spreading branches and forms rounded dense crown. The height of neem is commonly 10-25 metres. The bark is dark grey and longitudinally furrowed. Leaves are compound long and pendent, crowded at the ends of branches young leaves are pale, tender, green and tainted with rust. Their fresh green colour and shining surface give the tree a charming appearance. Flowers are small, greenish white and sweet scented. Petals-5, spathulate-oblong, faintly puberulous outside, minutely hairy along the margin. Fruits are yellow berries (one seeded drupe) about 1-2 cm long, usually one seeded or sometimes two seeded.

The panicle of white flowers appears during March to May and fruits ripen from June to August. Normally fruits collected from the ground are sown immediately, the seeds have short viability period of 8 to 10 days (Annon., 1983, Randhawa, 1983 and Nagveni *et al.* 1987).

**Silviculture characters**

*Light* : The *Azadirachta indica* is a light demander. Young plants are capable of standings a moderate amount of shade. In alley croppings system 3 mx3 m spacing are recommended and space between the two trees are filled by *Cajanus cajan* as fillars. This helps in improving fertility status of soil (Jha, *et al.*, 1990).

*Frost* : Its seedling and sapling stage are frost tender.

*Water* : It is not moisture loving trees. It can tolerate excessive drought. It does not stand water logging. They are intolerant of an excess of moisture in the soil, the tap root tending to rot (Troup, 1921).

*Coppice* : Its capacity to coppice is very well. Nearly 4-5 or more coppice shoots appear in a stump. It also produces roots-suckers.

*Establishment* : Seeds should be collected in July and August off the trees. Nagaveni (1987) reported that normally fruits collected from the ground are sown immediately, the seeds have short viability period of 8 to 10 days. This has been a limiting factor in its vigorous expansion over large tracts.

**Fig. 112 :** *Azadirachta indica.*

The viability of seeds can be improved and extended by adopted the following techniques. The greenish yellow fruits collected from the trees are depulped. The seeds are dried under shade to bring down the moisture level to around 5 per cent (which is normally obtained by keeping the seeds under shade for 2 days). These seeds gave germination up to 80 per cent for 4 months and the gemination decreased gradually to 50 per cent in about 6-7 months period. However, germination and viability after 7 months were very low.

Dry seeds when imbibed with water gave a characteristic garlic odour. As the viability decreases, the odour also decreases. This odouri-ferous character can also be taken as an index for the viability of seeds.

The seeds are having a thin papery testa. These are damaged early, when they are in the moist condition, especially when they are covered or enclosed with mucillaginous fleshy mesocarp, when the seeds are enclosed in the mucillage for 3 to 4 days period, it is likely that moisture is imbibed by the seed and the embryo becomes active progressing towards germination. These seeds when taken out and stored will not show viability and germination. The short viability could be accounted for the delay in producing and processing the seed before storage. So, it is critical that the greenish yellow fruits are collected, immediately depulped and dried to have good viability and germination for over six months.

Seeds are sown in July to August. Shading is not required in the nursery. The seedlings are watered throughout the hot weather. The seedlings are planted out as entire in prepared field, when these are 10-15 cm high, with balls of earth round the root system, with trimming the roots and stems. It does not stand winter planting. Sen and Jha, (1985) suggested that root and shoot cutting could be made from one year old seedlings. The stumps should be prepared with a 3 cm shoots and 23 cm of root.

Neem can also be used for the afforestation of usar lands by soil replacement technique, *i.e.*, by replacing a fairly large and deep plug of usar soil by good non-usar soil. A pit 0.6 m wide and 1.2 m deep is quite adequate for this purposes, as it helps to break the kankar pan, if any, and provides enough volume of normal soil for sustaining a moderate-sized tree. Depending on the nature of the soil, adequate quantities of farm yard manure and gypsum may also be applied in the pits with advantage. In the case of bad usars watering during the periods of drought and during the following summer is become necessary. The rate of growth is slow and a height of 6-12 m may be expected in 20 years, with a diameter of 20-26 cm. Under these conditions, neem is likely to be short-lived and an average life of 20-25 years may be expected. Such plantation may not be economically remunerative but would bring slow but steady improvement of the site, making it fit for moderate quality pasturage and poor wood production (Annon, 1983).

Mann and Muthana (1984) have reported performance of *Eucalyptus camaldulensis, Prosopis juliflora, Acacia tortilis, Tamarix articulata, albizia lebbeck* and *Azadirachta indica* on saline soils. In the experimental plot the surface soil texture has been loamy sand to loam with a sandy loam, sub-angular rocky sub-soil. The depth of soil varies from 7.5 cm to 9.0 cm at 0.1 per cent slope with pH varying from 8 to 9 and salinity ranging from 12,000 to 28,000 mmhos at 25°C. It has been observed that *Eucalyptus camaldulensis* has given the highest height-in-growth (299.0 cm) within a period of

**Fig. 113 :** *Azadirachta indica*

2 years followed by *Prosopis juliflora* Israel variety) (244.0 cm), *Acacia tortilis* (186.0 cm), *Tamarix articulata* (167.0 cm) and poorest by *Albizzia lebbeck* (94.0 cm) and *Azadirachta indica* (90.0 cm).

Under silvicultural system the percentage of seedling survival and growth in height of *Acacia tortilis* continued to exhibit higher (93 and 2.50 m) than *Azadirachta indica* (91 and 2.48 m), *Holoptelia integrifolia* (70 and 1.48 m), and *Albizia lebbeck* (62 and 1.34 m) respectively. Seeding with *Dichanthium annulatum* and *Panicum antidotale* continued to affect signifincantly on the seedling survival and growth in height of trees.

As regard the grass forage, mean maximum yield was recorded under *Albizia lebbeck* (22.4 q/ha) and followed by *Holoptelia integrifolia* (22.2 q/ha) and *Azadirachta indica* (21.5 q/ha) and least under *Acacia tortilis* (19.6 q/ha).

**Utility**

*Wood* : The sapwood is grey and heartwood dark-red. Wood is very hard. The wight of the wood at 12 per cent moisture content is 833 kg/m$^3$ (Annon., 1983). It is an easy wood to saw and work to a good finish. It is very popular wherever it is found and is favourable wood for panels, furnitures, boat buildings, toys, agricultural implements, etc.

*Fuelwood* : In agroforestry system, branches are pollarded regularly in order to facilitate the penetration of light. The branches are used as firewoods.

*Shelterbelts* : Neem tree has been used successfully in Rajasthan as shelterbelt. This tree has been recommended for central rows along with *Albizia lebbeck* and *Eucalyptus camaldulensis*.

*Soil improvement* : Neem is a tree ideally suited for agroforestry system and wasteland plantation. Its leaves and twigs are used as mulch and green manures.

*Fodder* : Neem foliage is a valuable supplement to grass. The foliage are nutritious. It is an excellent fodder for goats.

*Neem oil* : The seeds of this plant contain oil (margosa oil), which is in great demand for pharmaceutical, soap and disinfectant industry. Neem-tree seed oil industry can be developed as a rural cottage industry in areas where sufficient trees are available.

The seed cakes can be processed to produce biofertilizer thus biofertilizer industry in conjunction with oil extraction can also developed (Tiwari, 1983).

*Chew-sticks* : The soft tender portion of branches are used as chew-sticks.

*Repellent* : Neem contains several aromatic and odorescent principles which repel insects. Experiments conducted on the use of neem products as protectants against insect largely do not indicate whether the relief was due to repellent or antifeedent action (Attri, 1982).

*Attractant* : It also acts as attractant for many insect species. The nature of the attractant is not known.

*Insecticides* : Attri (1982) mentions that it is now amply clear that several neem products do have low to moderate toxicity particularly to soft bodied insects. Toxicity of water suspension of neem seed kernal to *Aphis gossypii*, *Urentius*, *enchinus* and *Saisetia nigra* was reported by Cherian and Menon (1944). Non-fatty alcohol extract of neem seed cake was found to possess appreciable aphicidal activity, the $LD_{50}$ value being 0.22% when tested against *Rhopalosiphum nymphae* L. (Goyal *et al.* 1971).

*Ropes* : The fibre of bark is used for rope making.

*Gum* : Its gum has medicinal value.

# 20. BAKAIN

*Melia azedarach* Linn.
(*Melia bukayun* Royle.)
(*Melia sempervirens* Sw.)

*FAMILY* - MELIACEAE

*Common names :* Bakain, bakayan, videshinim malla nim, mahanim, bakam, bokaiṇ, ghora nim, bakarjan, bevu, arebevu, sima veppu, pejri, puvempu, turka, makanim.

*Distribution :* The tree is native to India, China, Pakistan, Iran, Cuba, Malayasia, Philippines and Myanmar. In India it is common in Bihar, Gujarat, Bengal, Karnataka, Mizoram, Assam, etc. Trees are very common in cultivation.

*Climate :* In its natural habitat, the temperature varies from 8°C to 40°C. It thrives well in areas having 600-1000 mm average annual rainfall. It grows up to 2000 m altitude.

*Soil :* It is found growing under a wide range of soils. It can be grown on saline, alkaline, slightly acidic soils but it prefers alluvial sandy soils.

*Description :* A handsome deciduous with wide spreading branches. The height of Bakain is commonly 10-15 m and girth upto 2.5 m. The bark is, smooth greenish brown or dark grey and longitudinally shallow fissured. Leaves are bipinnate, sometime tripinnate, 20-50 cm long. Pinnae usually opposite. 3-7 leaflets are found in each pinna. The new leaves appear in March. The leaves are bright green, gives the tree a charming appearance. Flowers are small, liliac-blue. The flowers appears in the month of March. The flowers are bisexual in long peduncled axillary panicles. Calyx deeply 5 lobed. Petals 5, 5 to 6 cm long, linear-oblanceolate. Staminal tube purple, cyclindrical with 20-30 linear teeth, ovary 5 celled. Drupe about 1.3 to 1.5 cm in diameter.

**Fig. 114 :** *Melia azedarach.*

globose, generally 5-celled and 5-seeded, more or less dry, yellow when ripe, at first quite smooth, afterwards wrinkled (Kanjilal, 1927).

**Silvicultural characters**

*Light :* The *Melia azedarch* is a light demander. It requires full overhead light for proper growth. In young stage, it can tolerate moderate shade.

*Frost :* The seedlings require protection against excessive frost. The tree is frost hardy.

*Water :* It is not moisture loving. It tolerates excessive drought. Young seedlings are intolerant of an excess of moisture.

*Coppice :* Its capacity to coppice is not very well. It coppices only from tree having girth less than 1 m. Nearly 5-8 coppice shoots appear in a stump. It also produces root suckers. The tree is fast growing, can be grown in agroforestry system.

*Establishment :* It can be propagated by seeds, cuttings and root suckers. Fruits are collected off the trees or from the ground. Fruits are depulped and seeds are soaked in water for 2 days to enhance the germination per cent. The seeds are sown in nursery bed. Nursery bed is well prepared manured and soil is sprinkled with sand. Germinated seedlings are pricked out in Nursery bed when it attain 6 cm height and transferred in polythene bags. They are planted out in the planting sites on the out break of monsoons. Irrigation and weeding operation is essential to boost the growth of seedlings. Under agroforestry system a rotation of 5-7 years has been recommended in Bihar (Jha, *et al.* 1990)

The espacement usually adopted in Bihar is 3 m x 3 m. In South Bihar the tree attained the maximum height of 4.8 m and 5.2 m in 6th and 7 years, respectively.

**Fig. 115 :** Block plantation of *Melia azedarach.*

**Uses**

*Fuelwood :* A 6-8 years old plantation yielded 10-25 $m^3$ annual biomass per ha. In agroforestry system in order to reduce shed effect branches are pollarded regularly. The branches are used as fire wood.

*Wood :* The sapwood is yellowish-white; the heartwood is red, ageing to light reddish brown. It is lustrous with a dry feel and without characteristic colour or taste. It is moderately heavy (sp. gr; about 0.56) moderately hard, straight-grained, coarse and uneven textured. The weight at 12% moisture content in 577-609 kg/$m^3$. The wood saws and works with great ease, both on hand and machine tools. It can be brought to a moderately smooth surface, showing prominent grain making along the annual rings. The wood is extensively used for toys, small box, house building and ploughs (Anon, 1981).

*Soil improvement :* *Melia azedarch* is ideally suited for agroforestry system and wasteland reclamation. The foliage can be used as green manure or mulch to improve the fertility status of soil.

*Fodder :* The leaves and young shoots are lopped for cattle fodder. The fruits are consumed by goat, sheep and birds.

*Shade tree :* It has been introduced as shade tree in coffee and tea plantation.

*Repellent :* *Malia azedarach* leaves contains several aromatic and odorescent compound which is used as an insect-repellent and insecticide. This is mainly used against attacks of insects on dry fruits, clothing etc.

*Ornamental tree :* It is often cultivated as an avenue tree, but it is short-lived and cannot be relied upon for more than 20 years, by which time it starts to die off and, has to be replaced. It is also shallow-rooted and brittle and liable to be uprooted or broken by strong winds (Annon., 1981).

*Medical value :* Poisonous fruits yield a medicinal oil. The leaves and flowers are used as poultice and are said to cure the eruption of the scalp as well (Randhawa, 1983).

## 21. TOON

*Toona ciliata* M. Roem.
(*Cedrela toona* Roxb.)

*FAMILY* : MELIACEAE

*Common names :* Toon, tun, tuni, akil, mathagiri, kuruk, todu, drawi, agil, cevvagil, nandi, tunne, vembu, malarveppu.

*Distribution :* *Toona ciliata* is one of the common trees distributed in Bihar, Bengal, West Bengal, Assam, Jammu and Kashmir, Madhya Pradesh, Andhra Pradesh, Maharashtra, Kerala, Karnataka, and Tamil Nadu. It is found up to 1500 m. It is native also to Burma, Malaysia, Bangladesh and tropical Australia.

*Climate :* This species prefers annual rainfall between 1,100-4,000 mm. In Northern India this species grows in localities having rainfall nearly 750 mm. In the area where it has been growing, the absolute maximum shade temperature ranges from 37.5°-47.5°C and the absolute minimum from minus 1° to 17.5°C. The mean relative humidity in July ranges from 70-90 per cent and in January from 45-80 per cent.

*Soil :* The tree grows on variety of soil. It is at home on deep, rich, moist, sandy soils. It grows successfully in the area having good drainage.

*Description :* This is a fast growing tall elegant tree with numerous branches which give the tree a shady head. It usually attains a height of 15-20 m and a girth of 1.8-3.0 m. It has reddish-brown bark with shallow reticulate cracks. The cracks appears after middle age of rotation period. The leaves are pinnate, with a strong rachis and numerous leaflets on either side arranged in pairs. Leaflets are opposite or alternate, 5-15 cm x 2-6 cm, anceolate or ovate-lanceolate, acuminate, glabrous or pubescent and margin entire or wavy. The flowers are white, sweet-scented, in drooping open clusters at the ends of branchlets. Petals are 0.5 cm long, ovate-oblong, sub-acute, with ciliate margins. Stamens inserted on a round orange coloured disc. Fruits are an oblong, capsule dark brown, 2-2.5cm long; seeds numerous flat, with a membranous wing at each end (Annon., 1983 and Santapau, 1966). Haines (1925) recognised 4 varieties of this species (i) var. *typica* (ii) var. *hainsletii* (iii) *var. hasletii* (iv) *Var. multijuga*.

**Silvicultural characters**

*Light :* The tree is moderate light demander. It can however, stand moderate shade when young. Young seedling requires shade for proper growth.

*Frost :* After the first year this species can tolerate frost.

*Drought :* This is not drought resistant. In severe drought condition trees die. Regular watering is essential during drought period.

*Coppice power :* Due to it fast growing nature and enormous capacity to coppice it has great future as fuelwood and fodder.

*Establishment :* Ripe fruits of *Toona ciliata* are collected in May. The fruits are sun dried for 3-4 days, and broken to collect seeds. One kg of fruits will yield nearly 225 gm of clean seeds. As regard to storage, the seeds may, with care, be stored for one year as shown by results of tests recorded in West Bengal, Assam and Bihar. However, one year old seeds tested at the F.R.I., Dehra Dun failed to germinate. In Karnataka seeds

**Fig. 116 :** *Toona ciliata.*

are said to loose viability completely 3 months after collection. In Tamil Nadu, seeds are said to retain viability for 8 weeks, the germinative capacity falling to 50% by the 10th week. It is therefore advisible to sow them in the nursery as soon as possible after collection (Annon., 1983).

It can be propagated by direct sowing in the field, planting out entire seedlings and by stump planting. The seeds are sown in the nursery in May and June. The shading is essential till the seedling attains 10 cm height. The seedling will ready for plantation at the age of 4-5 months.

The stump planting is considered best method of propagation. The stump of 1.5-2.5 cm diameter proved better than stumps of 0.5-1.25 cm diameter. In agroforestry system espacement should be 2x2m or more. Stumps planted during the rains in July gave 90-100 per cent survival. In the Assam, plantation attain a mean height of 4.6 m in 3 years In U.P., the species showed an average height of 6.4 m with a girth of 27.9 cm at 6 years of age, and height of 10.4 m with a girth of 81.2 cm at 16 years of age. Toon gives a mean annual increment of 18.15m$^3$ at the age of 5 years and thus satisfies the criteria commonly accepted for a fast growing species (Annon., 1983).

**Utility**

*Wood :* The wood is a pinkish brick-red when freshly cut, toning to light brownish red on exposure. The wood is a light-weight, and usually straight grained timber. Its sp. gr. is 0.57 and weight at 12 per cent moisture content, 449-593 kg/m$^3$. The wood is widely used for furniture, tea-boxes, domestic building purposes (Trotter, 1982).

*Fuelwood :* Its branches are used as fuelwood. In agroforestry system the trees are managed under coppice system and provides effective method of producing high yield of fuelwood.

*Intercrops :* It has superficial root system. Farmers should be much careful in selection of inter crops in order to reduce the competition between the tree species and crops.

Its fodder is good supplement to pasture. The leaves are lopped in order to get fodder.

*Medicine :* The bark has astringent properties and is used in medicine Kanjilal, 1927).

## 22. BER

(*Ziziphus mauritiana* Lamk.)

*FAMILY*: RHAMNACEAE

*Common names :* Ber, bor, berii, boroi, yelchi, ilamda, badari, dadara, regureni, borkuli.

*Distribution :* It is native to India, China and Malaysia. This fruit tree species is ideal for agroforestry system. It is widely planted on the terrace and around the homesteads in Uttar Prades, Rajasthan, Madhya pradesh, Bihar, Assam, Haryana, Tamil Nadu, West Bengal and Maharashtra. In Himalayas this species grows up to 1800 m altitude.

*Climate :* It grows in different agro-climatic condition but does well under arid and semi-arid condition. Its yield is not influenced by excessive heat or cold. It does not grow properly in moist localities.

*Soil :* It can grow on a wide range of soils. It is at home on sandy loam soils, laterite and black cotton soils. This plant can be grown in the land having high pH. It can also be grown in the area having alkalinity.

*Description :* *Ziziphus mauritiana* is small to medium sized deciduous trees. It usually attains a height of 6-8 m. It has drooping branches, rounded crown and foliage are dark green. The bark is brown and deepely furrowed. The branches are covered with stipular prickles. Leaves are 1.8-6 cm x 2.5-7.5 cm in size and are covered beneath by white tomentosum.

The flowers are greenish yellow. Flowering occurs from April to October. The fruit is drup, stone is surrounded by fleshy pulp. The riped fruit turned in to reddish brown. Yadagni (1985) have reported that about 125 varieties are available in India, out of which a few have been classified as early, mid or late. The early ripening varieties such as Gola, Safeda selected, Seb and Sandhura Narnaul which ripen during February. The mid-season ripening varities include Kaithli, Sanaur-5, Reshmi and Muria Murhara, Desi Alwar and Banarsi Karaka. These varieties ripen between first and third week of March. The common late varieites are Umran, Katha phal, Ponda, Ilaichi and ZG 3. These varieties ripen between 4th week of March and middle of April.

**Silvicultural characters**

*Light :* *Zizyphus mauritiana* is a strong light demander and is intolerant to even partial shade condition.

*Frost :* This species is frost hardy.

*Drought :* This species is drought hardy.

*Fire :* Ber is fire hardy. It has power of recovery from fire injury.

*Coppice :* It coppices very well, the coppice shoots grows vigorously. It was recorded that 2.1 coppice shoots appear from one year old tree. It can coppice over a long period.

**Fig. 117 :** *Ziziphus mauritiana*.

*Pruning :* It has power to stand pruning and produces tender shoots. Pruning the ber tree is highly desirable, not only during its early years to build a strong framework but also in later years to obtain a profitable crop. The ber fruit is born in the axil of leaves on the young, growing shoots of the current year. Hence a regular annual pruning is necessary to induce a good healthy growth which will provide maximum fruit bearing area on the tree. For optimum yield of good quality of fruits, one-fourth of the one-year-old wood should be headed bock (Yamdagni, 1985).

*Establishment :* It can be propagated by direct sowing, planting out entire seedlings or by stump-planting.

Ripe seeds are collected in February or March and sun dried. Seeds are soaked in water for 48 hours and sown in the nursery bed. Regular watering and weeding is essential. The seedling will be ready for plantation or for stump planting at the age of 6-12 months.

Experimental results of transplanting of ber rootstock seedling at two, four, six and eight leaf stage of growth showed that the best success (82 per cent) was achieved when this operation was done at two leaf stage. The root stock raised from such seedlings attained buddable size in six weeks after transplanting. The budded plants so raised gave complete establishment in the field (Kanwar *et al.* 1980). The best time for budding in arid and semi-arid regions is from July to August.

The budding may be done either by shield or by the ring method. The ring budding may be done druing June, when the bark peels off easily and is full of sap. For ring budding, sticks of the scion trees should be of the same thickness as the stem of the stock which should be of pencil thickness, so that a ring bearing a bud fits in closely when transferred from the scion to the stock. Fresh buds should be used for higher success (Yamdagni, 1985).

**Utility**

*Wood :* The wood is moderately strong. It is used for cot-legs, poles, agricultural implements shaft, etc.

*Fuelwood :* The branches are pruned to get tender shoots every year. The branches are used as fuelwood and also produce good quality of charcoal. The sapwood of this species has high colarific value (4,878 calories).

*Fruit :* Ber fruit contains protein 0.8%, fat 0.1%, carbohydrates 12.8%, calcium 0.03%, phosphorus 0.03% and Iron 0.8%.

Generally the fruits are eaten fresh, candied, dried, smoked and pickled. Several products like ber butter, ber sgash or juice etc. may also be made (Yamdagni, 1985).

*Intercropping :* The intercrops can be grown in the ber orchard. The regular hoeing, weeding operation will facilitate the proper growth of the ber trees as well as intercrops. Thus total productivity of the land will increase.

## 23. LITCHI

*Litchi chinensis* Sonn.
(*Nephelium litchi* Camb.)

*FAMILY* : SAPINDACEAE

*Common names :* Litchi, safed guda, Ras guda fal.

*Distribution :* This species was introudeed in India from South China and is now widely cultivated in northern India. At present, it is also common in Southern Japan, Australia, Newzealand, Vietnam, Brazil, Taiwan etc. In Bihar Litchi orchard is common in Samastipur, Khagaria, Moonger, Muzaffarpur, Darbhanga, Bhagalpur, Katihar, Saharsa districts. In West Bengal, it is common in Hooghly, Murshidabad, Nadia and

24-parganas districts. Litchi orchard is also common in Uttar Pradesh, Punjab, Assam, Orissa and Tripura.

*Climate :* It grows well in moist subtropical climate with annual rainfall up to 200 cm. The temperature ranges from 20°C to 42°C. It is planted up to 800 m elevation.

*Soil :* It can be grown in a wide variety of soil but prefers soil with good drainage. It does best on deep, well-drained loam rich in organic matter.

*Description :* It is medium sized evergreen tree. The height of litchi is commonly 5-10 m or more. It has broad-spreading, much branched round canopy. The bark is greyish brown. Leaves are compound, with 4-8 leaflets, dark green above and greyish green on the undersurface, 6-9 cm in length. The inflorescence is a compound raceme developing both from terminal and axillary buds. Flowers are unisexual, bisexual or intermediate. The flowers of different sexes on the panicles do not open simultaneously. The mature litchi fruits are one seeded nuts, usually develop in bunches. The fruits are 2.5 to 3.5 cm in diameter and usually oval in shape. The pericarp is papillate like strawberry and turns pinkish red when the fruit is ripe. The edible portion in litchi which separates easily, lies under the pericarp and completely surrounds the seeds. It is fleshy, succulent, translucent, pearly white and soft in texture (Tiwari 1984, and Maiti, 1985). In India important litchi varieties are Bedana, Deshi, China, Seedless early, Seedless late, Rose scented, Gulabi, Purbi, Elachi late, Elachi early, Kasba etc.

**Silivicultural characters**

*Light :* It is a light demander and grows best in the open. In closely spaced trees in litchi orchard, fruit bearing capacity was recorded less in comparison to widely spaced trees in orchard.

*Frost :* It is frost tender. Frost in winter and dry heat in summer are limiting factors for its successful cultivation. Young trees especially requires protection against frost. When the trees are dormant, they are not as susceptible to frost as those in active growth (Maiti, 1985).

*Establishment :* Litchi is not propagated by seeds, because it produces inferior quality of fruits with very less edible portion. It is propagated by cutting, grafting, layering, and budding methods. Air layering is most widely accepted method. A terminal branch 45 to 60 cm in length is selected for air-layering. A ring of bark about 2.5 cm in length is removed from the branch where it is about 1.25 cm in thickness. The cambium layer is then rubbed off completely from the exposed wood. A rooting media consisting of two parts soil, one part sand and one part of leaf mould or cowdung manure or sphagnum moss properly soaked in water is used in the form of a ball. A layer of moist sphagnum moss is often placed on the rooting media to ensure supply of moisture and facilitates branching of roots. The rooting media is placed on the ringed portion covering about 2 cm from the upper end of the ring. A piece of polythene is wrapped around the rooting media which is tied properly at both ends. Within 40 to 60 days after layering the roots develop from the upper end of ring. After some time layer is removed by giving sharp cut about 5 cm below the lower end of the ring, preferably at 2-3 stage. The rooted layers are kept in partial shed in the nursery bed and they will be ready for planting during the next Monsoon (Maiti, 1985).

*Intercultivation :* The farmers can grow agricultural crop up to 5th years. The agricultural operations facilitate the growth of planted trees. The inter crop may be used as mulch to increase the nutrient status of soil.

Fig. 118 : *Litchi chinensis*.

**Utility**

*Fruit :* It is a sweet fruit. It contians 6.74 to 13.86 per cent sugar, 0.7 per cent protein, 0.3 per cent fat, and 0.7 per cent calcium and phosphorus. 100 gm pulp contains 64 mg Vitamin C.

It is also used as squah and pickles.

*Firewood :* The broken branches of Litchi (after fruit collection) are lopped and used as fuelwood.

## 24. MUNGA

*Moringa oleifera* Lam.
(*Moringa pterygosperma* Lam. and Gaertn.)

*FAMILY* : MORINGACEAE

*Common names :* Sahjan, Saijna, munga, sohjan, shajna, sondan, sunjna, moringa, morunga, sakhta, subunjna, saihan, sahjnao.

*Distribution :* *Moringa oleifera* is one of the popular hedge forming species, generally grown in the back yards in India. It has a wide range of distribution from tropical climate of sea cost to subtropical climate of sub-Himalayan tract. It is common in Bihar, West Bengal, Orissa, Uttar Pradesh, Gujarat, Assam, Mizoram, Jammu and Kashmir, etc.,

*Climate :* In its natural distribution the absolute maximum shade temperature may rise to 47.5°C and the absolute minimum from 2.5° to 3°C. The normal annual rainfall rages from 750 to 2,000 mm. It grows up to 1000 m altitude.

*Soil :* It prefers recent alluvial sandy land. It has been successfully grown in arid zone. Its growth is poor on shallow hilly soils.

*Description :* *Moringa oleifera* is a small to moderate sized deciduous tree. It attains height of 12 to 15 m and a girth of up to 2 m. The bark is pale brown, rough and deeply fissured. The leaves are 30 to 90 cm long, usually 3-pinnate, rachis slender, thickened and articulate at the base, pinnae and pinnules are opposite, deciduous. Leaflets are elliptic or obovate, 1.0 to 2.5 cm long, opposite, the terminal slightly larger, pale beneath. Flowers are 2.5 cm across, white strongly honey scented, in large puberulus axillary panicles. Flowering sarts from March and fruiting from May to June under South Indian conditions. Fruit capsule, greenish, 22.8 to 50.8 cm by 1.5 to 2 cm, pendulous, triangular, 9-ribbed, slightly constricted between the seeds. Fruits mature bearing white and light weight seeds during the months of June to August. A good tree yields as many as 1,000 fruits per year (Annon., 1980 Nautiyal and Venkataraman, 1987).

**Silvicultural characters**

*Light :* It is a strong light demander.

*Drought :* It is drought resistant, but seedlings are susceptible to drought condition.

*Frost :* It does not tolerate severe frost.

*Coppice :* The tree coppices very well. The observation recorded in Bangalore for one year old coppice shoots showed that nearly 4-8 coppice shoots sprouted from stump which attain the height from 7 to 10 m and girth 20 to 45 cm. The coppicing is preferable for wood production if it is meant for rayon grade and paper pulp. (Nautiyal and Venkataraman, 1987).

*Pollarding :* It pollards very well. The branches are lopped at the time of fruit harvesting. Many shoots appear after pollarding. Pollarding is preferred to harvest flower, fruits and fodder.

*Establishment :* It can be propagated by seeds and cutting. The seeds are collected in the month of September-October. Seeds are 3,000 to 9,000 per kg. The germination capacity is more than 50 per cent. Sharma and Raina (1982) reported that seeds sown at 10 mm depth showed the maximum germination percentage (66%). The seedlings can be raised in nursery for transplantation or can be directly sown in the field at the beginning of the rains.

Moringa can be propagated by stem cuttings. Two years old cuttings as such and treated with IBA at 50 ppm (24 hrs.), 150 ppm (24 hrs.) and with rootone have been found with good rooting behaviour. Spring season has been considered good for vegetative propagation (Sharma and Raina, 1982). Author experienced that in North Bihar this species could be propagated easily in any season by large branch cutting. Though, rainy season is preferable. It is better to plant large branch cuttings of 1 to 2.5 m. long directly in the field. Approximately half metre of cutting should be kept underground. It takes hardly 15 days time to sprout. Even in dry weather regular watering facilitate proper sprouting.

Nautiyal and Vankataraman (1987) have reported that in better silvicultural management practices the stem can attain a height of 10-15 m and a girth up to 60 cms at the end of 4th year. Though the specific gravity of wood is very low, it can achieve an yield of 30-60 tonnes dry weight at 4th year rotation. Coppice yield may be expected higher than seedling yield. Considering the rate as Rs. 400 per tonne, an income of Rs. 12,000 to Rs. 16,000 may be expected from 1 hectare apart from other usufructs (leaves, seeds and fruits, etc.).

**Utility**

*Vegetables :* The flowers tender pods are palatable and are used as vegetable.

**Fig. 119 :** *Moringa oleifera*

*Fodder :* The folliage are used as good green fodder. It is browsed by livestock. Leaves contain moisture 75.0%, protein 6.7%, fat 1.7%, carbohydrate 13.4%, fibre 0.9% and mineral matter 2.3%, calicum 440 mg, phosphorus 70 mg, iron 7.0 mg per 100 grammes, copper 1.1 microgram per gram and iodine 51 micrograms per gram. Leaves contain protein, vitamin-A 1130 *i.e.* vitamin $B_1$, 210 microgram, Nicotinic acid, 0.8 miligrams, as corbic acid 220 grams and tocopherol 17.5 microgram per 100 grams (Nautiyal and Venkataraman).

*Oil :* The oil extracted from the drumstick seeds are known as Ben oil and used as lubricant by watch makers. It is also locally used for edible purposes.

*Fertilizer :* The cake of meal left after extraction of oil from kernals has bitter taste. It is used as a fertilizer. It contians crude protein 58.93, Lime (CaO) 0.40, phosphporic acid ($P_2O$) 1.03 and potash ($K_2O$) O.80% (Nautiyal and Venkataraman, 1987).

*Living fences :* A row of this tree species makes a very effective living fences. This will last for many years without maintenance.

*Intercropping :* It is an excellent multipupose tree species. Farmers of North Bihar plant drumstick in their farm and grows agricultural crops/ vegetables in the interspace. In the initial years trees are managed by pollarding. After 8th year, trees are managed by coppice to get good yield of pods and to facilitate penetration of light.

*Medicinal :* All parts of the tree have considerable medicinal value. It is used in the treatment of ascites, rheumatism, venomous bites and as cardiac and circulatory stimulants (Annon., 1980).

# 25. ARJUN

*Terminalia arjuna* Bedd.

*FAMILY*: COMBRETACEAE

*Common names*: Arjun, arjuna, arjan, arjuna-sadra, kahu vellilaya, kakubhu, nirmatti, kaha, attumaruthu, vella-monda, sahado.

*Distribution*: Arjun is found in Bihar, Orissa, Maharashtra, Tamil Nadu, Punjab etc. It is also common in Myanmar and Sri Lanka. In Chotanagpur plateau it is found growing along streams.

*Climate*: In its natural distribution, the absolute maximum shade temperature may rise to 47°C and the absolute minimum varies from 1.5 to 4.5°C. The annual rainfall varies widely from 750 to 2000 mm. It grows upto 1200 m elevation.

*Soil*: It grows on a variety of soils but prefers fertile alluvial loam and deep sandy well drained soil. It also grows in saline and alkaline soil.

*Description*: Arjun is a moderate to large sized evergreen tree. In Bihar, it attains a height of 25 – 28 cm and a girth of 3 m or over. The leaves are oblong, opposite and sub-opposite The leaves on the lower surface usually carry a pair prominent glands close to the top of the leaf-stalk.

The flower are pale-yellowish-white, cup-shaped, small and crowded on a long axis. It flowers from March to June when its honey attracts swarms of bees. The fruit is a winged nut, the leathery wings are usually five in number and are closely veined, the veins not spreading horizontally but tending to curve upwards. The fruits is tan- coloured when dry (Randhawa, 1983).

**Silvicultural character**

*Light*: It is moderately light demander. Its growth is retarded in dense overhead shade.

Fig. 120 : *Terminalia arjuna*

*Drought* : It is moderately drought hardy. The seedlings are subceptible to drought condition.

*Frost* : It is frost tender only at seedling stage. The tree are frost hardy.

**Establishment**

It can be propagated by seeds, and stump planting. Sengupta (1986) have reported that raising *Terminalia arjuna* seedlings on a large scale has been successfully evolved by Central Tasar Research and Training Institute, Ranchi. The technique of raising the arjun plants involves dipping the seeds in water for 96 hours. After the treatment, soaked seeds are kept in heaps covered with moist gunny cloth in shade, till germination. This method gives a germination of about 85 to 90 per cent within a period of twenty days which permits to raise a stable planting stock material as against erratic and poor germination spreading for nearly two months under the convectional method.

Suitable rooting media for raising *T. arjuna* with seedlings in polythene bags has also been found to be farm yard manure, soil and sand in the ratio of 3:2:1 and the planting of three months old seedlings leads to about 90 per cent of planting success.

Experimental with stump-planting, conducted at Dehra Dun indicate that stumps of collar diameter 1.3 cm to 2.5 cm gave survival of 34 to 44 per cent at the end of 2 season. The best time for stump planting under Dehra Dun condition is the end of July (Annon., 1981).

**Utility**

*Wood* : The wood is hard strong and moderately heavy. Its sp.gr. ranges from 0.74-0.82. Wood is consider good for construction of furniture, agricultural implements etc.

Fig. 121 : Plantation of *Terminalia arjuna* (Three years old).

Fig. 122 : Plantation of *Terminalia arjuna* (One year old).

*Fuelwood* : The calorific value ranges from 5030-5128 Kcal per kg. It is a good fuelwood species.

**Tasar Worm rearing**

Tasar culture is an important part of our Indian forestry. It fulfils the outstanding objectives of generating income and employment opportunities, particularly amongst the weaker sections of Indian society. Block plantation of *Terminalia arjuna* is very common for Tasar worm rearing. participant gets employment starting from seed collection of host plant, planting of seeds, preparation of pits etc. Proper rearing of silk worms on arjuna provides jobs to the people. It has been proved that after 4 to 5 years of close plantation of arjuna in Wasteland under rainfed condition may provide per ha. Rs. 4 to 5 thousands per rearing seasons.

**Intercropping**

In the interspace of plants of arjun the cultivation of brinjal, tomato, lady's finger, ginger, gora paddy are very common in Bihar.

*Medicinal* : The bark has medicinal properties. It is used as tonic and astringent and is said to be useful in hypertension, fractures and ulcers. The fruit is used as tonic and deobstruent (Annon; 1981).

## 26. THE MYSORE GUM

*Eucalyptus tereticornis* Sm.
*(Eucalyptus umbellata* q.v.*)*

*FAMILY* : MYRTACEAE

*Common names* : Eucalyptus hybrid, Mysore gum, Nilgiri, Safeda.

*Distribution :* *Eucalyptus tereticornis* is an exotic species. It is found throughout the country. Large scale plantations of Eucalyptus have been raised in Bihar, Uttar Pradesh, Gujarat, Tamil Nadu, Andhra Pradesh, Karnataka, Kerala, Punjab, Haryana, Rajasthan and Madhya Pradesh.

*Climate :* In areas where it has been naturalised the maximum temperature varies from 30°C to 42°C. The range of average annual rainfall in the area of distribution varies from 500 to 1500 mm.

*Soil* : The tree grows on a variety of soils, *e.g.* lateritic soil as in the Bihar and Orissa; red sandy and loamy soils as in the Orissa, Bihar and Madhya Pradesh; black cotton soils as in the Karnataka, Gujarat and Maharashtra; recent alluvial soils as in Uttar Pradesh and Haryana and costal alluvial soils in the Tamil Nadu. It can grow on poor soils on wastelands. According to Kaushik *et al.* (1969) this species cannot grow on sites containing soil with pH exceeding 10, soluble salt content exceeding 0.7 per cent and possessing compact indurated sub-soil due to impervious pan, mostly kanker.

*Description* : A tall deciduous tree of quick growth reaching under good soil conditions up to 42 m or more in height. The bole is smooth and white. *Leaves* juvenile, simple, opposite, broadly lanceolate to orbicular and conspicuous intra-marginal vein. *Flowers* in umbels or heads. *Fruit* woody and hemispherical, seeds very minute. Flowering takes place usually twice in a year, first in May-June and again in October-December (Singh and Mathur, 1973).

**Silvicultural characters**

*Light* : It is a light-demander. Young plants are capable of standing a moderate amount of shade. For its best development the tree requires full overhead light. Patel (1985)

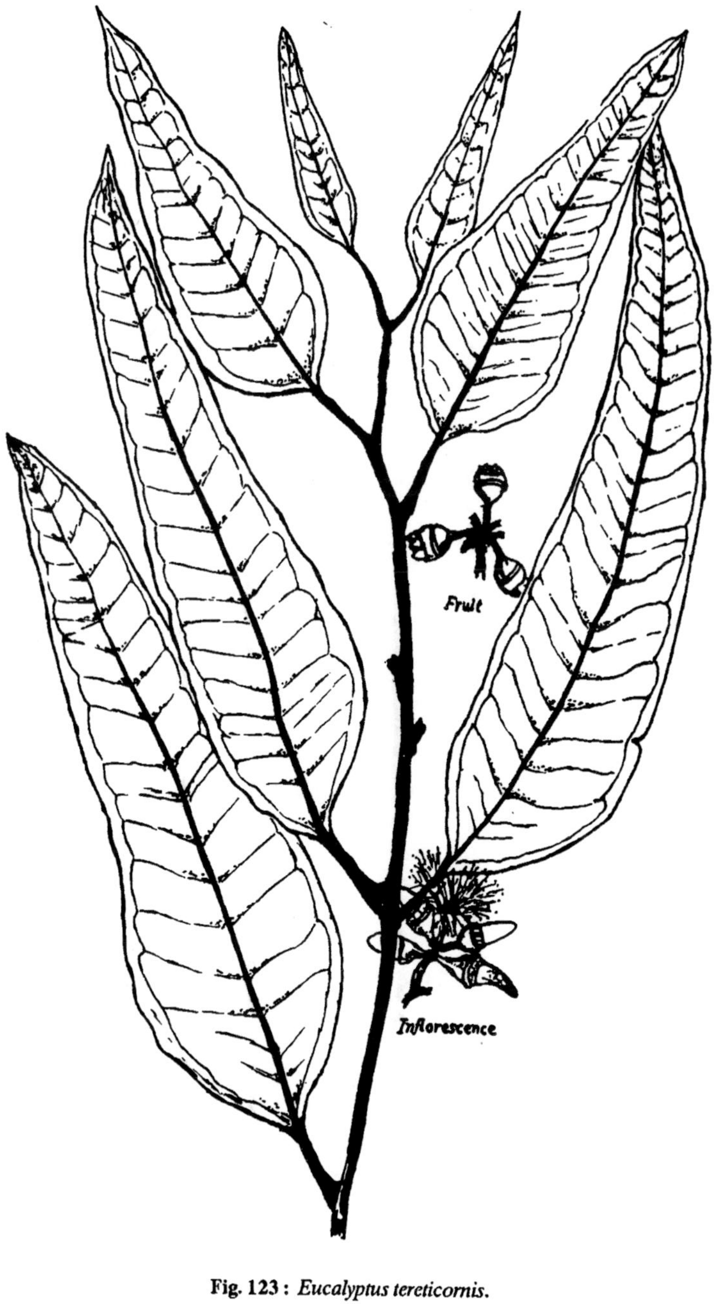


Fig. 123 : *Eucalyptus tereticornis.*

pointed out that by keeping 10 ft. distance between two plants, they get a lot of sun's rays. In winter, even when the sun is transcending the sky from south, the tree gets full sunlight up to 15 ft. below the top height. If the spacing is closer, the trees do not get sufficient sunlight and the trees in the inner rows are very thin. He has suggested to keep the rows in a east-west direction.

*Fire* : In farm forestry/agroforestry trees are managed on small rotations. Young trees are non-resistant to fire.

*Water* : There is unfortunately a wrong belief in our country that eucalyptus cause drying up of sub-soil water and receding of the water-table. The views of the policy and planning committee of Karnataka (1983) are as follows :

*Eucalyptus* hybrid is an efficient user of water compared to other tree species. But the consumption of water per hectare is high. Studies conducted by F.R.I. and colleges, Dehra Dun, shows that water consumed per gram of dry matter produced is only 1.41 mm per gram by Eucalyptus as against 2.59 mm per gram by rose wood and 3.87 mm per gram by chir (Shukla, 1980). In fact, however, one has to look at the total water balance including the impact of transpiration losses and the extent of percolation into the soil.

Regarding the lowering of the water-table. The fact is that the roots of *Eucalyptus* hybrid rarely go lower than 3-4 metres deep and usually do not spread out more than 1.5 metres. This means that Eucalyptus only consumes sub-surface seepage water, and it cannot tap subterranean ground water.

*Frost* : The species is moderately frost resistant.

Fig. 124 : Plantation of *Eucalyptus tereticornis* (10 years old).

Fig. 125 : Plantation of *Eucalyptus tereticornis* (12 months old).

*Nutrition* : Sinha (1973) reported the nutritional requirements of this species. N, P and K contents in the dominant, codominant and suppressed trees of two-three year old eucalyptus plantations were studied in different localities of Bihar. In dominant, codominant and suppressed trees percentage N (total) was 1.583, 0.940, 0.689; $P_2O_5$ was 0.620, 0.398 and 0.257 and $K_2O$ was 0.501, 0.383 and 0.215 respectfully. The difference between mean values of N, P and K for the three categories of trees are highly significant. The higher rate of growth has been associated with a higher level of all the three major nutrients.

**Litter Production and Nutrient return**

Studies on litter production and nutrient return through litterfall carried out in (5, 7 and 10 years old; stand density 1167, 1176 and 1133 trees/ha. respectively) plantation of *Eucalyptus tereticornis* have shown that these plantation produced a total litter of 3,377, 3801 and 6207 kg/ha respectively. The amount of various major nutrients returned to the soil in litter fall were found to be 30-59 kg/ha of Nitrogen, 2-4 kg/ha of Phosphorus, 15-31 kg/ha. of Calcium and 5-9 kg/ha of Magnesium of which the major proportion was contributed by leaf litter followed by twig and bark litter (George, 1979), Kushalappa (1987) has reported the annual nutrient return through litter on an unit area basis (kg/ha) from 6 years and 12 years old plantation. It was observed that the nutrient return had a direct relationship with the litter production and therefore, the maximum return of nutrients was also observed during premonsoon season, when the litter production was also maximum. The maximum nutrient return was Ca and N in both the plantations and were 38.9 kg/ha of Ca, and 32.8 kg/ha of N for 6 years and 37.4

kg/ha of Ca and 38.5 kg/ha of N, for the 12 years old plantations. The minimum return was P and was 0.28 kg/ha and 0.33 kg/ha for 6 and 12 years old plantations respectively.

Leaf litter contributed the maximum percentage of total nutrients returned and were, 97% of N, 89% of P, 93% of K and Ca and 92% of Mg in 6 years old plantation where as it was 95% of N, 85% of P, 85% of K and Ca and 87% of Mg in 12 years old plantations.

*Nursery Technique :* About 200 gm of seeds are sown per germination bed (10 m x 1 m x 10 cm) in a shallow furrow of about 1 cm width laid across the length of the bed and spaced about 5 cm from row to row. The seeds being very minute, are mixed with about an equal quantity of fine sand to prevent dense sowing in the furrows. A light covering of sieved fine sand is given on the sown seeds. Seeds are sown in October-November or in February-March. Germination of seed commences about 5 days after sowing. For planting naked root plants, the young seedlings are pricked out from the germination beds to the transplant beds at a spacing of about 10 x 10 cm. The beds are shaded for sometime till the seedlings get established (Shukla, 1980).

**Utility**

*Fuel wood :* In agroforestry system, short coppice rotations provide an effective method of producing high yield of fuelwood. Its calorific value is 4,800 Cal/gm.

*Eucalyptus wood :* The Eucalyptus hybrid timber is moderately heavy, moderately strong, tough and hard. Eucalyptus wood has been found suitable for a variety of uses. Eucalyptus wood is very well suited for furniture making. It has very good working and finishing qualities. Eucalyptus wood is also suitable for carving-Preliminary observations indicate that carving on

Eucalyptus wood can be done as easily as on sissoo (Annon, 1989).

*Match splints :* Eucalyptus wood are utilised for match splints. The processing involves giving a mild alkali treatment (1% NaOH) followed by a bleaching cycle (bleaching powder solution 5% followed by 1 per cent $H_2SO_4$ and finally by 1% Hypo). The match splints of normal light colour, proper wax penetration for adequate flame height and flame propagation coupled with good heat adherence are obtained (Annon, 1989).

*Pulp :* 6-7 years coppice rotation is considered best in view the pulp yield. It has been found to be suitable for rayon grade pulp also. We can produce different varieties of paper by proper change of pulping process and admixture of other pulp.

*Oxalic acid from bark :* Prabhu and Theagarajan (1977) have reported that bark can yield 40-50 per cent of oxalic acid by oxidation process. Its bark can be used profitably as a raw material for oxalic acid.

*Essential oil :* The leaves yield oil. It contain cineole. Small scale oil extraction units can be started by the farmers to increase their income.

## 27. LATAM

*Psidium guajava* L.

*FAMILY* : MYRTACEAE

*Common names* : Latam, amrud, guava, lal amrud, satam,

*Distribution* : It is distributed throughout the tropics and subtropics. Its cultivation has been attempted since early 17th century. According to Mitra and Bose (1985), the most important guava growing states are Uttar Pradesh (23,000 hectares), Bihar (6,850 hectares), Madhya Pradesh (3,738 hectares) and Maharashtra (2,700 hectares).

*Climate* : It occurs at elevations ranging from sea level up to 1500 m with a rainfall below, 12000 mm. It prefers only rainfall – between June to September. The mean annual temperature varies from 10°C to 40°C.

*Soil* : It occurs on variety of soils. It does best from alluvials to lateritic, well drained soil. It should hot be grown in alkaline soils.

*Description* : The most guava varieties are *Allahabad safeda, Banarasi, Lucknow-49, Chittidar, Baruipur, Harijha, Red Fushed, Seedless, Behat Coconut, and Apple Colour.*

Allahabad safeda is most popular variety of Uttar Pradesh. Trees are vigorous, medium sized, 5.8 to 6.2 m, branching heavy with dense foliage, tendency to produce long shoots. Crown broad and compact. Leaves 9.5 to 9.8 cm long and 4.8 cm wide, elliptical to oblong in shape. Fruits are medium, average weight 180 gm, roundish in shape, skin colour yellowish white in colour. The tree of *Banarsi* variety is medium to tall, 4.2 to 5.4 m in height, crown broad. Leaves are long, 10.0 cm to 10.4 cm in length and 5.4 cm in width, oblong in shape with acute apex and base obtuse. Fruits are round and dresden-yellow in colour. *Red fleshed* variety tree is medium sized 3.5 to 4.8 m in height with spreading branches. Leaves are 9.8 to 10.4 cm in length and 4.8 to 5.2 cm in breadth, elliptic-oblong in shape. Fruits are roundish-ovate in shape, saffron- yellow in colour, few red dots present on the surface. The flesh colour is down pink (Mitra and Bose, 1985).

**Silvicultural characters**

*Light* : It is light demander. For its best development the tree required full overhead light.

*Drough and cold* : The young plants are susceptible to drought and cold conditions. Although it requires dry

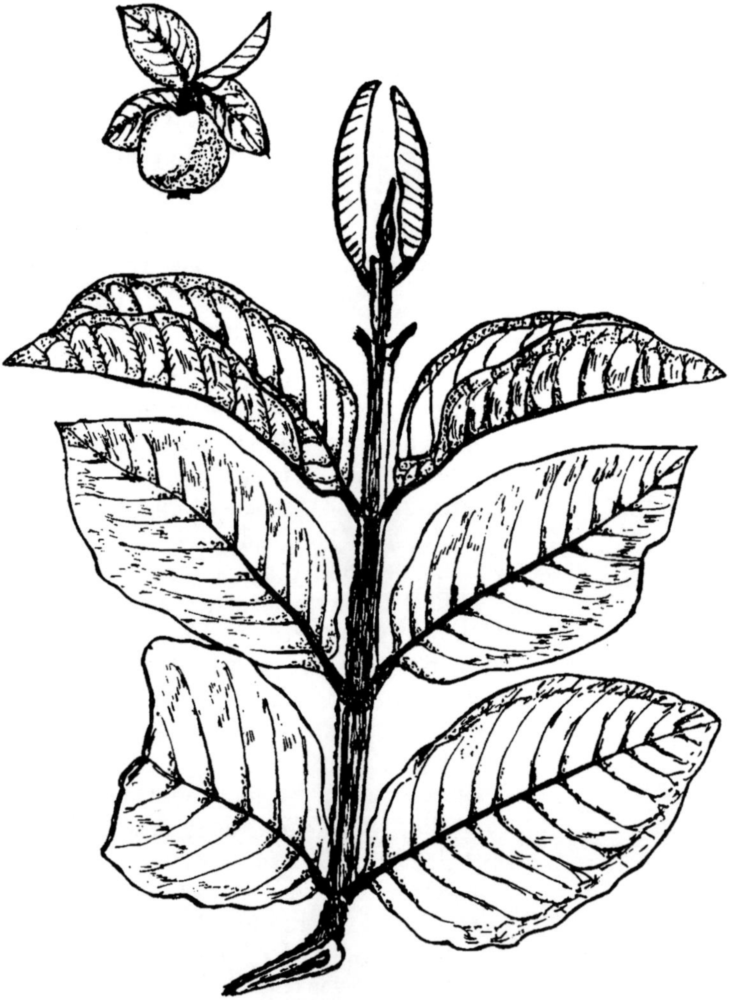

**Fig. 126 :** *Psidium guajava.*

atmosphere at the time of flowering and fruiting, high temperature at the time of fruit development causes fruit drop (Mitra and Bose, 1985).

**Establishment**

The tree bears fruits after 3rd years. Seedling can be raised in nursery by sowing nature seeds of current season in polythene bags. The seeds of fully matured fruits gives 80-85 per cent germination. Haq *et al.*, (1973) reported that soaking in cold water had beneficial effect, but treatment with hot water was harmful. Propagation through seeds has been the general practice in India. This species can be successfully propagated by grafting, cutting, budding and air-layering.

The green-wood cuttings taken from young seedlings and about 10-year old grafts of Allahabad safeda gave 100 and 90 per cent rooting respectively when etiolated and ringed for 39 days and treated with 5000 ppm of IBA at the time of planting (Bhandari and Mukherjee, 1969). Mitra and Bose (1985) have stated that air-layering is one of the most important commercial methods in practice for propagation of guava. Shoots selected for air-layering should be 1 cm in diameter and preferable from previous years growth. A ring of bark about 3 cm long is removed and covered with sphagnum moss, previously soaked in water and wrapped with polythene tape. The ends of the tape are carefully tied for rooting. It takes 30-40 days for rooting in the monsoon. It is reported that Forkert method of budding gives consistently better result than shield or patch-budding.

Planting is done without disturbance to the root-system in 1 m cube pits in the month of June or July. The soil is mixed with 10-15 kg Farm yard Manure. The young plants require

10 to 12 irrigation in a year. Singh (1981) has studied that spraying of 3 per cent urea along with calcium phosphate and muriate of potash each at 1 per cent significantly increased the yield of guava.

**Utility**

*Inter cultivation* : Intercropping with suitable agricultural crops can be practiced in guava plantation. Leguminous crops are considered safe for Guava orchard.

*Fruit* : The fruit contains 260 mg Vitamin C per 100 gm of fruits and pection. The fruits are used widely for salad, pudding, jelly.

*Medicine* : In some countries leaves are used for curing diarrhoea, and also for dyeing and tanning (Mitra and Bose, 1985).

# 28. JAMUN

*Syzygium cumini* (Linn.) Skeels
(*Syzygium jambolanum* D C.)
(*Myrtus cumini* Linn.)
(*Eugenia jambolana* Lamk.)

*FAMILY* : MYRTACEAE

*Common names* : Jamun, jaman, jam, jambu, jambo, phaunda, narala, neeram, naval, jambal, nava, nawar, neredu, black plum, Indian black cherry.

*Distribution* : It occurs either wild or is cultivated throughout India. It is particularly common in South India and is abundant in Chotanagpur plateau of Bihar. The jamun, is believed to be native to India. Outside India, it occurs in Central America, Australia, Philippines, Florida, Pakistan, Bangladesh and Myanmar.

*Climate* : Sub-tropical and tropical climates are the best for its growth. In its natural habitat, the absolute maximum shade temperature goes up to 17.5°C. It thrives in areas where the rainfall varies from 500 to 5000 mm. It grows up to 1800 m elevation.

*Soil* : It grows on most kind of soils, sand, limestone, alkaline as well as slightly acidic soils. While selecting a site for plantation one should avoid very heavy or light sandy soil. *Syzygium* can tolerate prolonged moist condition.

*Description* : *Syzygium cumini* is a moderate to large sized evergreen tree. It usually attains a height of 15-25 m and a girth of 3-4.5 m. The bark is light grey, fairly smooth, with shallow depression caused by exfoliation. Leaves are 7.6 to 15 cm by 3.8 to 6.3 cm, very variable in shape, ovate-oblong or oblong lanceolate, acuminate coriaceous, tough, smooth, shinning above except when very old. Flowers are dirty-white, fragrant, arranged in threes in lateral or terminal trichotomous panicles. Calyx tube is funnel shaped. Petals united into a calyptra and falling off in one piece. Petals are white and soon fall off. Fruit is juicy berry with a single stone. The skin and the pulp are at first green, and are then very astringent, later they turn purplish and finally dark purple or red or almost black. The wild fruit is seldom more than 15 mm long, 10 mm broad. However, cultivated varities have fruit over 3 cm long, 2 cm broad (Annon., 1981, Santapau, 1966 and Kanjilal, 1927).

**Silvicultural characters**

*Light* : It is a shed loving species. Seedling stage grows well under moderate shed condition.

*Forest* : It can tolerate frost.

*Fire* : It is moderately fire resistant.

*Drought* : It is drought resistant and can withstand prolonged drought period.

*Water logging* : It can tolerate water logging up to 2 to three months during the rainy season. This species can be tried in moderately waterlogged stretches.

*Coppice* : Its capacity to coppice is very well. Author has experienced that when felling was done at 25-30 cm above the ground in Bihar, it takes about 20-25 days to produce large number of coppice shoot. In six-months the coppice shoots attain an average height of 1.5 m.

*Establishments* : The species regenerates adequately in rainy season around the mother trees. Besides natural regeneration it can be easily propagated by seeds and stump planting, etc.

Bajpai and Chaturvedi (1985) have stated that at present, there are a number of seedling strains of jamun in India which provide good scope for selection of better varieties. While selecting a variety the following point should be kept in mind.

Oval or oblong shape of fruit, deep purple or bluish black colour big size, more pulp, small stone, juiciness, sweetness and early fruiting.

*Seed* : Fruits are collected from June to August from the ground. If the seeds are to be used for immediate planting, seeds are depulped by washing and rubbing and dried in shed condition. Under stored condition seeds loose its viability rapidly. The seed is sown in the nursery at 10 x 10 cm spacing or seeds may be sown directly in polythene bag. Regular weeding and watering are essential upto two years for proper growth. Two year old seedlings are best for transplantation. The best time of plantation is first fortnight of July.

*Inarching* : In the month of June-July one or two-year old rootstocks are inarched with the matching thickness of the scion. Root stocks are

**Fig. 127 :** *Syzygium cumini.*

watered, if necessary till the grafts are separated from the parent tree. The time taken for union is about six months (Bajpai and Chaturvedi, 1985).

*Stump-planting* : In the experiments at the Forest Research Institute, Dehra Dun, planting stumps, prepared from seedlings about 90 cm long and 1 year 3 months to 3 years old, gave encouraging result, as 96 per cent plants survived after one season. On larger scale, stumps have not proved successful (Annon., 1981).

**Utility**

*Wood* : The wood in heavy strong and durable. It is used for door and window frames, beams, cart-wheel, Agricultural implements and furniture.

*Fuelwood* : The wood is an excellent fuelwood. The Calorific value of the sapwood is 4,834 calories.

*Bark* : It is source of tanning materials in villages.

*Flower* : Its flower produces nectar. In agroforestry system farmers can also start apiculture in their farm land.

*Fruit* : Jamun fruit is eaten. it is a good source of iron. (Singh *et al.*, 1967) analysed nutritive value of the fruit at the Nutrition Research Institute Coonoor. The fruit contains moisture 28.2%, carbohydrates 19.7%, protein, 0.76, calcium, 0.02%, fat 0.1%, phosphorus 0.01%, mineral matter 0.4%, iron 1.0%, fibre 0.9%. It has calorific value 83 per 100 gm. It is used in preparation of jellies, jam, pickles, beverages, etc.

*Medicinal* : Its bark, fruit and seed have medicinal significance. Thapar (1958) reported that the vinegar prepared from juice extracted from slightly unripe fruit is stomachic, carminative and diuretic apart from having cooling and digestive properties. Jamun seed can be used as a concentrate for animals because it is rich in protein, carbohydrate and calcium.

Consumption of seed powder helps in reducing sugar quantity of urine.

## 29. AK

*Morinda coreia* Buch - Ham.
(*Morinda tinctoria* Roxb.)

*FAMILY*: RUBIACEAE

*Common names*: Maddi, ach, suranii, al, acchu, chayapattai, nuna, bartundi, hardi, haldi, manchapavatta.

*Distribution*: This species has a wide distribution. It occurs in dry regions of Bihar, West Bengal, Madhya Pradesh, Gujarat, Karnataka, Tamil Nadu and Kerala. In many states it is planted in clutivated fields.

*Climate*: In areas where *Morinda coreia* is indigenous the absolute maximum shade temperature ranges from 30° to 50°C and the absolute minimum from 1° to 8°C. The average annual rainfall in its habitat varies from 600 to 4,000 mm.

*Soil*: The tree generally tends to develope in localities having steep rocky hill slopes. It prefers sandstones and quartzite.

*Description*: *Morinda coreia* is a tree of variable size, usually small to medium sized reaching 10 m or more in height and a girth of 0.5 to 0.9 m. It has straight bole of 5-6 m.

The bark is pale brown with numerous deep, longitudinal cracks. Leaves are ovate or elliptic, abruptly acuminate at the tip, about 7.5 to 15 cm long and 2.5 to 6 cm broad. Stipules are membranous variable in size, ovate or acuminate, sometimes accrescent and foliaceous. Flowers are white and sweet scented. Fruit of each flower-head united in a compound succulent berry (Syncarpium), forming a pulpy mass (Annon., 1985). The white fragrant flowers appear from May to September and fruits ripen from January-February.

**Silvicultural characters**

*Light* : It is light demander.

*Drought* : It is drought hardy species.

*Coppice* : It has poor coppicing power.

*Establishment* : Information is lacking.

**Utility**

*Fuelwood* : As a fuelwood it is not much popular.

*Wood* : Sap-wood is not distinct from heart-wood. Wood is yellow or red with yellowish brown streaks. It is smooth and weight about 545 kg/m$^3$ at 12% moisture content. The wood is very durable. The wood is used for small furniture, frames, toys etc.

*Fodder* : Leaves and twigs are good fodder. The trees are lopped at regular interval to supplement the requirements of fodder during lean period.

*Dye* : The root are exploited as the source of dyeing material. These days use of root dye has been discontinued.

## 30. HARSINGHAR

*Nyctanthes arbortristis*

*FAMILY* : OLEAEAE

*Common names* : Harsinghar, harsingar, singhar, siharu, parijata, pavizhamalli, kaisar, kuri, pakura, sephalika, karsha, parijathmu.

*Distribution* : The tree is a native of India and commonly grown in the court-yard. It occurs in Bihar,

West Bengal, Madhya Pradesh, Orissa, Himachal Pradesh and Maharashtra.

*Climate* : In areas where *Nyctanthes arbortristis* is found the absolute maximum temperature ranges from 45.6 to 47.8°C and the absolute minimum from 2.8° to 8.3°C. The average annual rainfall in its habitat varies from 750 to 1,500 mm.

*Soil* : This tree grows on variety of soils in dry tropical habitats.

*Description* : Harsinghar is a tree of variable size, usually small to medium sized reaching up to 10 m height. It is deciduous tree. It has straight bold up to 5 m. The bark is brown, greenish or greyish brown in colour. Leaves are ovate, entire or somewhat toothed, acute or sub-acuminate, 7.5 to 15 cm by 3.8 to 8.9 cm densely pubescent beneath. Flowers are fragrant, small, arranged in trichotomous cymes, 3-7 in each head, 1.3 to 2.5 cm across, with a white salver-shaped bright corolla. The fragrant flowers appear from November to February the flowers open towards the evening and petals fall off in the morning, carpeting the ground beneath (Annon., 1985).

**Silvicultural characters**

*Light* : Harsinghar is a moderate light demander and can tolerant moderate shade.

*Frost* : It is frost-hardy species.

*Drought* : It is drought hardy species.

*Coppice* : It has good coppicing power and usually produce root-suckers.

**Establishment**

The mature seeds are collected in the month of January and February. The seeds are spread in the sun for drying. The seed have

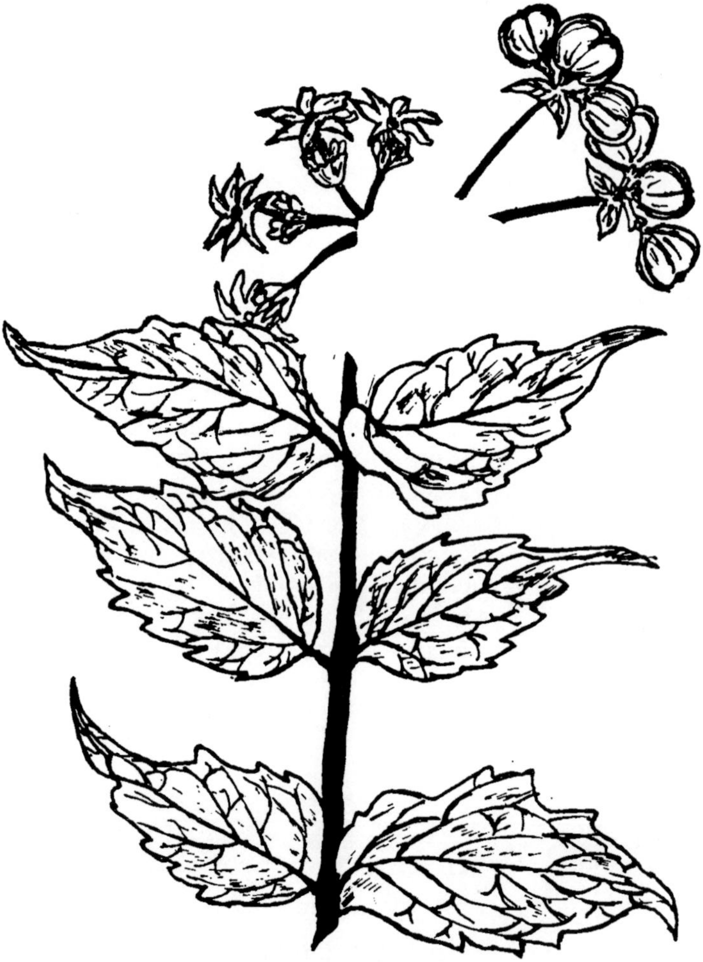

Fig. 128 : *Nyctanthes arbortristis.*

good viability. It can be raised successfully both by direct sowing and planting out entire plants. The seeds are sown in the nursery bed. Planting of container plants is a recent development and is being practised in Bihar. 12 months old container plants can be used for planting.

**Utility**

*Fuelwood* : As a fuelwood, it is excellent. This tree can be managed under coppice system in agroforestry system.

*Wood* : It is moderately hard, heavy and fine-textured. Its average weight is about 881 kg/m$^3$. The timbers is used for all types of agricultural implements.

*Acetic acid* : Warhadpande *et al.*, (1980). reported that this tree was found to give the highest yield of acetic acid (3-6%).

*Bark* : Its bark constitutes one of the tanning materials.

*Basket* : Farmers use young branches for making baskets.

*Essential oil* : The flowers, which on distillation yield an essential oil similar to that of Jasmine are used in perfumery (Annon., 1985).

*Dye* : The corolla tubes yield a beautiful buff or orange dye suitable for silk and other fabrics (Annon., 1985).

## 31. GAMARI

*Gmelina arborea* Linn.

*FAMILY* : VERBENACEAE

*Common names* : Gomari, gumhar, gumbar, shewan, sewan, shivani, kumula, pokki, kumilu, kumar, krishna-vritika, kumil, gumartek, white teak.

*Distribution* : *Gmelina arborea* has a wide range of distribution. The natural habitat of the tree is deciduous forests of the Indian sub-continent and the Andamans. It is now common in Uttar Pradesh, Bihar, Punjab, Bengal and Orissa.

*Climate* : In its habitat, the absolute maximum shade temperature varies from 30°C to 48°C and the absolute minimum from 11°C to 16°C. It grows satisfactorily on sites that receive normal annual rainfall from 750 mm to 4500 mm or more.

*Soil* : It grows on a wide range of soils, acid soils, calcareous loams and lateritic soils. It however, shows a preference for moist fertile valleys. It does not thrive where the drainage is bad, while on dry sandy or otherwise poor soil it remains stunted, and is apt to assume little more than a shrubby form owing to its being repeatedly killed back by drought. It does not thrive on heavy clay (Annon., 1974).

*Description* : *Gmelina* is a moderate sized tree. It develops straight trunk with low, heavy branches. It grows up to 15-25 m height with a clear bole of 6-9 m. Bark grey or greyish-brown, smooth, leaves 10-20 by 8-15 cm, broad-ovate, cordate with a shortly-cuneate glandular base, acute or acuminate, entire, finely tomentose and glaucous beneath. Inflorescence a terminal panicle with opposite decussate cymose branches. Corolla 2-lipped, oblique, brownish-yellow densely soft, tomentose out side. Stamens 4, didynamous, Ovary 4 called, style slender, unequally bified. Drup ovoid or oblong, pericarp leathery, shining, yellow outside when ripe, endocarp bony embedded in an aromatic pulp. Seeds 1 to 3, lenticular (Kanjilal, 1927).

**Silvicultural character**

*Light* : It is a light demander. Young plants are, however, capable of standing a moderate amount of shads. It creates heavy shade. Under agroforestry system wide spacing 3m x 3m are recommended in Chotanagpur plateau of Bihar (Jha *et al.*, 1989).

*Frost* : The species is moderately frost resistant.

*Water* : It is a moisture loving tree. It does not stand excessive drought as is evidenced in the case of plantations in Panch Mahals (Maharashtra) where plantation was raised on dry soil with a considerable proportion of sand. The trees have in many case either died out right or have died back and were kept alive only in a stunted busy form (Annon., 1974).

*Coppice* : Its capacity to coppice is very well. Nearly 5 coppice shoots appear in a stump. It does not produce root-suckers.

*Establishment* : Seeds should be collected in May off the ground. Before collection, ground under mother trees should be cleared and seeds collected frequently so as to avoid collection of seeds already damaged by insects. Brown fruits are the best; black and green ones should be discarded. The seeds should be allowed to rot, or better still, fed to cattle in a pan and the excreted stones are collected, washed and dried in the sun. Seeds thus treated gives better germination. Seeds can be stored in a dry and well ventilated shed after removing the pulp, but these are usually sown in the year of collection. If stored for a year, seeds can be sown in March and the seedlings will be over 30 cm high by the break of the rains, ready for stump-planting. Dried seed can be soaked in water for one or two days to accelerate germination (Homfray, 1937).

Seeds are sown in June and the seedlings planted out as entire in prepared field, when these are 7.5 to 10 cm high, with balls of earth round the root system, without timming the roots. Plants up to 1.8 m high have been planted out with success. It does not stand winter planting ; most winter transplants die and even those that live grow very slowly first. Starting from seed in June, 1923, the few surviving winter transplants were only 0.9 m high in December, 1925, while rain transplants were 2.4 to 2.7m high (Howard, 1926).

It can also be propagated by stump planting. Jha *et al.* (1989) suggested that stump planting is better method of raising it in agroforestry system. It is desirable to grow crops from second year. Under agroforestry system coppice rotation should be kept 12 years for timber and 4-5 years for fuel.

**Utility**

*Fuelwood* : In agroforestry system, short coppice rotations provide an effective methods of producing high yield of fuelwood. If trees are managed under long rotation period, the branches are cut to facilitate the light penitration. Branches are used as firewoods. It burns quickly. The calorific value of sapwood is nearly 4,800 kcal per kg.

*Wood* : Wood is light strong, pale-yellow or pinkish buff in colour. It is fairly close textured wood, odourless and with a smooth feel. It is moderately strong wood. It is an easy wood to saw and work to a good finish. It is very popular wherever it is found and is favourable wood to local crafts men for furniture, panelling, boxes, etc. It is very suitable for camp furniture (Trotter, 1982).

*Plywood* : The wood is suitable for class I commercial plywood (I.S.I., 1951).

Fig. 129 : Plantation of *Gmelia arborea* in agroforestry system.

Fig. 130 : Block plantation of *Gmelina arborea*.

*Pulp* : Superior quality of papers are produced from its pulp. Different varieties of paper can be produced by change in pulping process and admixture of other pulp.

*Honey* : It produces ample quantity of nectar. Farmers can start apiculture in the agroforestry farm.

*Medicinal use* : The juice of tender leaves is demulcent. Mixed with cow's milk and sugar, the juice is useful in gonorrhoea, catarrh of the bladder, cough, etc. The root is a bitter tonic, stomachic, laxative, and galactogogue. It is used as a nerve tonic in insanity and epilepsy. The roots and bark are used in snake bite (Annon., 1974).

## 32. AONLA

*Emblica officinalis* Gaertn.
(*Phyllanthus emblica* Linn.)

*FAMILY* : EUPHORBIACEAE

*Common names* : Amla, amali, Aonla, Tam, Lisirikai, nelli, amalakamu.

*Distribution* : The plant is native of India, China, Malaysia, Sri Lanka. It grows vigorously in North Eastern States Mizoram, Assam, Tripura etc. It also grows vigorously at lower altitudes.

*Climate* : In its natural habitat, the temperature ranges from 40 to 46°C. Moderate warm climate favour for the initiation of floral buds. It occurs upto an altitude of 1,500 m. In Mizoram it grows upto 1,200 m elevation.

*Soil* : It does best on in well-drained fertile loamy soil. It thrives well in light as well as heavy soils. It is found even on moderately alkaline and acidic soils.

*Description* : Aonla is a small or medium sized deciduous tree. The height of aonla is commonly 5-8 m, but ranged from 3 to 10 m. The bark is greyish-white. Leaves are small, 10-13 mm long, 2-3mm broad. Flowers are pale green, usually in small dense clusters below the leaves. Fruit is 1.5-2.5cm diameter, flesh, roundish, rather indistinctly marked into 6 lobes (Singh and Srivastra, 1989).

Bajpai and Shukla (1985) have stated that the varieties mainly classified according to their colour are green-tinged, Red-tinged, Pink-tinged, White-streaked and Bansi Red. Teotia *et al.* (1968) described physio-chemical characteristics of *Banarsi*, *Chakaya*, *Hathijhool* and *Deshi* varieties.

**Silvicultural characters**

*Light* : It is a light demander. Young plants are capable of standing a moderate amount of shade. For its best development the tree required full overhead light.

*Fire* : Young plants are non-resistant to fire. In Mizoram in slash-burn cultivation, new buds appears just before monsoon rains.

*Frost* : The species is sensitive to frost. Gangwar *et al.* (1975) reported that older plants have been found damaged with frost and plants were also rejuvenated by T-budding.

*Coppice* : It coppices well upto a considerable size, stumps up to 1 m in height have been found to give a large number of shoots.

*Establishment* : The tree bears fruits after 6 to 8 years. Seedlings can be best raised in nursery by sowing mature seeds directly in polythene bags. Two fruits are dibbled in each bag. The seeds are collected in the month of April or May. Freshly collected fruits stored in tins gives 70 to 80 per cent germination till three months. Watering to keep the soil moist must be provided until the seedlings have put forth four or five leaves. Bhujbal (1975) obtained

Fig. 131 : *Emblica officinalis.*

the highest percentage of germination (92.5) and the best root system with stones soaked in 500 ppm $GA_3$ for 24 hours. The aonla tree riased by seedlings yield small sized fruit. It is only used as root stock.

Propagation though seeds has been the general practice in India. However, vegetative propagation of selected high yielding varities is desirable. Of the various methods of vegetative propagation, budding has been found to be most practical and shield budding is the commercial methods of aonla propagation under North India conditions (Singh, 1974).

Planting is done without disturbance to the root-system in 1 m cube pits in the month of June or July. The soil is mixed with 5-10 kg Farm Yard manure. In order to protect the attack of termite 5% B.H.C. powder is mixed in the FYM. Pruning of the bearing plants should be done after the end of fruit bearing season. The young plants require watering during months at fortnightly interval, particularly till they are fully established. Watering of mature, bearing plants is also necessary from April to June at bi-weekly interval to secure higher fruit set and reduced fruit drop. Irrigation during October to December at 20 days intervals helps in better development of fruits (Bajpai and Shukla, 1985).

**Utility**

*Intercultivation* : Intercropping with suitable agricultural crops can be practised in aonla plantations. In Mizoram vegetables and crops are grown in the interspace. Farmers should always keep in mind that grain farming timing of the crop should not coincide with flower bud differentiation in aonla trees. The North-Eastern part of India is considered natural

Fig. 132 : Block plantation of *Emblica officinalis*

Fig. 133 : Silvihorticulture (*Emblica officinalis* tree to support climber vegetables).

home of aonla. Leguminous crops may be considered as safe for aonla orchard.

*Industrial use :* It is occasionally used as fuelwood and for making small farm implements and small furniture. Due to its durability under water it is used in well work. Its wood also yields good charcoal. Bark and leaves are used in tanning industry (Tiwari, 1983).

*Fruit :* The fruit contains Vitamin C and polyphenols. The fruits are used widely for murabba, jellies, tophies, pickles etc.

*Medicine* : Aonla is valued as antibiotic, antiscorbutic and diuretic. It is also used to control diarrhoea, bronchitis, cough, fever, jaundice, etc. The fruits are one of the three constituents of the well-known Indian ayurvedic medicine *Triphala* the other two constituents being Bahera and Harra (Singh and Srivastra, 1989).

## 33. KATHAL

*Artocarpus heterophyllus* Lamk.

*FAMILY* : MORACEAE

*Common names* : Kathal, phanas, jackfruit, barafal.

*Distribution* : *Artocarpus heterophyllus* is widely distributed almost throughout India. It is indigenous to western Ghats. Besides India it is common in Pakistan, Myanmar, Malaysia and West Indies.

*Climate* : In its natural habitat the temperature ranges from 0°C to 46°C. It grows satisfactorily on sites that receive annual rainfall up to 1250 mm. It grows up to 1400 m altitude.

*Soil* : It grows on most kinds of soil, it does well on deep, alluvial soil, but adapted to acidic, alkaline, laterite and sandy soils as well. It does not tolerate waterlogged situation.

*Description* : Kathal is a middle sized to large sized evergreen tree. Bark is dark brown, almost black and fissured. The leaves are elliptic in shape, acute in apex, narrowed in the base, leathery and stiff in structure, smooth and shinning above and deep green, rough and pale beneath. Leaves are 10-20 cm long. Flowers occur at the end of branches or on the trunk and branches away from the leaves ; the two sexes are usually separate in different inflorescence. Male heads 2-10 cm. long, more or less cylindrical roughly about the size and shape of human thumb ; individual flowers have a tubular perianth or corolla 1-2 mm long, hairy ; anthers are minute and yellowish, the stalk of the male heads is stout, expanded into a ring just below the head. Female head ; flower very numerous, the styles free for about 2mm ; the real fruits of this tree are not the jack-fruit, but the smaller pieces into which the jack fruit is split, roughly the edible portion of the fruit, the outer part of perianth is fleshy and edible with sweet odour ; the seeds are over 2.5cm long and 1.6cm thick (Santapau, 1966).

Samaddar (1985) has stated that being cross polinated and mostly seed propagated, the jackfruit has innumerable types or forms based on the fruit characteristics. The types differ widely among themselves in density of spike on the rind, bearing size, shape, quality and period of maturity. Innumerable variations in sweetness, acidity, flavour and taste are also observed in jackfruit growing areas. He has pointed out that cultivated types are broadly classed into two groups by consumers, *viz*.

(*i*) *Soft flesh* : When fully ripe the fruit yeilds to the thrust of a finger easily. The pulp is very juicy and soft. The taste varies from very sweet, sweet-acid to insipid.

(*ii*) *Firm flesh* : The rind does not yield to the thrust easily. The pulp is firm and crispy. The taste is variable in degree of sweetness.

**Silvicultural characters**

*Light* : It is a light demander, but tolerate shade during late years.

*Frost* : It can tolerate frost, but severe and regular frost is harmful for its growth and also influence over fruit bearing capacity of trees.

*Drought* : It is capable of withstanding drought condition.

*Water* : Rise in water table or flood causes damage to the tree. Young tree may die.

*Establishment* : It can be propagate by seeds, cuttings, grafting, air layering and budding.

Propagation by seed is most popular method. Direct sowing of seed is common, but seedlings are also raised in nursery. Direct sowing in trenches is done at the beginning of the rainy season. Freshly extracted seeds have more than seventy per cent germinative capacity. Stored seed gradually looses its germinative capacity. Sanmugavelu (1971) reported 100 per cent germination by soaking seeds for 48 hours in GA, up to 500 ppm compared to 80 per cent under control. Seedling growth was also better. IAA, NAA and CPA were less effective.

Dhua *et al.* (1983) observed 90 per cent rooting success from shoots which were etiolated and ringed for 30 days and then treated with IBA at 3,000 ppm in combination with ferulic acid at 2,000 ppm.

**Fig. 134 :** *Artocarpus heterophyllus.*

According to Gunjate *et al.* (1982) epicotyl grafting with mature, pulp, terminal scion shoot on germinating jackfruit seedling of about 8 to 10 days by wedge method during April-May gave 50 to 90 per cent success and the successful epicotyl grafts attained saleable size within a year.

**Utility**

*Wood* : The timber is yellowish-brown in colour and is used for a wide range of colour, etc. It has good demand locally for making good quality of furnitures, window, frames etc. Its wood has a sp. gr. 0.59.

*Vegetables* : Unripen fruits and seeds of ripen fruits are good vegetables. Ripe Jackfruit is also referred as 'poor-mans food. It contains, minerals, Vitamin A and C.

*Fodder* : Its foliage is nutritious and palatable and is browsed by livestock. The skin of the fruit is used as a cattle feed in Bihar.

*Fruit* : Ripes fruit contains moisture 77.2%, carbohydrate 18.9 gm, protein 1.9 gm, fat 0.1 gm and total mineral matter 0.8 gm. It has calorific value 84 (Annon., 1979). The ripe fruit is largest testy fruit.

*Intercropping* : In its natural habit farmers grows agricultural crops, vegetables in the interspace between the rows of Kathal trees.

## 34. COCONUT

*Cocos nucifera* Linn.

*FAMILY* : PALMAE

*Common names* : Narial, coconut, palm coconut tree, nariker, nariyl, naral.

*Distribution* : Coconut is an exotic, believed to be native of South or Central America. In India it has naturalised for centuries. It has been suggested that fruits were carried all over the pacific by sea currents, and finally brought to India. In India it has been naturalised in the coastal part of India, Assam, Kerala, Tamil Nadu, Orissa, West Bengal. Coconut orchard is now common in Bihar and Uttar Pradesh. In Bihar few districts like, Saharsa, Darbhanga, Purnia, Katihar are recognised as suitable for Coconut farming. Planting of coconut tree in the farm land is very common is most of the area.

*Climate* : The optimal mean annual temperature for its proper growth should be 27°C with a diurnal range of 6-7°C. In its natural habitat the temperature varies from 20 to 32°C. A warm and humid condition is preferred by coconut. Persistence of highly humid conditions throughout is, however, not conducive for two reasons. One is that it reduces transpiration and their by limits nutrient up take, secondly, it favours rapid spread of dreadful diseases like bud rot (Thangaraj and Muthuswami, 1985). With in its habitat the annual rainfall varies from 1300 to 2300 mm. It is found at elevations up to 900 m.

*Soil* : It occurs on a variety of soils. It can be grown in area having soil pH between 5.0 to 8.0. It grows well on alluvial soils, red loamy soils, lateritic soil, peaty soil, marine and costal soils.

*Description* : It is a tall, unbranched tree, 8.25 m tall, straight or slightly curved, marked with ring leaf scars. The stem does not grow in thickness in the way of, but remains more or less uniformly thick 40-70 cm throughout its length. The leaves 2-4 m long, feather like in appearance consisting of a stout axis on which the numerous leaflets are fixed ; the leaflets from the green part of the leaf, they are over 100 in number, each being 60-80 cm long and very tough, green in colour and shining. Flowers come out from a boat-like spathe or bract which is 60 cm or more long, male flowers are scarcely 12mm across, female flowers may be about 25 mm across and long. Fruits, the coconuts of commerce, are 20-30 cm long, slightly less broad, somewhat three sided. The outermost part of the fruit is green and shining when fresh, being almost entirely water-proof and very hard. Beneath this, there is a thick layer of stout fibres, the layer being at times over 25mm in thickness. There follows a hard, stone like coat or layer, about 6mm thick. The inside of this stony layer is lined with a fairly thick coating of soft, milky-white flesh, the cavaty inside the flesh is filled with a watery fluid (Santapau, 1966).

There are two distinct varieties of coconut, namely the tall and dwarf. The varieties differ in size, shape, colour and quality of nuts, shape of crown, leaves, flowers and also trunk character (Thangaraj and Muthuswami, 1985).

**Silvicultural characters**

*Light* : It is strong light demander. The coconut tree requires full sunlight for its proper growth. It does not grow well under shade.

*Frost* : It is moderately frost hardy. In North Bihar this tree grows well, even in severe frost.

*Establishment* : A ripe coconut should be collected from mother palms having following characteristics :

* The trees should not be older than 10-15 years.
* The trees bearing barren nuts should not be selected as mother palm tree.
* The tree producing round or spherical,

medium size sound nut should be selected as mother tree.

* The mother tree should be selected from palm orchard instead of growing close to human habitat.

A ripe coconut can be made to germinate if planted with its husk in a bed of leaf mould in a moist and shady place ; germination may be hastened by leaving the coconut in water until the new roots begin to sprout out of the shell and its covering fibres. Seedling needs plenty of water during the first few years (Santapau, 1966). The best soil for a nursery is sandy soil near a water source which will provide effective drainage and prevent termite attack. The area should be open and free of shade. If sandy areas are not available, it is better to remove 0.3 to 0.5 m depth of soil and fill with sand (Thangaraj and Muthuswami, 1985). The seedlings can be planted at 8 x 8 x 8 m to have 74 palms per acre (180 palms/ha or 7.6 x 7.6 x 7.6 m to have 80 palm/acre or 200 palms/ha. (Perera *et al.* 1989). Traditionally, husks have been used in pits for moisture conservations. Husks acts like a sponge and can absorb and retain about six times their own weight of water, which will be conserved for dry periods. Husks breaks down slowly. Thus, its moisture conservation ability will be available for a period of 3 to 4 years. Burying of husks induced an increase in coconut yield in both sandy and gravely soils but the benefits were more in sandy soils. Husk and Coir dust should not be burried in areas subjected to water-logging (Annon, 1989). Husks have manurial and other properties. The main constituent of husk is potash. On an average about 100,000 husks contain potash equivalent to 1MT of muriate of potash.

**Fig. 135 :** *Cocus nucifera.*

The Coconut Research Institute of Sri Lanka recommended two general fertilizer mixtures to provide the nitrogen, Phosphorus, and Potassium requirement of coconut. They are : the Young Palm Mixture (YPM, 13-12-17) for seedling/Young coconut and the Adult Palm Mixture (APM, 12-6-32) for adult coconut. In addition, application of dolomite at the rate of 1.5 - 2.0 kg/ palm once every three years is also recommended to provide the magnesium requirement of coconut (Annon., 1989). Jayasekera (1989) has stated that boron deficiency in young coconuts may be observed in localised with high rainfall where the soil is acidic and in lands that are intensively intercropped. He has suggested that boron deficiency in young coconut could easily corrected at the incipient stage. In Bihar, an integrated farming model comprising, tree fodder, agricultural crops/ vegetables and cattle is in practice. Similarly in Assam farmers grows paddy, vegetables etc., along with palm trees and rear animals near their home. Dung and urine of the cattles are used in the agricultural field. This reduce the cost of inorganic fertilizer. In addition the cost of weeding is also reduced under agri-silvi-horticulture system. This system helps in increasing the productivity of land and reducing the cost of coconut production. Tayasundara and Marasinghe (1989) have stated that at Ratmalagara Estate, Madampe a model was developed to demonstrate the feasibility of integrating legume based pasture, tree fodder and cattle with coconut and supplementing inorganic fertilizer with cattle manure for the plants. During the year, palms in the integrated system showed an increase of 17.5 per cent of nut and 11.4 per cent in copra yields over monoculture system. In the integrated system, leaf nutrient levels of palms were above the critical level and soil organic matter, total nitrogen and available phosphorus level was significantly higher than in the monoculture system. In the first lactation period which commenced during the year, cattle produced 7-10 liters of milk per day, providing an additional income.

**Utility**

*Nariel pani* : The coconut water contains, water 95.5%, protein 0.1%, fat 0.1%, mineral 0.4%, Carbohydrate 4%, Calcium 0.02%, Phosphorus, 0.01%, Iron 0.5% mg/100gm (Thangaraj and Muthuswami, 1985).

It is good drink. It is prescribed patient having problem like diarrhoea, dysentery, urinary antiseptic.

*Endosperm* : The wet carnel and dried coconut carnel are an excellent product and are used for preparation of delicious dishes. Coconut flour is used in bakery and confectionary preparations.

*Toddy* : The juce tapped from coconut tree is used in the preparation of alcohol, vinegar, etc.

*Nariel oil* : The coconut oil is used as edible oil, hair oil, and soap making.

*Oil cake* : It is used to feed poultry and cattle.

*Coconut fibre* : The coconut fibre is used for making ropes, mats, nets, etc.

*Leaves* : Leaves are used for thatching purpose and brooms are made from the leaf ribs of coconut tree.

*Coconut Shells* : The empty coconut shell is used for making hookah, decorative items, utensils etc.

*Coconut charcoal* : Breag and Josheph (1989) have reported that the traditional method of production of charcoal has an yield of 25 to 30 per cent of the dry weight of shells used. The charcoal produced by this method is of variable quality, and is often contaminated with extraneous matter and soil. During the process, over 50 per cent of the gross heat content of the shells is lost and at the same time, a large volume of abnoxious smoke is evolved to the atmosphere. In view of this, the operation of the traditional pit is either restricted or not permitted at all in several areas in Sri Lanka.

The control carbonisation of shells in the Waste Heat Recovery Unit (WHV) converts shell into charcoal and gas evolved during the process is subsequently burned in heat exchange system. This process, therefore, replaces either imported fuel oil or locally felled wood, and subsequently reduce deforestation and associated environmental problems. In addition, the WHV both maximises the utilisation of shell feed stock for charcoal production and gives the processor a greater degree of self-sufficiency.

## REFERENCES

Amin, R.S. (1978). Curr. Sci, 47: pp. 468-469 (C.R.).

Anderson, D.M.W. (1985). Contributed papers involving one NFT Genus, *Nitrogen Fixing Tree Research Reports*, Vol. 5, pp. 12-13.

Annon (1984). *Shisham*, issued on the occasion of Doon environment week, F.R.I and College and conservator of Forests, Shiwalic circle, Dehra Dun, pp. 1-4.

Annon. (1935-1936). *Annual Adm. Rep.* Ajmer-Merwara Forest Department for the year 1935-1936, (C.R.).

Annon. (1948). *The wealth of India raw materials, vol. 1*, C.S.I.R., New Delhi. (C.R.)

Annon. (1974). *Indian timbers, Gamari*, Forest Research Institute and Colleges, Dehra Dun, pp. 1-14.

Annon. (1979). *Extension Bulletin No. 22*, Indian Institute of Horticultural Research, Bangalore, (C.R.).

Annon. (1980). *Firewood crops, shrub and tree species for energy production*, National Academy of Sciences Washington, D.C. pp. 110-111.

Annon. (1980). *Firewood crops, shrub and tree species for energy production* National Academy of Sciences, Washington, D.C., pp. 106-107.

Annon. (1980). *Firewood crops, shrub and tree species for energy production,* National Academy of Sciences, Washington, D.C., pp. 152-153.

Annon. (1980). *The Silvicultural of Indian Trees,* Vol. III, Joshi, H.B. *et. al.* (eds.), The Controller of Publication, Delhi, pp. 300-303.

Annon. (1981). *Ann. Rep. All India Coordinated Fruit Imp. Project* Rajasthan Agril. University, pp. 30-31.

Annon. (1981). *Silviculture of Indian Trees,* Vol. V, Joshi, H.B. *et al.*, (eds.), The Controller of Publication, Delhi pp. 95-100.

Annon. (1981). *Silviculture of Indian trees, Vol. V, Joshi, H.B. et al.*; (eds.), The Controller of Publications, Delhi, pp. 268-276.

Annon. (1981). *The Silviculture of Indian Trees,* Vol. III, Joshi, H.B. *et. al.*, (eds), The Controller of Publications, Delhi, pp. 169-174.

Annon. (1983) *The silviculture of Indian trees, Vol. IV* Joshi, H.B., *et al.* (eds.) The Controller of Publications, Delhi, pp. 199-203.

Annon. (1983). *The silviculture of Indian trees, Vol.IV,* The Controller of Publications, Delhi 110054, Joshi, H.B. *et al.* (Eds.), pp. 208-222.

Annon. (1983). *The Silvicultural of India Trees.* Vol.III, Joshi, H.B., *et al.*, (eds.), Controller of Publication, Delhi, pp. 76-79.

Annon. (1983). *The Silviculture of India trees,* Controller of Publications, Delhi, Joshi, H.B. *et al.*, (Edited), Vol. IV, pp. 57-59.

Annon. (1983). *The Silviculture of Indian Trees,* Vol. III, Joshi, H.B. *et al.*, (eds.) Controller of Publications, Delhi pp. 79-81.

Annon. (1983). The *Silviculture of Indian Trees,* Vol. III, Joshi, H.B., *et al.*, (eds.) Controller of Publications, Delhi, pp. 188-196.

Annon. (1983). *The Silviculture of Indian trees,* Vol. III, Joshi, H.B., *et al.*, (eds.), The Controller of Publications, Delhi, pp. 147-152.

Annon. (1983). *The silviculture of Indian trees,* Vol. IV, Joshi, H.B. *et al.* (Edited), The controller of publications, Delhi pp.76-80.

Annon. (1983). The Silviculture of Indian trees, Vol. IV, Joshi, H.B. *et al.* (eds.), Controller of Publication, Delhi, pp. 204-205.

Annon. (1983). *The silviculture of Indian trees,* Vol. IV, Joshi, H.B., *et al.*, (eds.), Controller of Publication, Delhi, pp. 326-332.

Annon. (1983). *The silviculture of Indian trees, Vol. IV,* The Controller of Publications, Delhi, (Joshi, H.B. *et al.*, Edited), pp. 80-83.

Annon. (1983). *The Silviculture of Indian Trees, Vol. IV,* The Controller of Publications, Delhi, (Joshi, H.B. *et al.*, Ed.). pp. 126-137.

Annon. (1985). *Troups the silviculture of Indian Trees, Vol. VI,* Revised and enlarged by Mondal, R.K., Chaudhari, D.C. and Chauhan, R.S., The Controller of Publication, Civil Lines, Delhi pp. 37-39.

Annon. (1985). *Troups the Silviculture of Indian Trees,* Vol. VI, Revised and enlarged by Mandai, R.K.; Chaudhuri, D.C. and Chauhan, R.S., The Controller of Publications, Civil lines, Delhi, pp. 12-131.

Annon. (1987). All India Co-ordinated Research Project on Agroforestry, *Annual Report,* Division of Forestry, S.K.University of Agricultural Sciences and Technology, Srinagar, J & K p. 52.

Annon. (1989). *Annual Research Report,* (1988-89), Dept. of Silviculture and Agroforestry, U.H.F., Solan, (C.R.).

Annon. (1989). *Firewood crops. Shrub and tree species for energy production*, National Academy of Sciences, Washington, D.C., pp. 36-37.

Annon. (1989). New "Adult coconut" Fertilizer mixture, *Coconut bulletin*, Vol. 6, No. 2, pp. 13-14.

Annon. (1989). Use of husks and coir dust in coconut land, *Coconut Bulletin*, Vol. 6, No. 2, pp. 10-12.

Annon. (1989). *Utilisation of Eucalyptus Wood*, F.R.I., Dehra Dun, pp. 1-12.

Annon. (1990). *Annual Research Report*, (1989-90), Dept. of Silviculture and Agroforestry, U.H.F., Solan.

Annon. (1992). *Acacia nilotica*, NFT Profile, *M.P.T.S. News letter*, Hedge, N, Daniels J. and Dhar T., (Edited), BAIF Development Research Foundation, Pune, pp. 6-7.

Arya, S.P. (1988). Studies on the improvement of tree leaf fodder, *Annual Research Report (1988-89)*, Department of S.A.F., U.H and F., Solan, pp. 35-37.

Arya, S.P. (1988). Studies on the leaf fodder, fibre and fuelwood yields from *Grewia optiva* in relation to its girth class, *Annual Research Report, (1988-89)*, Dept. of S.A.F., UH and F, Solan, pp. 37-38.

Arya, S.P. (1989). Substitution of concentrate by *Grewia optiva* leaves in the ration of lactating cows, *Annual Research Report* (1989-90), Dept. of S.A.F., UH & F., Solan, pp. 107-110.

Attri, B.S. (1982). Neem (*Azadirachta indica*) as a source of pest control material. *In : Cultivation and utilization of medicinal plants*, Atal, C.K. and Kapur, B.M., (eds.), Regional Research laboratory, C.S.I.R., Jammu-Tawi, pp. 680-684.

Ayensu, S. Edward. (1980). *Firewood crops and tree species for energy production*, National Academy of Sciences, Washington, D.C. (C.R).

Ayensu, S. Edward. (1980). *Firewood crops—shrub and tree species for energy production*. National academy of Sciences, Washington, D.C. (C.R.)

Bagchi, S. and Emmanuel, G.J.S.K. (1984). Analysis of variability for pod length and seed count in *Albizia lebbeck*, Benth ; *Myforest*, Vol. 20(2), pp. 85-90.

Bajpai, P.N. and Charturvedi, H.S. (1985). *Aonla*, In : Fruits of India : Tropical and sub-tropical, Bose, T.K. Eds., Naya Prokash, Calcutta, pp. 581-590.

Bajpai, P.N. and Shukla, H.S. (1985). Aonla, In : *Fruits of Indian tropical and Subtropical*, Bose, T.K., (eds.), Naya Prokash, Calcutta, pp. 591-600.

Bakshi, B.K. (1976). *Forest pathology, Principles and practice in Forestry*, Manager of Publications, Delhi, pp. 1-250.

Balasundaram, M. and Mahamed Ali, M.I. (1987). Root nodulation potentialities of *Leucaena leucocephala* in Kerala, *K.F.R.I. Report* Peechi, Kerala, pp. 1-22.

Banerjee, A.K. (1973). Plantations of *Acacia auriculiformis* (Benth), A Cunn. in West Bengal, *Indian Forester*, Vol. 99(9), Dehra Dun (C.R.)

Bhandari, K.R. and Mukherjee, S.K. (1969) *Curr. Sci.*, Vol. 38, pp. 197-198 (C.R.).

Bhimaya, C.P. & Kaul, R.N. (1966). Role of seedling root habit on initial seedling survival in some desert tree species, *Sci & Cult.*, Vol. 32(4), Calcutta, (C.R.).

Bhujbal, B.G. (1975). *Res. Jour*, Mahatma Phule Agri. Univ., Vol. 6, pp. 73-75.

Breag, G.R. and Joseph, P.G. (1989). Coconut shell Carbonisation/waste heat recovery technology in coconut processing in Sri Lanka, *Coconut Bulletin*, Vol. 6, No. 2, pp. 21-23.

Brewbaker, James L. and Mark Hutton, E. (1987). *Leucaena versatile tropical tree legume*, 10, pp. 206-260.

Brown, R.S. (1929). Revised Working Plan for the Nilambur valley Forests, 1928-29 to 1937-38, C.C.F. Kerala (C.R.).

Chaturvedi, A.N. (1984). Water consumption and biomass production of some forest trees. *Common-wealth for. Rev.* 63 : 3 (C.R.)

Chaudhari, K.N. (1957). Afforestation technique for the laterite zone, *West Bengal Forest Bull.,* No. 5 Supt. Govt. Printing, Alipore.

Cherian, M.C. and Menon, G.E.R. (1944). *Madras Agril. J.*, pp. 10-11 ; (C.R.).

Dayama, O.P.(1985).Effect of sucrose on the growth, nodulation and nitrogen content of *Pongamia pinnata, Nitrogen Fixing Tree Research Reports*, Vol. 3. pp. 9-10.

Dhua, *et al.* (1983) Cited from jackfruit, Samaddar, H.N., Fruits of India ; Tropical and sub-tropical Naya Prokash, Calcutta, p. 491.

Diangana, D. (1985). Recherche dun traitement d avant semis capable d accelerer la germination des graines de *Acacia mangium, Albizia falcataria, Calliandra calothyrus* et al. *Leucaena leucocephala Nitrogen Fixing Tree Research. Reports,* Vol. 3, pp 2-3.

Duguma, B. Kang, B.T. and Okali, D.V.V. (1988). Effect of pruning intensities of three woody leguminous species grown in alley cropping with maize and cowpea on an alfisol, *Agroforestry system,* 6, pp. 19-35.

Economic and Planning Council for Karnataka (1983). *Report of Panel discussion on the ecological and socio-economic impact of the cultivation of Eucalyptus*, 7th Sept. 1983, F.R.I. Press, Publicity and Liaison Branch, pp. 1-5.

Edwards, M.V., Badhwar, R.L. and Dey, A.C. (1952). The vegetative Tannin Materials of India and Burma, *Indian Forest Rec.*, (N.S) Chemistry and M.F.P., 1(2), Manager of Publication, Delhi.

Fay emilihin, A.A. and Kang, B.T. (1988). Note on compative early growth of *Cassia siamea, Gliricidia sepium, Leucaena leucocephala* in an alfisal, *Nitrogen Fixing Tree Research Reports*, Vol. 6, pp. 18-19.

Gamble, J.S. (1922). *A manual of Indian Timbers*, Sampton low Marston and Company Ltd., London.

Gangwar, R.P., Singh, D. and Chundawat, B.S. (1975). *Haryana J. Hort. Sci.*, Vol. 4, pp. 150-151.

George, M. (1979). *Organic productivity and nutrient cycling in Eucalyptus hybrid plantation*, Abstract of Ph.D. Thesis, Meerut University.

Gill, A.S. (1986). A fuel/fodder/production system using *Sesbania grandiflora, Nitrogen Fixing Tree Research Reports*, Vol.(4), pp. 14-15.

Goyal, R.S., Gulati, K.C., Sarup, P., Kidwai, M.A. and Singh, D.S. (1971). *Indian J. Ent.*, Vol. 33(1), pp. 67-71. (C.R.).

Gunjate, R.T., Kolekar, D.T. and Salvi, M.J. (1982). *Proc. Nat. Sem. on Pl. propagation*, Calcutta, p. 87, (C.R.).

Gupta, A.K. and Jindal, S.K. (1985). Variability for quality traits in *Prosopis cineraria and Acacia senegal, Nitrogen Fixing Tree Research Reports*, Vol. 3, pp. 15-16.

Gupta, R.K. and Balara, G.S. (1972) Comparative studies on the germination, growth and seedlong biomass of two promising exotics in Rajasthan desert (*Prosopis juliflora* (Swartz) DC and *Acacia tortilis* (Frosk) Hayne ssp. *tortilis*, *Indian Forester*, Vol. 98(5), Dehra Dun (C.R.).

Haines (1925). Flora of Bihar and Orissa.

Halliday, J. and Somasegaran, P. (1983). Nodulation, nitrogen fixation and Rhizobium strain affinities in the genus Leucaena, In : *Leucaena research in Asia-Pacific region*, Proc. of a workshop held in Signapora, 23-26 Nov. 1982.

Haq, F., Khan, M.S. and Faridullah, I. (1973). *W. Pakistan J. Agric. Res.*, Vol. 11; p. 121, (C.R).

Homfray, C.K. (1937). *Nursery and plantation notes for Bengal*, Govt. Printing, Alipore, Bengal (CR).

Howard, (1926). Cited from *Indian timbers Gamari*, F.R.I. and Colleges, Dehra Dun, pp. 1-14 (C.R.).

Husain, A., Arshad, M., Hussain, A., Ghugtai, F.A. and Manzur, R. (1986). Effect of fertilizer *on Albizia procera, Nitrogen Fixing Tree Research Reports,* Vol. 4, pp. 32-33.

I.S.I (1951). *Specification for commercial and moisture proof plywood*, Indian standard Institution, Delhi, pp. 303 (I.S.).

Jha, L.K. (1989). *AICRP Agroforestry Annual Report*, B.A.U. Ranchi, pp. 1-25.

Jha, L.K. Jha, G. and Sah, R.B. (1990). *Annual report-O.R.P. Extension programme*, B.A.U. Ranchi, pp. 1-25.

Jha, L.K., and Jha, G. and Toppo, S. (1989). *A.I.C.R.P., Agroforestry, Annual Report,* Faculty of Forestry B.A.U. Ranchi, pp. 1-29.

Jha, L.K., and Sen-Sarma, P.K. (1988). *Annual Report, Lab to land programme,* Faculty of Forestry, B.A.U. Ranchi, pp.1-30.

Jha, L.K., Jha, G., Toppo, S. (1989). *Agroforestry ; A.I.C.R.P., Annual Report*, Faculty of Forestry, B.A.U., Ranchi, pp. 1-35.

Jha, L.K.; Jha, G. and Toppo, S.(1989) *AICRP Agroforestry Annual Report,* Department of Social Forestry and Extension, B.A.U. Ranchi, pp. 1-39.

Jha, L.K. ; Jha, G. and Toppo, S. (1990) A.I.C.R.P. On Agroforestry, *Annual report* Deptt. of Social Forestry and Extension, B.A.U., Ranchi.

Joshi, *et al.* (1983). *Troup's the silviculture of Indian trees, Vol. IV, (Edited), Leguminosae*, The Controller of Publication, Delhi, p. 96.

Joshi, H.B. *et al.* Eds. (1983). *The Silviculture of Indian Trees,* Controller of Publications, Delhi, Vol. IV, pp. 3-6.

Joshi, H.B., *et al.*(Eds.) (1983). *Troup's, the Silviculture of Indian Trees*, Vol.IV, published by the Controller of Publications, Delhi, pp.332-334.

Joshi, H.B; Ed, (1983). The Silviculture of Indian Trees, Vol. V, Controller of Publication, Delhi, p. 122.

Kandaswamy, M.C. and Venkata Row, A. (1956). A note on the extraction of seeds from the pods of *Prosopis juliflora, Madras Agric. J.*, Vol-43, Coimbatore.

Kanjilal, U. (1927). *Forest flora of the chakrata, Dehra Dun and Saharanpur forest divisions, Uttar Pradesh*, Periodical expert, Book Agency, Delhi, pp. 186.

Kanjilal, U. (1927). *Forest flora of the Chakrata, Dehra Dun and Saharanpur Forest Division, U.P.*, Third Edition by Gupta, B.L. Periodical expert book agency, Delhi, pp. 78-79.

Kanjilal, U. (1927). *Forest flora of the Chakrata, Dehra Dun and Saharanpur forest divisions*, U.P., Revised by Gupta, B.L., Periodical Expert Book Agency, Delhi, pp. 78-79.

Kanjilal, U. (1927). Forest Flora of the chakrata, Dehra Dun and Saharanpur Forest Division, U.P., Revised and Enlarged by Gupta, B.L., periodical Expert Book Agency, Delhi pp. 101-102.

Kanjilal, U. (1927). Forest Flora of the Chakrata, Dehra Dun and Saharanpur Forest Divisions, U.P., Revised and enlarged by Gupta, B.L., Periodical Expert Book Agency, Delhi, pp. 240-241.

Kanjilal, U. (1927). Forest flora of the chakrata, Dehra Dun, and Saharanpur forest Division, Uttar Pradesh, Revised and enlarged by Gupta, B.L., Periodical Expert Book Agency, p. 100.

Kanjilal, U. (1927). *Forest flora of the chakrata, Dehra Dun, and Saharanpur Forest Division, Uttar Pradesh*, Revised 3rd edition by Gupta, B.L. Periodical expert Book agency, Delhi pp. 373-374.

Kanjilal, V. (1927). *Forest flora of the Chakrata, Dehra Dun and Saharanpur forest divisions,* U.P., I.B.D. Dehra Dun, p. 208.

Kanjilal, V. (1927). *The forest flora of the Siwalik and Jaunsar forest division*, U.P. Third revised edition by Gupta, B.L., Periodical Expert Book Agency, Delhi pp. 200-202.

Kanwar, J.S. ; Singh, S.N. and Gill, R.P.S. (1980). *Punjab Hort. J.*, 20, pp. 167-177 (C.R.).

Kaul, R.N. (1965). Chemical and mechanical scarification for breaking dormancy of seeds of tree species, *Agric. Research*, Vol. 5(1), New Delhi, (C.R.).

Kaul, R.N. (1965). Chemical and mechanical scarification for breaking dormancy of seeds of tree species, *Agric. Research*, Vol. 5(1), New Delhi, (C.R.).

Kaul, R.N. and Ganguli, B.N. (1965). Trials in the introduction of Acacias in the Arid zones of Rajasthan I. Seed studies, *Indian Forester*, Vol. 91(8), Dehradun (C.R.).

Kaushalappa, K.A. (1989). Biomass studies in *Calliandra* in Karnataka, *Myforest*, Vol. 25(4), pp. 325-329.

Kaushik, R.C. (1969). Suitability of soils for *Eucalyptus* hybrid (Mysore gum, syn, *E. tereticornis*) in Haryana and Punjab; *Indian Forester*, Vol. 95(6); Dehra Dun.

Krishna, S. and Ramaswami, S. (1932) Calorific values of some Indian woods, *Indian Forest Bull.*, 79, chemistry, (N.S) Government Publication Branch, Calcutta.

Krishna, S. and Ramaswami, S. (1932). Calorific values of some Indian woods, *Forest Bull*, No. 79, (N.S) Chemistry, Govt. of India, Central Publication Branch, Calcutta.

Krishna, S. and Ramaswami, S. (1932). Calorific values of some Indian woods, *Indian forest Bull* (N.S.), 79 Chemistry, Govt, Publication Branch, Calcutta.

Krishnaswami, V.S. and Mathauda, G.S. (1954). Phenological behaviour of a few forest species at New Forest Dehra Dun *Indian Forest Rec.* (N.S), Silviculture 9(20), Manager of Publication, Delhi.

Kushalapa, K.A. (1988). Comparative biomass of *Acacia auriculiformis* and *Leucaena leucocephala* trees from moist region of Karnataka, *My forest*, Vol. 24(1) pp. 12-15.

Kushalappa, K.A. (1987). Productivity of Mysore Gum (Eucalyptus hybrid) under different ecosystem in Karnataka, *My forest*, Vol. 23(4), pp. 217-226.

Lal, D. ; Krishnan, J. ; Negi, S.S. ; Goel, G.C. and Callar, J.F. (1985). Effect of partial replacement of concentrates by roughage in the diet of Angora rabbits on wool production, *Ind., J. Anim*, Vol. 55, p. 152.

Mac Donald, K.R. (1942). Farm and Forest - 11(1) (C.R.)

Mackay, J.H. (1952). Notes on the establishment of Neem plantation in Bornu Province of Nigeria *Farm and Forest*, Vol. II, Forest Deptt. Nigeria, cited from the *Silviculture of Indian Trees*. Vol. III, p. 148 ; Controller of Publications, Delhi.

Maiti, S.C. (1985). Litchi, In : *Fruits of Indian tropical and subtropical*, Bose, T.K. (eds.,) Naya, Prokash, Calcutta, pp. 388-408.

Mann, H.S. and Mathuna, K.D. (1984). *Arid zone forestry, with special reference to the Indian Arid zone, C.A.Z. R.I., Jodhpur, pp. 1-43.*

Mann, H.S. and Muthana, K.D. (1984). *Arid zone forestry (with special reference to the Indian Arid zone)*, Central Arid Zone Research Institute, Jodhpur, pp. 1-48.

Mann, H.S. and Muthana, K.D. (1984). *Arid Zone Forestry*, C.A.Z. A.R.I., Jodhpur, p. 23.

Meelink, W.H.H. (1989). Management of trees for fuelwood and fodder production on small farms, In : *Promotion of fodder and fuelwood trees*, Hegde, N.G., Relwani, L.L. and Kelkar, V.D. (eds.) BAIF, Pune.

Mishra, C.M. and Singh, S.L. (1985). Salt stress studies upon germination of *Sesbania grandiflora, Nitrogen Fixing Tree Research Reports*, Vol.3, p. 22.

Mishra. (1973). *Albizia procera* planting in saline and alkaline area. Cited from (The silviculture of Indian trees Vol. IV, Revised and enlarged by Editorial Board F.R.I. Dehra Dun, 1986) p. 112.

Mitra, S.K. and Bose, T.K. (1985). Guava, In : *Fruits of tropical and subtropical*, Bose, T.K., (eds), Naya Prokash, Calcutta, pp. 277-298.

Muhammad, S. (1984). Agroforestry practices - In : *Proceeding of social forestry training programmes*, Muhammad, S. (eds), B.A.U.Ranchi, pp. 45-54.

Muthana, K.D. (1988). Gum arabic the backbone of kordofan region, *Myforest*, Vol. 24(2), pp. 95-98.

Muthana, K.D. and Arora, G.D. (1973). Study of *Acacia tortilis* (Frosk). A fast growing tree species for arid zones, *Forestry conference*, F.R.I. Dehra Dun.

Muthana, K.D., Meena, G.L. Bhate, N.S. and Bhatia, O.P. (1984).

Nagaveni, H.C. ; Ananthapadmanabha, H.S. and Rai, S.N. (1987). Note on the extension of viability of *Azadirachta indica*, *My forest*, Vol. 23(4), p. 245.

Nautiyal, B.P. and Venkataraman, K.G. (1987). *Moringa* (Drumstick)—An ideal tree for Social forestry, Growing conditions and uses-part I, *Myforest*, Vol. 23 (1), pp. 53-58.

Patel, K. (1985). *Tree cultivation,* Rameshbhai Patel, Adarsh Farm Vatra, Ahmedabad, pp. 1-24.

Pathak, P.S and Gupta, S.K.(1985). Seed germination and seedling growth of *Hardwickia binata* Roxb, *N.F.T.R.R.* Vol.3. pp. 25-26.

Pathak, P.S. and Gupta, S.K. (1987). Soil amelioration role of Leucaena in the agroforestry system, *Proc. Int. Workshop on Agroforestry for rural needs,* New Delhi.

Pathak, P.S. and Patil, B.D. (1985). Growth of *Leucaena leucocephala* compared to four native species in India, *Leucaena Research Report*, 6 : 46-47.

Pathak, P.S.; Patil, B.D.; Gupta, S.K. and Deb Roy, R. (1982). Leucaena cultivation improves soil fertility, *Leucaena Research Report,* 3 : 37.

Perera, M.R.L.A., Liyanagama, P.S. and Mahindapala, R. (1989). Makandura seed Garden, *Coconut bulletin*, Vol. 6, No. 2, pp. 6-8.

Prabhu, V.V. and Theagarajan, K.S. (1977). Utilisation of Eucalyptus hybrid (Mysore gum) bark for production of oxalic acid) *Indian forester*, 103(7), Dehra Dun.

Puri, D.N. (1990). Production potential of degraded lands with multipurpose tree species, In : *Multipurpose tree species research for small farms, strategies and methods*, pp. 184-187.

Puri, S. and Shamet G.S.(1988). Propagation of *Robinia pseudacacia* by seeds and stem cuttings in the Himalayas, *N.F.T.S.*, Vol. 6, pp. 71-72.

Raizada, M.B. and Chatterji, R.N. (1954). A diagnostic key of the various forms of introduced mesquite (*Prosopis juliflora* D.C.), *Indian Forester,* Vol. 80(11) Dehra Dun.

Randhawa, M.S. (1983). *Flowering trees*, National Book Trust, India, p. 136.

Randhawa, M.S. (1983). *Flowering trees*, National Book Trust, India, pp. 150-151.

Randhawa, M.S. (1983). *Flowering trees*, National Book Trust, India, pp. 165-167.

Randhawa, M.S. (1983). Flowering trees, National Book Trust, India, pp. 169-170.

Randhawa, M.S. (1983). *Flowering Trees*, National Book Trust, India, pp. 93-94.

Randhawa, M.S.(1983). *Flowering trees*, National Book Trust, India, pp.107-109.

Root system of desert tree species, *J. Arid Environments*. Cited from NFTRP, Vol. 3, Page-28 (C.R.).

Samaddar, H.N. (1985). *Jackfruit*, In : Fruits of India Tropical and sub-tropical, Naya Prokash, Calcutta, pp. 487-497.

Sanmugavelu, K.G. (1971). *Madras Agric. J.*, 58 : pp. 97-103, (C.R.).

Santapau, H. (1966). *Common Trees*, National Book Trust India, pp. 82-86.

Santapau, H. (1966). *Common trees*, National Book Trust, India, pp. 100-102.

Santapau, H. (1966). *Common Trees*, National Book Trust, India, pp. 11-13.

Santapau, H. (1966). *Common Trees*, National Book Trust, India, pp. 112-113.

Santapau, H.(1966). *Common trees,* National Book Trust, India, pp. 91-92.

Saxena, G.K. (1963). *The Allahabad Fmr*. Vol.37, pp. 28-31.(C.R.).

Sekhar, A.C. and Rawat, B.A. (1960). A note on mechanical propeties of *Prosopis juliflora* D.C., *Indian Forester,* Vol. 86(8), Dehra Dun.

Sen Gupta, (1937). Cited from the *Silviculture of Indian Trees*, Vol. III, Controller of Publication, Delhi, p. 80.

Sen, P.P. and Jha, L.K. (1985). Lab to land extension programme extension report, Faculty of Forestry, B.A.U. Ranchi, pp. 1-15.

Sengupta, K. (1986). *Tasar Silk Industry in India*, Central Tasar Research and Training Institute, Central Silk Board, Govt. of India, Ranchi, pp. 8-9.

Shamet, G.S. and Kumar, S. (1988). Rooting studies of *Punica granatum and Delbergia sissoo* cuttings under controlled phyto-environment conditions, *The Indian Forester,* Vol.114(6) pp. 331-334.

Shankar, G. (1985). Phalsa, *Fruits of India tropical and subtropical*, Bose T.K. (eds.), Nayo Prakash, Calcutta, pp. 559-565.

Sharma, G.K. and Raina, (1982). Propagation techniques of *Moringa oleifera,* In : improvement of forest Biomass. Khosla, P.K., eds., Ind. Society of Tree Sicentist, HPKVV, Solan, pp. 175-181.

Sharma, S.K.Verma, K.S. and Mishra V.K. (1988) Natural grass yield as influenced by various fodder tree species under pastoral system, Annual Research Report 1988-89. Deptt of S.A.F., U.H. & F., Solan, pp. 83-84.

Shukla, S.K. (1980). *Eucalyptus for planting,* Extension series-4, F.R.I. and Colleges, Dehra Dun, pp. 1-12.

Sihha, J.P. (1973). A study on nutrient level of Eucalyptus hybrid, *Proc. Forestry Conf*; Dehra Dun (Cross Ref.)..

Singh, A. (1980). *Fruit, physiology and production*, Kalyani Publisher, pp. 323-326.

Singh, B and Mathur, R.B. (1973). Natural regeneration of *Eucalyptus hybrid, Proc. Forestry Conf*; Dehra Dun.

Singh, M.P. and Srivastra, J.L. (1989). *Medicinal plants of Chotanagpur*, Directorate of Extension Education, B.A.U. Ranchi, pp. 59-60.

Singh, M.P. and Srivastra, J.L. (1989). *Medicinal Plants of Chotanagpur*. Directorate of Extension Education, B.A.U., Ranchi, pp. 1-78.

Singh, M.P. and Srivastra, J.L.(1989). *Medicinal plants of Chotanagpur,* Directorate of Extension Education, Birsa Agril. University, Ranchi, pp. 15-16.

Singh, R.N. (1974). *Indian Hort.*, Vol. 19, pp. 17-18.

Singh, S., Krishnamurthy, S. and Katyal, S.L. (1967). Fruit culture in India, ICAR, New Delhi pp. 255-259.

Sinha, M.P. (1984). Experiments on agroforestry at *Mandar*, (Ranchi), In : (Muhammad, S. Ed.), *Proceeding of social forestry training programme,* B.A.U, Ranchi, pp. 14-28.

Solanki, K.R., Jindal, S.K., Muthana, K.D. and Kackar, N.L. (1985). Performance of *Acacia senegal* and *Prosopis cineraria* in western Rajasthan, *NFTRR*, Vol. 3, pp. 27-28.

Sugur, G.V. (1988). Coppicing ability of natural stand of *Prosopis juliflora, My forest,* Vol. 24(2), pp. 129-132.

Teaotia, S.S. ; Singh, D.B. ; Singh, R.D. and Singh, R.N. (1968). *Punjab Horti. J.*, Vol. 8, pp. 241-244.

Thangaraj, T. and Muthuswami, S. (1985). Coconut, In : *Fruits of India, Tropical and Sub-tropical*. Bose, T.K., (eds.), Naya Prokash, Calcutta, pp. 345-385.

Thapar, A.R. (1958). *Farm Bull.* No. 42, ICAR, New Delhi, (C.R).

The Silviculture of Indian trees, Vol. 5 (1984). Revised and enlarged by Joshi, H.B., The Controller of Publications, Delhi, pp. 953.

*The silviculture of Indian trees*, Vol. IV, Revised and enlarged by editorial board Joshi H.B. *et al.*, (1984), Controller of Publication, Delhi, p. 267.

Tiwari, K.M. (1983). *Social Forestry for Rural Development*, International Book Distribution, Dehra Dun, pp. 94-95.

Tiwari, K.M. (1983). *Social Forestry for rural development*, International Book Distributor, Dehra Dun. pp. 91-92.

Tiwari, K.M. (1983). *Social forestry for rural development,* International Book Distribution, Dehra Dun pp. 74-76.

Tiwari, K.M. (1983). *Social Forestry for rural development*, International Book Distributors, Dehra Dun, pp. 79-80.

Tiwari, K.M. (1985). *Social Forestry for rural development*, Int. Book Distributor, Dehra Dun, pp. 77-78.

Tiwari, N.K. (1987). Productivity of Litchi in North Bihar, *Tech. New*, Muzaffarpur pp. 1-2.

Trolldenier, G. (1970). The influence of potassium and nitrogen nutrition of wheat on bacterial colonization of the rhizosphere. *Butchof Abstract,* pp. 18-20 (C.R).

Troller, H. (1982). *The common commercial timbers of India and their uses*, The Controller of Publications, Delhi, pp. 47-48.

Trotter, H. (1982). *The common commercial timbers of India and their uses*, reprint Govt. of India Press Nasik, Controller of Publication, Delhi, pp. 118-119 (Reprint).

Trotter, H. 1982. The *common commercial timbers of India and their uses*, The Controller of Publication, Delhi, pp. 84-86.

Trotter, H.(1982). *The Common commercial timbers of India and their uses*, the Controller of Publications, Delhi, pp. 120-124.

Troup (1921). Cited from the *silvicuture of Indian Trees, Vol. IV*, Controller of Publication, Delhi, p. 144.

Troup's the silviculture of Indian trees (1983), Revised and enlarges by the Editorial Board, Joshi, H.B. *et.al* F.R.I and College, DehraDun, pp. 296-299.

Varma, D.S. and Dey, A.C. (1942). Ash content of some Indian fuelwoods, *Indian Forest leaft* ; No. 28, Chemistry, Manager of Publication, Delhi.

Warhadpande, U.R. ; Mene, P.S. and Saletore, S.A. (1950). Destructive distillation of some Central Provinces Woods, *J. Scient. Ind. Res.*, Vol. 9(9), New Delhi.

Yamdagni, R. (1985). Ber, In : *Fruits of India : Tropical and Subtropical*, Bose, T.K., (eds), Naya Prokash, Calcutta, pp. 520-536.

17

# *Relevance of Research, Training and Extension in Successful Transfer of Agroforestry Technology*

Many poverty removal programmes along with sound technology have been transferred in rural areas to tackle the problems of the peasants. However, it must be admitted that most of the programmes were less effective than what they could have been otherwise. We cannot blame that programmes were less effective due to its demerits. The failure of programmes are mainly due to lack of realistic approach, want of sense of dedication in the workers; non-participation of the target group, weak co-ordination among the different agencies involved in programmes etc.

Inherent problems of Indian agriculture such as absence of land reforms, fragmentation and subdivision of holdings (in states where consolidation of holdings has not been done/completed), deforestation and acceleration of erosion and rise of river beds causing floods, frequency of floods, waterlogging of fields due to poor drainage, increase in the frequency and intensity of droughts, limited facilities of irrigation (30.7% in 1980-85), high erodability of soils in about 48 per cent of our land surface, degradation of lands due to salinity and alkalinity, desertification, etc. continue to exist. They have continued not only from one plan period to other, most of them have increased progressively. For example, in spite of about 32 million hectares of land having been treated with soil conservation measures in 30 years, there are about 125 million hectares of land eroded and subject to further erosion. The eight plan may not solve all problems or overcome the evils, something tangible can be done, *e.g.*, halting deforestation and expediting afforestation as community enterprises, in addition to direct efforts of Govt. departments. The same can also be done in the field of soil conservation and land reclamation. Without 'Peoples participation, Govt. efforts alone have not been as effective, as desired. Technology and financial support compounded with 'Peoples participation' can combat many man-made evils and natural hazards in a more significant way (Mandal, 1989).

Most of the on-going programmes are not comprehensive enough in their approach and coverage. These programmes have been concerned with those having land. The vast majority rural population living below poverty line belong to marginal, sub-marginal farmers and agricultural labourers. Under the present circumstances emphasis should be given on programmes in which all sections of rural population can get regular employment.

Adoption of agroforestry programmes on large scale can play pivotal role in villagers life which can solve most of their problems.

* Farmers will get regular employment. The migration of rural population will be stopped; because this system is labour intensive. In this programme women will take more interest, because this system meet subsistence needs of fruits, fuelwood and timbers and men may take more interest in growing cash-crop along with trees for short term benefit and planting trees will provide benefits in long term.
* Even wasteland can be fruitfully utilised under agroforestry system. If wastelands/jhum lands are allotted for settle cultivation, farmers will adopt agroforestry system with proper motivation and training.

Development of agroforestry system requires and integrated multidisciplinary approach for which existing institutions, except the Agricultural Universities, are ill equipped. Compartmentalisation of governmental agencies into agriculture, forestry and animal husbandry is a serious impediments in the development and popularisation of agroforestry systems. Lack of trained manpower at various levels and institutional financing is other serious handicaps. Even the village level workers (Govt.) are trained in agriculture and/or animal husbandry and never in agroforestry. Therefore, message on the benefits of agroforestry system cannot reach to the farmers. Extension network for agroforestry is non-existing. Even international funding agencies do not provide separate funds for the development of agroforestry under the social forestry programme.

If we consider about the main constraints of successful implementation of agroforestry programmes, the major constraint is not technology, rather it is sociological. There has been weak linkages between research, training and extension programmes.

In the recent decade ICAR, State Agricultural Universities, volunteer agencies and other had established on-station agroforestry demonstration plots. During transfer of technology emphasis on proper training and motivation of farmers is not given.

Thus, successful implementation of any programmes depends upon:

* On-station research and proper training to different level of functionaries and motivation of farmers.
* Transfer of need-based, location specific, appropriate technology with farmer's participation.

## NEED OF RESEARCH IN AGROFORESTRY

In the present decade research work on different components of agroforestry systems are in progress at Agricultural Universities and ICAR regional Research stations. However, more emphasis on on-station research needs to be given in each agroclimatic zone, to develop demonstration models at district level, so that farmers can be motivated, trained after demonstration at the research station.

Once scientists are satisfied with their model/design only than technology should be transferred in the farmers field in collaboration of farmers and extension workers. After implementation of programme, extension worker should analyse the percentage adoptibility and constraints faced by the farmers, etc. and if required model should be modified as per farmers need. Under this condition transferred technology will create permanent impact over the villagers and programme will be a success.

In general research on agroforestry will help in knowing the

* How and which trees and crops can be grown better than what they are growing now (Under agroforestry systems in specific biophysical condition) ?
* Which agroforestry system will be more economical ?
* Which system will improve the productivity of site and can maintain the fertility status of soil and reduce the soil erosion problems and conserve more moisture ?

Above problems can be solved by conducting a strong and time bound research work through various research agencies in each agroclimatic zone.

**Research priorities under agroforestry to produce maximum fuelwood, fodder, timber and crops**

* To study the available genetic resources and genetic improvement of suitable trees and crops/grasses for agroforestry system.
* To study tissue culture of trees and its biotechnological application in rapid propagation and improvement.
* Identification and evaluation of multipurpose tree species having deep root system, nitrogen fixing ability, coppicing ability and nutritious palatable leaves for each agroclimatic zone.
* To study the best combination of trees and grass species under silvipastoral system and trees and crops under agrisilviculture system.
* To study the best design for rehabilitation of hill slopes, with maximum soil conservation and increase in water retention capacity.
* To establish the simplest design/model under agroforestry system (*e.g.* Alley cropping, etc.), which may be followed by the farmers with minimum efforts.
* To study nutrient recycling, hydrological balance, cost benefit analysis, etc.
* To study the role of mycorrhiza and non symbiotic nitrogen fixing bacteria in relation to the productivity of crops and trees under agroforestry system.

**More emphasis on on-farm research trial**

* To test on-station design in the different edaphic and psycho-socio-economic condition of farmers.
* Modification in on-farm trial/design as per psycho-socio-economic and biophysical constraints of the farmers and introduction of modified model by the farmers under supervision of scientist/extension workers. This will create desired socio-economic impacts on the farmers

life. These village laboratories will be a continuous features on long term basis.

* To study the pyscho-socio-economic condition and pattern of migration to the remote forest area or migration to urban areas, by the farmers before and after implementation of agroforestry models.

## IMPORTANCE OF TRAINING

Farm practices are being revolutionised. Agroforestry is a complex system. It requires an integrated and multidisciplinary approach. This system is an challenge for extension workers and farmers, who need to be updated in technical know-how. To make above programme successful we have to bring change in the attitude of extension workers and farmers. Allport (1935) has defined an attitude as a "Mental and neutral state of readiness, organised through experience, exerting a directive or dynamic influence upon the individuals response to all objects and situation with which it is related". Thus, to change in the attitude, it is essential to give more importance to the training programme. This will help in mobilizing the human resources and help in analysing and solving the real problems of concerned community or person.

Srivastava (1989) has stated that farmers training and education is distinct from motivational communication with farmers. Much of the extension work during the fifties was of the latter category. Motivation may be generated but cannot be sustained by communication and exhortations, whether audio-visual or direct. In case of farmers and other primary producers the most effective motivation is the economic motivation. The prospect of getting higher yield and income is sufficient motivation for seeking the skills required for using the new improved technology and participating in the new programme. The skills are being refined by research scholars from year to year, some-times from month to month. This fast pace of technology, therefore, not only makes training of their producer imperative, but also requires constant alertness of new methods on the part of those responsible for training. Training thus is a direct response to the production process implied in a programme; in fact, it has to be part and parcel of this process. Bigger and more resourceful farmers are not only motivated but have also better access to the skills necessary for the adoption of new technology than the small farmers. The small farmer is not wanting in motivation but he has no access to new technologies and inputs. *Thus, production based farmers training* becomes a national necessity.

In the present situation there is large gap between demand and supply of food, fuel, timber and fodder, etc. This is only due to lack of knowledge of latest technology. Villagers, village level workers are ignorant about the scientific achievements. The gap between the latest technology known and the traditional practices used by the farmers may be bridged by imparting training to the farmers which will help in changing the attitude of farmers. Thus in the present circumstances training needs to communicate.

**Meaning and aims of training**

In training, a person increases knowledge, skill which influence change in hehaviour and increase in creative ways of thinking. When we say training of extension workers means we want to update knowledge, skills, attitude, etc. by educating them, so that a extension worker fits in his job and becomes efficient in communicating the desirable knowledge to the farmers. (See Chart No. 9).

**Steps involved to achieve the aims of training**

In Chart No. 10 mentioned schematic diagram various steps are shown which are essential to achieve the aims of training.

**Suggestion for trainers**

* A trainer should know in details about the community and must be careful in selecting beneficiaries from the community we wanted to work for.
* Selection of right group and proper training will create awareness. Utimately, this section will be example for whole community.
* We should never select person who is convinced that he cannot put into practice what is suggested by trainers. If we force them to learn, he may consider it an issue of prestige and always behave that he cannot learn. In 1st phase of training programme we should not select such type of beneficiaries from the group.
* In many areas trainees know better than the trainers. We should always impart the image of equality. The two ways

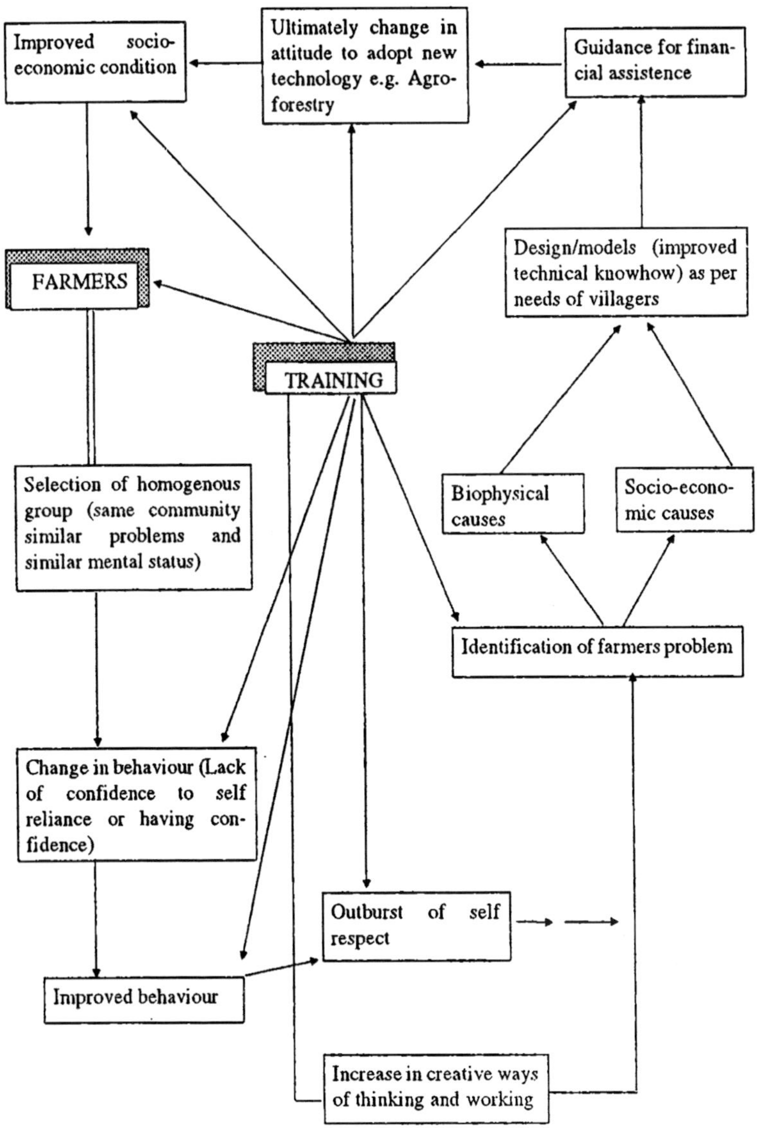


**Chart No. 9 :** A schematic diagram showing the aims of training in transfer of technology in village.

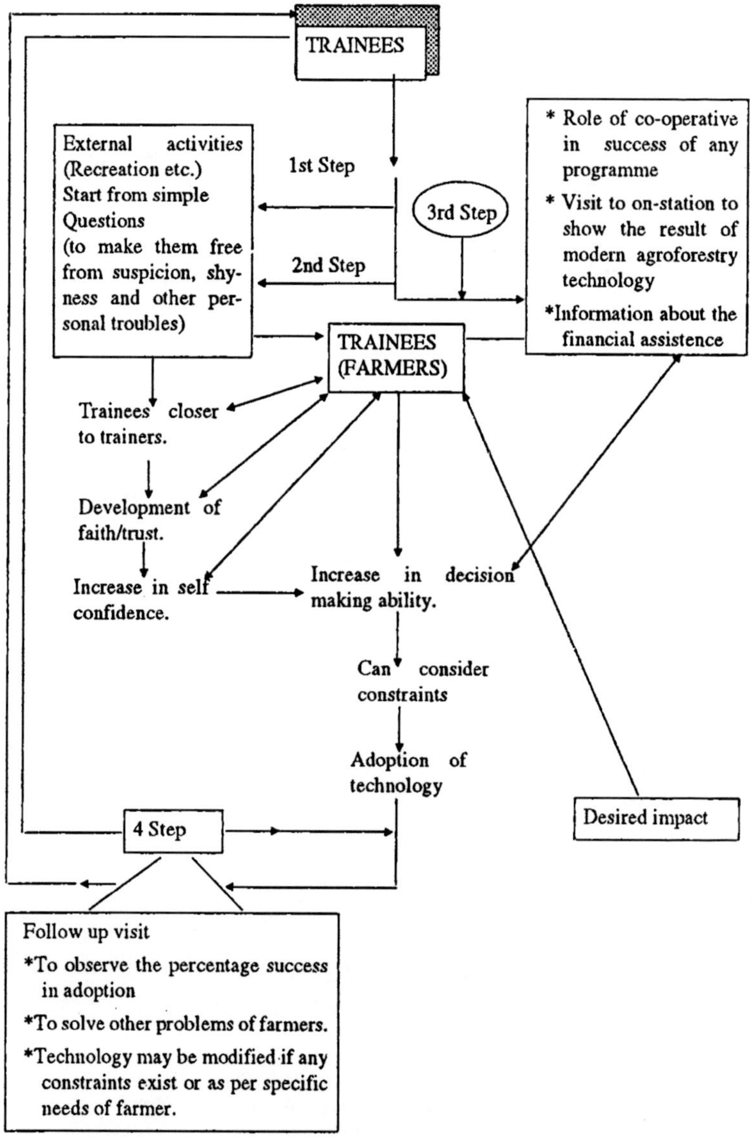


**Chart No. 10 :** A schematic diagram showing various steps which are essential to achieve the aims of training.

interactions between trainers and trainees aims in developing self confidence, faith in themselves and help in modification of on-station trial as per requirements of trainees and transferred technology will be accepted forever.

* Trainers should never doubt over the ability of the target group. They want to be convinced that they can.
* Trainers should always be co-operative in participant's thinking process.
* Trainer should explain many times in order to convince the participants. For example how agroforestry technology can solve their problems ?
* Trainers should give more emphasis over psychological motivation. In general people have ability in themselves the power and resources to achieve progress. If they feel the need to adopt any technology, economic help will not make much difference. They will try to adopt it to fulfil their own needs as well as that of society.

**Principles of extension training**

* Training objective should be clear.
* A good training creates feeling of requirements of training among the trainees, so that actual willingness and desire to tackle the problems comes from the trainees mind.
* A good training provides effective learning situation and ability to change the attitude by change in the knowledge, skills, etc.
* Good training should be designed also to evaluate the results for future modification in the programme/ technology as per needs and constraints faced by participants.

**Nature of training**

The nature of training depends upon the training needs of the particular locality. Training needs can be determined by finding out the difference between what is being done now and what ought to be done in near future. The gap gives clues to the kind and amount of training needed (Srivastava, 1989).

**Types of Extension training**

(A) Training for personnel involved in extension work.

(B) Training for the members of the target group.

**Training for personnel involved in extension work**

Training methodology, contents, technical know-how, details are developed for different level of extension worker as per their mental ability, depth of knowledge, etc.

(*i*) *For top level extension worker (trainers) :*

For top level of extension workers (training for trainers) subject matter are discussed in depth. For example, if we have to train top level extension worker regarding the agroforestry technology all aspects of the technology must be taught. (See Fig. 136)

* Scientist may acquaint themselves with different types of agroforestry system, best combination of tree species, crops or tree species, fodder or tree species, agricultural crop and fodder or tree species and fruit crops etc., for each agroclimatic zone, under different edaphic condition.
* Impact of tree species on crops/fodder etc., may be discussed at length (supported by relevant data).
* The way agroforestry technology improves the nutrient status of soil, reduces soil erosion and increases moisture retention capacity of soil and also makes other micrometrological parameters conducive for growing crops in conjunction with trees must be taught supported by relevant research data, slides and field trip.
* Comparative analysis of different agroforestry models should be discussed in details to find out the best model in specific edaphic and climatic condition.
* Cost benefit analysis as studied during on-station trial.

(*ii*) *For middle level extension workers (trainers) :*

* Only about the best combination of tree species and crops or fodder or fruits for their locality.
* Best suitable design/models for their area.
* About the effect of agroforestry system on the nutrient status of soil, soil erosion/water retention capacity of soil, etc. There is no need of discussion about the comparative study of all experimental design.

Fig. 136 : Orientation course for senior extension staff.

Course contents are increased or decreased as per the level/rank of functionaries, *e.g.* orientation training course, short range courses, periodical meeting and conferences are organised for different level of functionaries (Extension workers) by different state government organisations, state agricultural universities, ICAR and other organisations.

**(B) Training for the members of the target group (farmers)**

We can divide training in two phase :-

1st phase —Before implementation of programme

2nd phase—After implementation of programme.

**1st phase**

* Course contents of training programme are designed as per need of the beneficiaries (farmers).
* Discuss only about the best suitable models or best improved technologies for their land.
* Discuss only how relevant technology will solve their biophysical and socio-economic problems.
* Evaluation of impact of training programme on their attitude.

**2nd phase**

* Transfer of technology in the farmers land under supervision of extension workers.
* Discuss regarding various problems faced by farmers during adoption of improved technology.
* If necessary, modify the model/technology to solve the constraints and fulfil the needs of farmers.
* Evaluation of the impact of training (motivation) programmes on the adaptability of the programme.

Krishi Vigyan Kendras, farmers training centres, etc., play important role in conducting training to change the attitude of farmers adoption of improved technology.

**Training centres at District/Block level to motivate the farmers to adopt agroforestry system**

Special agroforestry training camp should be organised at District level/Block level as per need of the farmers.

* Short term course on agroforestry for 3 days/7 days/15 days should be organised firstly at village level than at the research station to demonstrate the model. Again training programme should be organised at village level to help them in adoption of agroforestry technology.
* Leader among the farmers community may be trained under different professional farmers training programme *i.e.* one month/2 months/4 months/6 months/1 year course.

**Few knotty points to be cleared up for successful training programme :**

* Trainers do not have clear concept about the techniques of training, and how to motivate or create the desire/willingness among the trainees ?
* Training programme will be only effective when it is designed as per needs and biophysical and socio-economic condition of participants.
* Trainees should never be selected from different socio-economic status groups.
* Training should be supported by research finding and actual visit to demonstration plots.
* Training should be conducted at regular intervals to the same group of beneficiaries to create awareness, willingness and finally to assess the adaptibility per cent.
* There should be sound communication and feed back.

**Mobilisation of Villagers**

In transfer of technology in farmers land, mobilisation is first step for proper training/communication. Motivation is the process of initiating a conscious and purposeful action.

In village, community peoples are associated with social group and institutions which influence the life style of the individuals. A group consists of many individuals for example—Family group, neighbourhood, informal groups (play groups, chhopal group), formal group (societies etc.), religion and cast group. The same individuals may be a member of different groups. Each member has his own motives and attitudes when he enters into interaction process; each group also influence interaction process due to its rule. Any change in the member due to interaction process, changes characteristics of the group. Trainers have to mobilise the particular section of rural community in order to introduce certain latest technology. Trainers should know the difference between groups and institution. Generally groups are unorganised pattern of association of persons; while in institution people's dynamic interest are expressed in organised and stable pattern to achieve the goal.

The rural groups are characterised by the following characters :

1. Sufficient good numbers of members.
2. Tendency to live together at one place.
3. Common interest.
4. Leaders and follower system prevails.
5. Internal and external conflicts prevail.
6. Group possess a sense of belonging (I belong to this community), pride and satisfaction and feel secure.
7. Cohesiveness (*i.e.* strength of the forces that keep the group together).

**Principles of group mobilisation**

Kuppuswamy (1985) has stated that motivated behaviour is the behaviour in which a person engages in, as he strives to reach a goal. Motivated behaviour is good-directed and goal-oriented behaviour. The behaviour continues till the goal is reached, or some other motivated behaviour intervenes. It is organised around a goal. Motive is a state of the organism in which the bodily energy is mobilized and directed to attain a goal. Motivation implies a state of energy mobilization and a direction towards a goal. Thus, motive is a concept that joins together a state of energy mobilization and a goal. State of energy is linked by learning.

For successful implementation of any programme, first step is group mobilisation and in order to mobilise the group extension worker should carefully consider points mentioned below :

* In rural community group systems comprise of leader and followers system. Identification of leadership pattern in a group is essential step for group mobilization. The group leader draws its power from the group. This is the reason, his decision or interest in regard to any programme also influences other members of the group. Leader is to coordinate the activities of other members.

* Extension worker should not only depends upon leader's decision but also involve all members of a group in any decision making programme. Due recognition to all members should be given to make programme successful.
* Co-operative decision creates the feelings among the members that execution of the programme is their responsibility and then all members along with leader will participate in the execution of programmes with full co-operation under the supervision of the extension worker.
* The group leader should always create environment to sustain the continuous interest in the programme.
* Leader has to plan and find out the ways and means by which the group can attain its objectives.

Besides above points, extension worker should also keep in mind few basic elements while motivating the villagers.

* The socio-economic and biophysical causes of particular villages.
* The villager himself.
* Past device created by villagers to improve themselves and community (*e.g.* formal group, religion, etc.)
* Use of any communication media.

**Specific problems of rural tribes and other weaker sections of society**

During motivation training camp, extension workers should always keep in mind problems of rural tribes or other weaker sections.

The problems can be classified as follows :

(1) *Economic problems* :

a. Land use pattern.
b. Land ownership.
c. Economic condition.
d. Labour migration pattern.
e. Marketing of the agriculture and forest produce.

(2) *Biophysical causes* :

a. Topography of land.
b. Fertility status of soil.

c. Soil erosion problem.

d. Water conservation problem.

(3) *Infrastructural problems* :

a. Drinking water facility.

b. Road linkage.

c. Transport facility.

(4) *Education*

a. Poor education facilities due to absence of school and non-availability of teachers.

b. Language problem.

(5) *Special problems* :

a. Problem of drinking habit.

b. Rigidness towards traditional practices.

c. Never thinks about the future work plan.

Here it is worthwhile to mention that when a project is accepted by the leader and members of group, we think that members will actively participate, but this is not so. Each member is also influenced by various other factors which play important role in their participation. For good communication, extension workers/trainers should identify and eliminate the problems.

**Communication and its role in transfer of technology**

According to Kuppuswamy (1985), human interaction is based on communication. The behaviour of one person, the message that he sends by speaking are received by the other person and he responds to the message by another message which the first person receives and so on. Sharing of information will make information the common property of the whole group, and thus enhances the cultural life of the group.

Thus, communication is an exchange of information, knowledge between two or more persons. Interpersonal communication refers to the transmission of message from one person to another. Thus, this process involves two persons—the transmittor and the receiver, and the message (information packages), which is understood and approximately responded by the receivers.

Attitude of a personal changes, when he receives information by proper communication media.

**Communication model**

Hovland Janis (1959) has reported a useful model of attitude change that includes many stages and variable.

In communication process, sender sends desired message having clear objective and gives special treatment for transmission over a channel to a receiver. The receiver responds to the message and interprets before taking desired action. The following figure illustrates the model in a very simplified form.

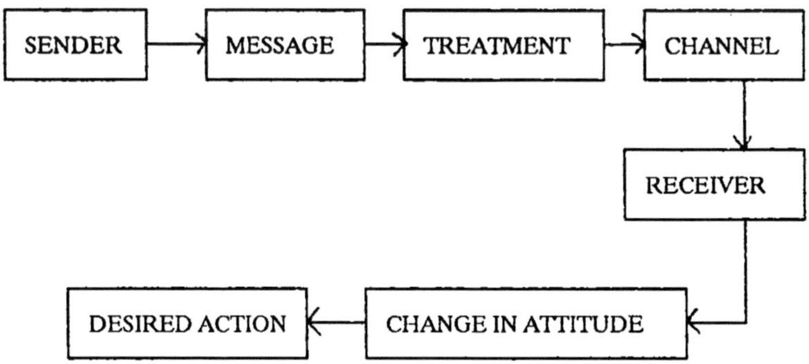


**Chart No. 11.** A schematic diagram showing the communication process.

The sender, message, treatment, channel and receiver are the elements of the communication process.

**Sender**

In the communication process senders are speaker (Trainers/Extension workers). Trainers have to decide about the message to communicate and channel (Lecture/slides/V.C.R./Booklets etc.) to be used for successful transmission to create desired impact on the receiver. If sender selects poor choice his communication cannot create desired impact to achieve the goal.

**The message**

Message is information package, *e.g.* for agroforestry system details technical know-how, design, choice of species etc., for improving productivity and economic status of the farmers. Message may be in the form of booklet.

**The treatment of message**

To make the message more effective, realistic to the receiver, proper treatment is given before placing it on the channel.

**The channel**

The channel is medium of communication *e.g.* microphone/radio/newspaper/lecture during training programme, etc. choice of channels varies as per objective.

**The receiver**

Under communication process, information is trasmitted by the sender to another person or group of people. When information package communicated by the sender is effectively received by receiver, the message is understood and at the same time receiver responds in regard to message.

Thus in communication process there are three phases :

* Expression
* Interpretation
* Response.

In proper communication process, expression always remains clear and thus the interpretation will be accurate and the response will be maximum.

**Communication approaches to make extension programmes successful**

On on-station trial, development of desired technology by the scientist and training of the extension workers, farmers by suitable communication approaches are the tools to increase the effective transfer of technology.

Srivastava (1989) has stated that an extension worker must first know what methods are available to him, secondly, he should know when to use a given method and thirdly, he should become efficient in using each method.

Communication approaches may be grouped into three :

(*a*) *Individual approach*

The farmers are approached by personal visit in his house in order to motivate to adopt latest technology under supervision of experts.

(*b*) *Group approach*

In this process psycho-socio-economic conditions and biophysical causes are studied by the extension workers in 1st phase. Leaders are identified. In consultation of leader as well other member of groups training programmes are organised as per their need and constraints. Group discussion and field tour to on-station research trials are conducted to show them on-station model. Finally technology developed at on-station is transferred in the farmer's land under supervision of experts.

Again, group discussion and interviews are conducted to know the impact of transfer of technology. (See Fig. 137)

(*c*) *Mass approach*

Under this communication approach we consider problems and needs of mass population. Information about the latest technology are communicated through T.V., Radio, Newspaper, film show, Kisan Mela, etc. Farmers from different group, community, participate in the Kishan gosthi of the Mela to discuss about their problems. Extension workers communicate about the solutions to tackle their problems.

In my opinion combination of more than one methods will be more effective in motivation programme.

**Conclusion**

It is apparent that Indian farming is subsistence farming. They only produce grains, etc. to fulfil their own requirements. We can increase productivity of marginal, sub-marginal and waste lands by introducing sound technology, like agroforestry system. Technology should be transferred in the farmer's field after creating awareness among the villagers by proper training/learning process. When desires to achieve some standard of excellence emanates from farmer's side they will accept agroforestry technology on large scale as a permanent features.

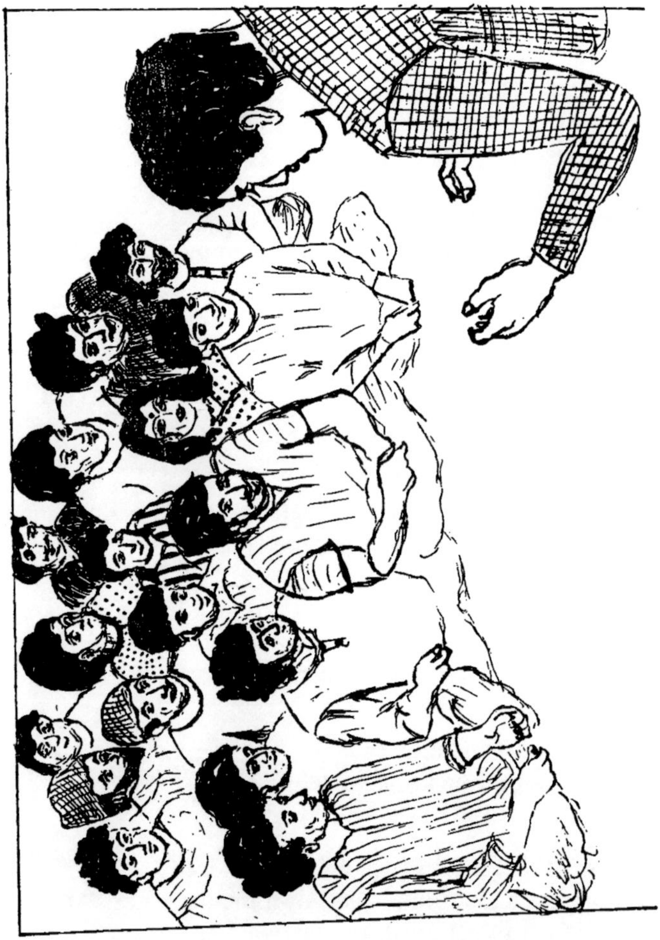

Fig. 137 : Regular contact programme to motivate the farmers to adopt latest technology.

## REFERENCES

Allport, G.W. (1935). Attitudes, In : *Handbook of Social Psychology,* Murchison, (ed.), Clark Un. Press, Mass.

Kuppuswamy, B. (1985). *Elements of Social Psychology*, Vikas Publishing House, Pvt. Ltd, New Delhi, pp. 1-194.

Mandal, S.C. (1989). Eight Five Year Plan and Agricultural Production, *Grow more Food Souvenior,* On the occasion of VI National seminar-cum-workshop on Krishi Vigyan Kendras, Ramkrishna Mission Ashram, Morabadi, pp. 18-24.

Srivastava, R.M.(1988). *Extension Education and Communication,* Lecturer notes. Faculty of Forestry B.A.U.,Ranchi.

Srivastava, R.M.(1989). Training for development, paper presented in the *workshop on training programme for the voluntary agencies of the C.N. and S.P. of Bihar,* organised at Faculty of Forestry, B.A.U. Ranchi

18

# *Incentives to Make Agroforestry Programme a Grand Success*

Rapid promotion of agroforestry cannot be obtained by the participation of limited conscious farmers. Involvement of mass in general seems absolutely necessary to attain this objective. A large number of our population, particularly belonging to weaker section suffers from poverty, ignorance, conservatism and fatalism. Thus, for the active participation of mass all possible incentives are to be provided in our country.

(*a*) *Expected gains*

Incentives are to be provided to the participants by making them aware of the expected gains of agroforestry.

Agroforestry is increasingly accepted and promoted as an alternate approach to land use which generates steady income for subsistence farmers. Large areas of uplands which are only partially cultivated can be made to yield high return per unit area under a combination of suitable crops/legumes/fodder with trees. This system can generate several positive environmental impacts like improvement in hydrological balance, improvement of physical properties of soil favourable to improved infiltration, retention capacity and indepth percolation of surface runoff and reduction of sedimentation of rivers and streams etc., recycling of nutrients, creation of favourable microclimatic condition conductive to higher food production, maintaining balance in oxygen-carbon dioxide, atmospheric temperature and relative humidity. It is a labour-intensive programme and provides job, particularly to weaker section of society. If the gains are fully understood by the participants. They will get special incentive for intensive efforts.

(*b*) *Techniques and design*

The foremost incentives is to present detailed information regarding the techniques of agroforestry to the participants. This dual system of production needs latest research based informations.

The technology is to aim at ensuring a higher return which is within the reach of the farmer and which can be achieved with reasonable amount of certainty instead of the illusive maximum (Subarna, 1990). The establishment of agroforestry research centres throughout the country in large numbers will provide informations to the participants as to which kind of combination of tree and fodder/crops/legumes/tubers/fibres/forage/vegetable will be suitable in a particular area. The formation and functioning of a net work of agroforestry research centres throughout the country will incur heavy expenditure, but that will definitely provide good incentive to the farmers on agroforestry and the expenditure will amply be compensated by the higher income generated through production of crops and woody material.

(*c*) *Demonstration plot*

Formation of ideal demonstration plot in each Block seems desirable to provide incentive to participants. Farmers will personally visit and derive incentive to adopt agroforestry in their lands. Further the associated advantages of agroforestry compared to crop cultivation may also be fed into their minds so that they are able to take decision for a viable alternative (Boraiyan, 1985).

(*d*) *Training centres*

In a democratic country like India, any programme/project which does not aim at educating the people to secure their participation and involvement cannot finally succeed and can hardly be expected to produce desired results. In other words education/training of people is a primary and major factor to intimately involve them in accomplishing the agroforestry programme. Training of the participants should be an important part of the programme. Proper training of farmers is assumed to be the base for success of programme like agroforestry (Sagwal, 1985).

Thus, training centres should be established at Block level. An incentive should be given to the participants in the form of free lodging and subsidised food.

(*e*) *Superior quality of physical inputs*

The success of this programme depends largely on the speed and efficiency with which procurement of seeds and planting material is organised. Any default in this work may jeopardise entire plantation programme, successful arrangements of all other necessities notwithstanding (Sen Sarma and Jha, 1990). Provision of superior quality of physical inputs like seeds, seedlings/saplings, fertilizers, pesticides, farm equipment, etc. at a concessional rate, may be considered an important incentive for the participants in our country. Utmost care is to be taken by the Government so that superior quality of inputs may be made available to the participants adequately and timely even in the remotest corner of the country without any harassment (Jha and Sen, 1991). A single window delivery system facilitate timely distribution. (See Figs. 138 and 139)

*(f) Distribution of land amongst poor family (Co-operative farm)*

As the land holdings are small and scattered, agriculture practice is unproductive and unscientific. Owning of small pieces of land and cultivating them independently may give psychological satisfaction to the cultivator, but this arrangement can certainly not reap the benefits of better water management and better tree, crop and animal husbandry management. Therefore, if 'big push programme' like agroforestry is to be organised, small and scattered holdings must be made compact (Srivastava, 1985 and Jha, 1986).

In tribal belt to provide an incentive, the degraded forest areas lying in the neighbourhood of villages may be distributed amongst the poor families in the form of co-operative farms (not less than 2.5 ha). In such cases the allottees may be induced to adopt agroforestry in order to meet the expanding and ever growing fuel, food, fodder, needs of people and cattle.

*(g) Promotion of cottage industries*

Agroforestry can play an important role in cultivating host plants of tasar silkworm and Lac insect. The major host plant of tasar silk in Chotanagpur is arjun (*Terminalia arjuna*) for which bush plantation is resorted to maximise rearing of tasar silkworm.

It has generally been accepted that a steady decline of lac host plants has taken place during the past few decades due to shortage of fuelwood, extension of agriculture etc. In agroforestry system, lac cultivation can be boosted by planting conventional lac hosts such as palas (*Butea monosperma*), ber (*Zizyphus mauritiana*) and kushun (*Schelichera oleosa*). In addition bhalia (*Moghania macrophylla*), Galwan (*Albizia lucida*) and poutri (*Croton oblongifolia*) are also good hosts. Because of their bushy nature, the cultivation in agroforestry system in easy.

Figs. 138 & 139 : Timely supply of superior quality seedlings.

All inputs should be made available by the government required for the functioning of these cottage industries. Proper training facilities are to be provided to the participants. Sericulture, lac and forest Department in different states in collaboration with national scheme of training of rural youth for self employment may fulfil the task easily.

**Special incentive for Taungya cultivators**

Taungya system (a type of agroforestry system) has been successful in many places. This system may be made more popular by motivating poor persons, specially belonging to the tribal belt. They may adopt this system in the wastelands of the government lying in the neighbourhood of their villages. Taungya system should be popularised in the plain areas of the country, wherever suitable conditions prevail. Persons involved in taungya cultivation, belong to weaker section and hence they may be provided special inducement in the shape of interest free credit and subsidised inputs. Attention must be taken that taungya cultivators should surrender the possession of the land after the expiry of the period (Jha and Sen, 1991).

(*h*) *Village agroforestry co-operative society*

Extensive deforestation, indiscriminate land use and use of fertilizer etc., to boost production have created global atmospheric change. Unless precautionary measures are taken right from now, a perilous situation for us will exist as we will enter the 21st century. To overcome such a situation, adoption of agroforestry in large scale by developing agroforestry Co-operative societies in each village under direct involvement and supervision of one or more Government Departments to raise production and quality life of the people is essential (Jha and Nath, 1992).

Thus, for promoting agroforestry, special incentive may be provided in the form of legal and financial aids for the formation of joint Co-operative agroforestry society in villages. If majority of land-owner of a village decide to adopt agroforestry in a particular area, consisting of marginal or sub-marginal lands they should be encouraged to form agroforestry co-operative society. Small number of recalcitrant land-owners will ultimately be obliged to join such society in their own interest. The successful functioning of such societies will accelerate the tempo of development in agroforestry.

In the Republic of Korea the functioning of village forestry association is contributing significantly for the promotion of social forestry programme. In India formation of agroforestry co-operative society in villages will serve this purpose (Jha and Sen, 1991).

(*i*) *Institutional credit*

An incentive will be provided to the participants by providing cheap and adequate amount of institutional credit *i.e.* credit from Banks, Co-operative societies and other agencies of the government in proper time. Agroforestry programme cannot be implemented rapidly with the help of private source of credit, where rate of interest is very exorbitant. The participants, particularly, the poor, require credit for purchasing physical inputs, farm implements, fencing of the plot, etc. Physical labour may be done by several participants, but for the maintenance of their families, till the crops are harvested, provision of consumption credit is absolutely required. Moreover, institutional credit is to be provided for the establishment of cottage industries based on the raw materials of agroforestry.

Institutional credit may be misutilized by some participants. For example, instead of purchasing physical inputs, the poor participants may spent the amount for purchasing household requirements. This credit may be provided, as far as it can be practicable in the form of physical inputs and not in cash. Moreover, inspection must be done by the appropriate agency to check misutilization of institutional credit. Credit should be realised in easy instalments and at suitable intervals. Utmost vigil is required by the government and social agencies in our country so that the ignorant and poor participants may not be forced to grease palms of the personnels, who provide institutional credit to them (Jha, 1986).

(*j*) *Marketing facilities*

Products of agroforestry system are often sold at a very cheap price. The unhappy situation of tribal haats, can be averted by suitable marketing arrangement. Formation of joint co-operative marketing societies may be much helpful to tackle the problems. Moreover, if marketing of products is done through these co-operative societies, it will be an easy task for the Government to realise loans from the participant (Jha and Sen, 1991). (See Fig. 140). If such agencies are not operating , prices of the commodities must be made known to the participants clearly and distinctly and strict supervisory control should be exercised so that no trader is able to cheat. Likewise, under no circumstances, dual pricing system, advocating different rates for fair price shops and open markets, as is prevailing in urban areas, should be introduced in the rural tracts, as it is a potential destroyer of fair marketing and creater of blank marketing system (Srivastava, 1985).

Hence the government must provide incentive to the participants by managing suitable marketing facilities. Fluctuations in the price of the products of agroforestry will be disincentive to the producers. This problem can be solved by announcing a support price by government or wherever necessary by arranging for purchase of such wood even by wood-based industries directly, at predetermined rate.

(*k*) *Crop Insurance*

In our planned economic development of the country, crop insurance scheme is being implemented throughout the country so that progress with stability can be achieved in agriculture. In the field of agroforestry, the implementation of insurance scheme will provide an additional incentive for active participation. At several times due to the absence of insurance scheme, the participants suffer badly. For instance, if larvae of tasar silk are abruptly infected by some harmful disease, the participants are bound to suffer enormous loss. Such a loss can be easily compensated, if insurance scheme is implemented in this field. The implementation of insurance scheme in agroforestry is not an easy task. It may be implemented gradually on the basis of findings of experts. Needless to say, the adoption of the scheme in due course will provide obvious incentive to the participants (Jha, 1989).

**Infrastructure**

In several areas there is lack of better infrastructure. Therefore, some sizeable investment would have to be made to enable the farmers to take his produce in the existing market. Provision of infrastructure is basic incentive for raising production in any field. Hence, by providing appropriate infrastructure, incentive will be available to the participants.

Fig. 140 : Local market pithoria (Ranchi)

**Policy and Laws**

Due to ignorance some people think that if plantation in a large area is done on private land, it will be nationalised in due course.

Several matters relating to rules and regulations concerning harvesting, transport and marketing of wood also need attention with a view to simplifying the procedures to eliminate harassment and delays in obtaining permission. This would require amending the timber transit rules for exemption of species most commonly grown by farmers under various agroforestry programmes.

Raising tree plantation over larger area under social/ agroforestry system with a view to improving environment and maintaining ecological balance can be achieved through well-to-do farmers if laws relating to land reforms are suitably amended to exclude land under such forest tree crops from the limits of land holdings. Income tax and wealth tax exemptions may also have to be thought of to induce well-to-do farmers particularly to take to more and more tree planting in the overall interest of maintaining ecological balance (Rao, 1989).

**Avenues of promotion**

Extension workers perform arduous task in the execution of agroforestry programme in imparting technical knowledge to the participants. They are to be provided incentive in the form of better salary and avenues of promotion. If some members of the department perform excellent works, they must get special pecuniary and non-pecuniary rewards. Persons guilty of dereliction of their duties must be punished. Failure of such action adversely affects the incentive to other employees for performing their duties sincerely (Jha and Sen 1991).

Promotion of agroforestry is now an international programme. In India we will have to incur heavy expenditure for providing incentives in various forms. Several efforts have been made by the government on this score, but the pace of work cannot be considered upto the mark as yet. We may expect that concerted efforts will be taken to provide all possible incentives to the participants during the eight five year plan so that nation shall be able to tackle this herculean task to a large extent.

## REFERENCES

Boraiyan, M.T. (1985) Social Forestry, a way of life, *Yojna*, Vol. 29 (19), pp.6-8.

Hegde, N.G. (1987) *Hand-book of wastelands Developments,* The Bharatiya Agro-Industries Foundation (Kamdhenu), Senapati Bapat Marg, Pune, pp. 1-102.

Jha, L.K. (1986). Incentives for social forestry in India, *My forest*, Vol. 12(2), pp. 122-128.

Jha, L.K. and Sen, P.P. (1991) *Social Forestry*, Himalaya Publishing House, Bombay, pp. 1-199.

Jha, L.K. and Nath K.P. (1992) Population: Impact on demand and supply of resources and environment-Areas of relevance to tackle and the problem with emphasis on agroforestry land use system, *Vol. of abstract*, I.I.D.S., Int. Seminar, (Calcutta), pp. 78-79.

Kerketta, V. ; Lakra, V. ; Bara, N. and Tirkey, K. (1990) Women's participation in agricultural development of tribal region of Bihar, Workshop-cum-Training Programme, Director of Extension Education, B.A.U., Ranchi, pp. 1-5.

Rao, A.L. (1989) Social forestry: Basic requirements for making it a peoples movement, *Training-cum-workshop on community participation in Social forestry*, Tata Institute of Social Sciences, Bombay, pp.1-10.

Sagwal, S.S. (1985) On popularising agroforestry, *Yojna*, Vol. 29(19), pp. 11-12.

Sen Sarma, P.K. and Jha, L.K. (1990) Social Forestry for improved land management, *K.V.K. Souvenir,* Ranchi, pp. 40-48.

Srivastava, J.L. (1985). Tribal Man *'Why thy fat is not prosperity'?* Forest Deptt. Doranda, Ranchi, pp. 1-6.

Subarna, M.C. (1989). Importance of socio-economic and cultural attitude in planning for agriculture development of the tribal area, *Annual Research Conf.*, B.A.U. , pp. 1-8.

# 19

# *The Role of Different Sections of Indian Society in Agroforestry Programme*

A combination of appropriately selected woody components and herbaceous crops essentially contributes to both productivity and sustainability of the farming system on marginal and submarginal lands by maintaining soil fertility, reduction of water and wind erosion and creating a microclimate favourable for associated crops and livestock. Further, integration of farmers into tree management processes has a salubrious effect in making the farmers conscious of the importance of the forest cover in sustainability of the farm production. Thus, peoples participation in afforestation in real sense can be achieved through agroforestry. Agroforestry programme is labour intensive and consequently large employment opportunities are created. Small and marginal farmers, agricultural labour, landless persons particularly females get special preference in getting employment in it. Agroforestry is aimed to alleviate directly the economic condition of Indian community, particularly rural community by providing different amenities (Kerketta and Jha, 1989).

This programme will be the mainstay for greening the country and bringing at least 33 per cent of land area under forest cover in the coming decades for which a major portion of financial allocation in the forestry sector is made by the government under social forestry. Farm/agroforestry can be financed under the Jawahar Rozgar Yojna on the lands belonging to scheduled caste, and scheduled tribes, emancipated bonded labours, etc. Therefore, the usual financial constraints for the progress of social/agroforestry is not envisaged. However, only the government agencies cannot make this programme a success on account of massive nature of the problem that the country faces. Agroforestry programmes are executed by the different sections of the society in which the role of agriculturist is conspicuously dominant. The non-government organisations (NGOs) must play a very important and decisive role in this.

**Extension worker**

The role of extension worker in agroforestry is outstanding. He is the field level functionary who is supposed to translate the core objective of ensuring spontaneous and willing co-operation and participation from the people in social/agroforestry (Boraiyan, 1985). It becomes the duty of extension staff to examine the socio-economic conditions of farmers and motivate to adopt agroforestry, so that their income may rise rapidly. The extent of participation of farmers in agroforestry depends on the nature of motivation, by the extension staff. The extension staff must explain clearly the agroforestry programme, so that the farmers may be able to know answers to following questions.

1. What is agroforestry ?
2. How interaction of trees along with crops improves the fertility status of soil ?
3. What are the techniques of plantation of the important species ?
4. What are their economic and other gains ?
5. What are government helps to the participants ?
6. How can products of agroforestry be sold at reasonable prices ?

Regular discussion of the problems of agroforestry grass root level is absolutely necessary. It can create spirit of popular enthusiasm and voluntary participations.

Tribals are at time harassed by the Government personnels while obtaining the benefits granted by the Government for the promotion of social/agroforestry. Most of the extension staff in tribal belt consists of non-tribals. Attempts should be made to include at least 30 per cent tribal employees in the extension staff of the Government working in tribal belt, for developmental programmes of social/agroforestry. Non-tribal extension staff, working in a particular tribal area, must have basic ideas of local language, culture and customs.

**Voluntary organisations**

Voluntary organisations work with local people to promote people's involvement and participation in developmental tasks. The underlying objective is to develop people's abilities and resources, so that they could take over and manage village development tasks by establishing their own organisations. A voluntary organisation is thus conceived as a catalyst institution which is expected to develop the community organisational competence and ability to mobilise resources for development (Narayana, 1991).

Voluntary organisations have played a pioneering role in rural development right from the last century to this day. They have demonstrated the qualities of flexibility of service, personal touch and closeness to the client groups. In addition, they have demonstrated the capacity to initiate and experiment with new programmes, stimulate the interest of the community, and showed ingenuity to meet difficult problems that cannot easily be tackled by state Government. Most of them profess Gandhian ideology. According to them, a Voluntary organisation should be a trailblazer, experimenting with new methods and strategies. The state may profit from experiences and experiment in larger geographical areas with wider resources at its disposal. A small number of organisations which function as independent bodies have experimented successfully and established new strategies for rural development. A large number of voluntary organisations have played supplementary roles to the efforts of the state (Mishra, 1989).

Some of the voluntary organisations are doing laudable works. Among them could be listed Kasturbagram Krishi Kshetra (Kasturbagram), Gayatri Parivar (Khargon district) Ramkrishna Mission Ashrama, etc.

Krishi Kshetra is engaged in farming dairying and plantation activities. Agroforestry and social forestry are two of the schemes of Krishi Kshetra. In 1980 the Krishi Kshetra launched a massive silvicultural scheme on the Tekri. Under this scheme many trees of timber, fuel, fruits, and fodder were planted on the slopes of the hillock. Nearly, 1,35,000 saplings of various trees like *Emblica officinalis, Terminalia bellerica*, *Pongamia glabra*, *Dendrocalamus strictus*, and *Cassia* spp., were planted with the help of the Forest Department of the state Government of Madhya Pradesh. After the rainy season, the plants were irrigated in the first year with buckets and cans. Watering was not necessary in the second year. At present, the hillock is a combination of silvipastoral development area. At the foot of the hillock where the land is plain, different species of *Leucaena leucocephala* have been planted with the idea of developing it into an energy plantation. After almost 10 yrs. of persistent hardwork and planting, the barren hillock has been turned into an economically viable pasture land. The trust is an organisation for women. All activities of the trust, therefore, are for women and to a large extent by women.

There are several Missions and Ashrams which function as Non-Government organisation in the country. They may offer their Co-operation, because they are close to the people they should be able to mobilize people, especially women, for government programmes (Murthy, 1990).

**Large farmers**

There are approximately 2.5 crores of large farmers in the country. It is apparent that the active participation in agroforestry programme by large farmers would increase their income and employment to the weaker sections of the society. Concerted efforts are required to motivate them for more active participation. The role of large farmers in agroforestry programmes may attain new dimension due to the following reasons :

* Several of them have enough land and hence they can afford land for agroforestry. Many of them are aware of the gains of agroforestry and hence their inclination towards active participation may be considered as natural.
* The incentives provided by the government like institutional credit for purchasing inputs, etc. attract them, to adopt this dual cropping system.

Presently we find that rich farmers of Bihar, Haryana, Punjab and Uttar Pradesh have adopted agroforestry. In Bihar some large farmers of Saharsa district have planted *Mangifera indica, Litchi chinensis*, and *Dalbergia sissoo*, etc. in their farm land and in the interspace vegetables/paddy are produced. They also plant sissoo along the boundaries. They can harvest timber after 15 year, but fuel wood is obtained from the trees after 3rd years by lopping the branches. In summer the green leaves of sissoo are mixed with the dry fodder and fed to cattle.

In Uttar Pradesh some large farmers plant poplars which remain without leaves in winter and in the interspace wheat cultivation is done. A few large farmers of South Bihar have made block plantation, in their marginal lands of *Terminalia arjuna* at a distance of 1.5 x 1.5m. Farmers also grow agricultural crops, vegetables in the interspace. Tasar worm rearing is not possible by sporadic plantation and hence large farmers of plateau region require constant guidance for plantation.

**Small and marginal farmers**

Small and marginal farmers constitutes nearly 12 crores, who possess land less than 2 hectares. If majority of small and marginal farmers of a village decide to adopt agroforestry, they should be encouraged to form agroforestry Co-operative Society. Owning of small pieces of land and cultivating them independently may give psychological satisfaction to the cultivator, but this arrangement can certainly not reap the benefits of better tree crop and animal husbandry management. Therefore, small and scattered holdings must be made compact.

In the Kumharia village a small nursery was developed in the year 1984, by the marginal farmers in their own land, under technical and partial labour assistance by Deptt. of Extension and social forestry, B.A.U., Ranchi. After great pursuation, 24-marginal farmers offered 17 acres wastelands where silvipastoral system was introduced by the farmers. The farmers formed a society to protect the silvipastoral plot from the stray cattle and due to their pressure the villagers gradually gave up habit of uncontrolled grazing and started stall feeding (Social fencing). In the year 1987, 40 farmers of the same village voluntarily offered 50 acres of land for agroforestry under overall supervision of scientist of Birsa Agril. University, Ranchi. The participants whose lands were adjacent agreed to form common boundary with mutual understanding in order to protect the entire area from the biotic pressure collectively and to share produce among themselves as per their land area. In the year 1989, nearly 100 acres of wasteland of marginal farmers were put under agroforestry system (30 acres under wide alley cropping system and 70 acres under intensive alley cropping system). In this way, marginal and submarginal farmers of a single village could cover 210 acres of wastelands under agroforestry.

The cost incurred by the small and marginal farmers are far less than that incurred by larger farmers, as small and marginal farmers themselves participate in all type of works with greater enthusiasm and dedication.

**Agricultural labour**

There are approximately 14 crores of agricultural labour. After independence, several measures have been adopted to ameliorate their socio-economic condition. Till now, the participation of the poor masses in development programmes may at best be compared with the participation of the bullock in ploughing done by a farmer. Interestingly, it has got reflected in the Seventh Plan (1985-90) document which observed, "The objective of removal of poverty can be fulfilled in the measure in which the poor themselves become conscious, and assert themselves". But nothing tangible has been accomplished in this direction yet and the poor still remain just onlookers without having any role of participation in the poverty-alleviation programmes. To enable poverty-afflicted men to gradually exercise their birthright the development workers and the poor people together will have to gain an understanding of the situation of the community in terms of its needs, resources, constraints, relations etc., and plan a strategy of action for removal of poverty on a family and also a community basis (Swami Prabhananda, 1989). Low standard of living of agricultural labourer is the consequence of the vicious circle of poverty which starts from the point of low production. Obviously the low productivity affects the income, saving and investment and the standard of living also.

All these problems of agricultural labour come out from the basic problem of farm resources, and therefore proper allocation of resources base may be the solution of the problem (Rajgopal, 1986).

The agricultural labour has been given homestead land throughout the country. Due to enforcement of land ceiling act a limited percentage of agricultural labour received small area of land. The Govt. of West Bengal and Bihar has provided land to each tribal family for afforestation purpose. The Jawahar Rozgar Yojna also urges the government agencies to encourage small, marginal farmers and to adopt social agroforestry. The profits of the agroforestry should accrue to those poor who are given tree pattas or tree planting permits. These beneficiaries may be permitted to raise loan from financial institutions by mortgaging the tree pattas/tree planting permits. Farm forestry can only be financed under the Jawahar Rozgar Yojna on the lands belonging to scheduled caste, and scheduled tribes, and emancipated bonded labours.

The agricultural labourer of plateau region in the country, particularly belonging to scheduled castes and scheduled tribes, are still in more deplorable conditions. Many of them still pilfer wood from traditional forests and sell them for mere subsistence. They deserve sympathetic consideration. Firstly, they have to be provided jobs under social forestry/agroforestry programme, so that they may subsist without pilfering wood from the forest. Secondly, they require special inducement in the form of liberal subsidies, training, allotment of land, etc. The denuded forest obtained by the government after abolition of Zamindary system, may be distributed among the agricultural labourers for farm forestry agroforestry practice. If there is full harmony and sense of co-operation among the recipients of such lands, joint co-operative tree plantation society may be formed with the active assistance of the oppropriate government agencies. This will promote income and employment opportunities among the poorest section of society. Agricultural labourers plant, tend and protect trees themselves and as a result the cost of work is much less in comparison to the works done by the departmental agencies.

Undoubtedly continuous efforts are needed to motivate them by offering comparative large quantum of subsidies and effective guidance so that they may fully participate in the agroforestry programme (Jha & Sen, 1991).

**Women participation**

India has experienced unequal distribution of benefits between men and women from many development schemes, despite efforts to reverse this trend. Women have had limited access to extension services, their enrolment in producer-oriented organisations has been low, and they have suffered from a lack of training programmes, which would allow them to upgrade their technical and managerial skills (Sharma *et al.*, 1989, Chen *et al.*, 1986 and Sreen *et al.*, 1992).

The programme for rural women still continues as a separate exercise within the sectoral programmes with marginal attention, resources and inadequate monitoring. Therefore, a fresh look needs to be given for the identification of trades and activities which may gainfully be taken by the women. Participation of women in forestry is now being recognised as essential. Forestry is increasingly seen as essential to the advancement of women in rural communities, where life is directly dictated by the nature and quality of available natural (and renewable) resources.

Realistic opportunities for participation can be created only when a programme is designed to accommodate and correct disadvantages and to build upon existing interests and advantages for each given group in the concerned community. Women and forestry programmes are no exception to this rule (Siddiqui, 1992). Dr. Anoja Wickramasinghe has stated that the stereotype thinking about male/female and the male dominance in the forestry sector are greatly hampering progress in community forestry management development. Womens equity should be ensured in policy, planning decision making and proper information flow. The rural women can be gainfully engaged in the agroforestry activities. Much can be achieved by the organising women at the local level. The forestry sector provides regular employment to women. The women should be trained in nursery raising, processing of various, minor products, marketing skill etc. Tribal society has a tradition of both men and women working on an equal footing whether in agriculture or in other vocations. They perform almost all agricultural operation including animal husbandry, sericulture, lac culture, bee keeping, vegetable cultivation etc. Women are considered more efficient than men in the mulbery plots. An acre of land optimally requires 10 labourers out of which 7 would be women. The active participation of women in agroforestry programme is only possible by massive training programme, development of infrastructure facilities at Panchayat level, policy support in respect of marketing and liberalised funding to women co-operatives and empowerment of women through women's organisation (Kerketta, *et al.*, 1990). There is need for integration of women in the main stream of development programme.

The rural branch of the Kasturba Vanavasi Kanya Ashram at Niwali have done considerable work in organising women, creating social awareness among them, and harnessing their energies towards social improvement. Mahila mandals with 10-12 members have been formed in each of the 60 villages surrounding the Ashram. The girls in the Ashram are trained in nursery operations. Members of the mahila co-operative had planted fruit trees around their homes. Mahila mandals are also organised by Aditi in Chotanagpur plateau of Bihar. They are actively engaged in social forestry / agroforestry programmes. (See Figs. 141 and 142)

Mobilization of women to popularise social forestry/ agroforestry can succeed in involving those who have time and resources. The poor women, who have to earn their living, cannot

**Figs. 141 & 142 :** Mobilization of women to pupularise social forestry/ agroforestry.

afford to participate in these schemes unless they can directly benefit from their involvement (Murthy, 1990). (See Fig. 143)

Thus, the women can definitely play a dominant role in agroforestry programme. It is apparent that the active participation by women would increase adoption of agroforestry farming system. Concerted efforts are required to motivate them for more active participation.

**Intellectuals**

A significant percentage of this section owns land in rural areas. They may adopt agroforestry farming system in their suitable land and can avail the gains of agroforestry. They can create awareness towards agroforestry in the country and thus help in the socio-economic promotion of the country.

**Government Departments**

It is apparent that in agroforestry programme, collaboration and coordination of all the three departments *i.e.* Agriculture, Forests and Animal Husbandry are essential.

The personnel of the Forest, Agriculture and Animal Husbandry departments should chalk out developmental plans, models, arrange credit and inputs to introduce agroforestry farming system in private sector. In the present set up, in each state Forest Departments is executing the programme according to the direction given and funds allocated by the Planning Commission. Several efforts have been made by the government on this score, but the pace of work cannot be considered up to the mark as yet. Active involvement of all the concern department seems absolutely necessary to attain the objective.

**National Development Council and Planning Commission**

Lastly, it must be accepted that our present achievement towards social forestry (agroforestry) is due to the favourable attitude and decision of National Development Council and Planning Commission of India. Needless to say that such favourable allocation of resources becomes the basis of rapid development of social Forestry/agroforestry (Jha and Sen, 1991).

Fig. 143 : Kisan Gosthi to motivate the farmers.

## REFERENCES


Boraiyan, M.P. (1985). Social forestry a way of life, *Yojna*, Vol. 29(19), pp. 6-8.

Chen, M., Mitra, M., Athreya, G., Dholaki, A., law, P. and Rao A. (1986) *Indian Women : A study of their role in the dairy movement*, Shaki Books, New Delhi.

Jha, L.K. and Sen, P.P. (1991). *Social Forestry*, Himalaya Publishing House, Bombay, pp. 1-199.

Kerketta, R. and Jha, L.K. (1989). The role of different sections of Indian society in social forestry programme, *Souvenir, Social Forestry and environment amelioration*, seminar, B.A.U. Ranchi, pp. 1-10.

Kerketta, V., Lakra, V., Bara, N. and Tirkey, K. (1990) Women's participation in agricultural development of tribal region of Bihar, *Workshop-cum-training programme*, Directorate of Extension Education, B.A.U. Ranchi, pp. 1-5.

Mitra, H.R. (1989). Rural transformation in tribal Bihar, *Souvenir*, Sixth National Seminar-cum-Workshop as Krishi Vigyan Kendras, Ramakrishna Mission Ashrama, Ranchi, pp. 29-32.

Murthy, N. (1990). NGO and Women's involvement in wasteland development : A case study of Kasturba Trust and Nagaziri experiment, In : *Studies on social forestry in India*, Shingi, M.P., (*eds*)., RAPA Publication, Bangkok, pp., 161-167.

Narayana, E.A. (1991). Managing voluntary organisations with limited finances and unskilled personnel. *The Indian Journal of Social Work*, Tata Inst. of Social Sciences, Bombay, pp. 361-368.

Poonam Smith-Sreen and John Smith-Sreen (1992). Constraints faced by women dairy farmers in India and strategies to overcome them, *The Indian Journal of Social Work*, Vol. III (1), T.I.S.S., pp. 105-114.

Rajgopal (1986). Agricultural labour in tribal areas, *Kurukshetra*, Vol. XXXIV (4), pp. 17-18.

Sharma, J., Rao, R.J. and Srinivasan, V. (1989). *Dairying, Training Manual for Women*, Women's Development Division, National Institute of Public Co-operation and Child Development, New Delhi.

Siddiqui, N. (1992). Opportunities and constraints for equitable participation of women in Forestry, Summary, *Asia-Pacific Community Forestry Newsletter*, Vol 5(1), pp. 1-2.

Swami Prabhananda (1989). Swami Vivekananda on alleviation of poverty, *Souvenir*, Seminar-cum-workshop on Krishi Vigyan Kendras, Ramkrishna Mission Ashrama, Ranchi, pp. 1-10.

Wickramasinghe, A. (1992). Role of women and children, *workshop on sustainable and effective management systems for community forestry*, RECOFTC, Bangkok, 15-17 Jan, 1992.

# Indexes

## *General and Species (Crops, Fodder, Trees and Pathogens)*

## II. Genera and Species (Insects)

## *III. General and Species (Birds)*

# IV. General Index

T